T0314296

Condition Monitoring with Vibration Signals

Condition Monitoring with Vibration Signals

Compressive Sampling and Learning Algorithms for Rotating Machines

Hosameldin Ahmed and Asoke K. Nandi
Brunel University London
UK

Registered Offices
John Wiley & Sons, Inc., 111 River Street, Hoboken, NJ 07030, USA
John Wiley & Sons Ltd, The Atrium, Southern Gate, Chichester, West Sussex, PO19 8SQ, UK

Editorial Office
The Atrium, Southern Gate, Chichester, West Sussex, PO19 8SQ, UK

For details of our global editorial offices, customer services, and more information about Wiley products visit us at www.wiley.com.

Library of Congress Cataloging-in-Publication Data
Names: Ahmed, Hosameldin, author. | Nandi, Asoke Kumar, 1976- author.
Title: Condition monitoring with vibration signals : compressive sampling
    and learning algorithms for rotating machines / Hosameldin Ahmed and
    Asoke K. Nandi, Brunel University London, UK.
Description: Hoboken, NJ, USA : Wiley-IEEE Press, 2019. | Includes
    bibliographical references and index.
Identifiers: LCCN 2019024456 (print) | LCCN 2019024457 (ebook) | ISBN
    9781119544623 (cloth) | ISBN 9781119544630 (adobe pdf) | ISBN
    9781119544647 (epub)
Subjects: LCSH: Machinery–Monitoring. | Machinery–Vibration–Measurement.
    | Rotors–Maintenance and repair.
Classification: LCC TJ177 .N36 2019 (print) | LCC TJ177 (ebook) | DDC
    621.8028/7–dc23
LC record available at https://lccn.loc.gov/2019024456
LC ebook record available at https://lccn.loc.gov/2019024457

Cover Design: Wiley
Cover Image: © PIRO4D/2442 images/Pixabay

Set in 10/12pt WarnockPro by SPi Global, Chennai, India
Printed and bound in Singapore by Markono Print Media Pte Ltd

10 9 8 7 6 5 4 3 2 1

To

My parents, my wife Intesar, and our children – Monia, Malaz, Mohamed, and Abubaker Ahmed.

Hosameldin Ahmed

My wife, Marion, and our children – Robin, David, and Anita Nandi.

Asoke K. Nandi

Contents

Preface

As an essential element of most engineering processes in many critical functions of industries, rotating machine condition monitoring is a key technique for ensuring the efficiency and quality of any product. Machine condition monitoring is a process of monitoring machine health conditions. It is incorporated into various sensitive applications of rotating machines, e.g. wind turbines, oil and gas, aerospace and defence, automotive, marine, etc. The increased level of complexity of modern rotating machines requires more effective and efficient condition monitoring techniques. For that reason, a growing body of literature has resulted from research and development efforts by many research groups around the world. These publications have made a direct impact on current and future developments in machine condition monitoring. However, there is no collection of works, including previous and recently developed methods, devoted to the field of condition monitoring for rotating machines using vibration signals. As this field is still developing, such a book cannot be definitive or complete. But this book attempts to bring together many techniques in one place, and outlines the complete guide from the basics of rotating machines to the generation of knowledge using vibration signals. It provides an introduction to rotating machines and the vibration signals produced from them at a level that can be easily understood by readers such as postgraduate students, researchers, and practicing engineers. The introduction introduces those readers to the basic knowledge needed to appreciate the specific applications of the methods in this book.

Based on the stages of the machine condition monitoring framework, and with the aim of designing effective techniques for detecting and classifying faults in rotating machines, a major part of the book covers various feature-extraction, feature-selection, and feature-classification methods as well as their applications to machine vibration datasets. Moreover, this book presents the latest methods, including machine learning and compressive sampling. These offer significant improvements in accuracy with reduced computational costs. It is important for these to be made available to all researchers as well as practitioners and new people coming into this field, to help improve safety, reliability, and performance. Although this is not intended to be a textbook, examples and case studies using vibration data are given throughout the book to show the use and application of the included methods in monitoring the condition of rotating machines.

The layout of the book is as follows:

Chapter 1 offers an introduction to machine condition monitoring and its application in condition-based maintenance. The chapter explains the importance of machine condition monitoring and its use in various rotating machine applications, machine maintenance approaches, and machine condition monitoring techniques that can be used to identify machine health conditions.

Chapter 2 is concerned with the principles of rotating machine vibration and acquisition techniques. The first part of this chapter is a presentation of the basics of vibration, vibration signals produced by rotating machines, and types of vibration signals. The second part is concerned with vibration data acquisition techniques and highlights the advantages and limitations of vibration signals.

Chapter 3 introduces signal processing in the time domain by giving an explanation of mathematical and statistical functions and other advanced techniques that can be used to extract basic signal information from time-indexed raw vibration signals that can sufficiently represent machine health conditions.

Chapter 4 presents signal processing in the frequency domain, which has the ability to extract information based on frequency characteristics that are not easy to observe in the time domain. The first part describes the Fourier transform, the most commonly used signal-transformation technique, which allows one to transform the time domain signal to the frequency domain. In addition, this chapter gives an explanation of different techniques that can be used to extract various frequency spectrum features that can more efficiently represent machine health conditions.

Chapter 5 introduces signal processing in the time-frequency domain and gives an explanation of several techniques that can be used to examine time-frequency characteristics of time-indexed series signals, which can be figured more effectively than the Fourier transform and its corresponding frequency spectrum features.

Chapter 6 is concerned with vibration-based machine condition monitoring using machine learning algorithms. The first part of this chapter gives an overview of the vibration-based machine condition monitoring process, and describes fault detection, the problem diagnosis framework, and types of learning that can be applied to vibration data. The second part defines the main problems of learning from vibration data for the purpose of fault diagnosis and describes techniques to prepare vibration data for analysis to overcome the aforementioned problems.

Chapter 7 presents common, appropriate methods for linear subspace learning that can be used to reduce a large amount of collected vibration data to a few dimensions without significant loss of information.

Chapter 8 introduces common, suitable methods for nonlinear subspace learning that can be used to reduce a large amount of collected vibration data to a reduced amount without loss of information.

Chapter 9 introduces generally applicable methods that can be used to select the most important features that can effectively represent the original features. Also, it provides an explanation of feature ranking and feature subset selection techniques.

Chapter 10 is concerned with the basic theory of the diagnosis tool decision tree, its data structure, the ensemble model that combines decision trees into a decision forest model, and their applications in diagnosing machine faults.

Chapter 11 is devoted to a description of two probabilistic models for classification: (i) the hidden Markov model (HMM) as a probabilistic generative model, and (ii) the logistic regression model and generalised logistic regression model, also called multinomial logistic regression or multiple logistic regression, as probabilistic discriminative models, and their applications in diagnosing machine faults.

Chapter 12 begins with a discussion of the basic principles of the learning method known as artificial neural networks (ANNs). Then, the chapter describes three different types of ANNs (i.e. multi-layer perceptron, radial basis function network, and Kohonen network), which can be used for fault classification. In addition, the applications of these methods in diagnosing machine faults are described.

Chapter 13 presents the support vector machine (SVM) classifier, by giving a brief description of the basic idea of the SVM model for binary classification problems. Then, the chapter explains the multiclass SVM approach and the different techniques that can be used for multiclass SVMs. Examples of their applications in diagnosing machine faults are provided.

Chapter 14 describes recent trends of deep learning in the field of machine condition monitoring and provides an explanation of commonly used techniques and examples of their applications in diagnosing machine faults.

Chapter 15 provides an overview of the efficacy of the classification algorithms introduced in this book. This chapter describes different validation techniques that can be used to validate the efficacy of classification algorithms in terms of classification results.

Chapter 16 presents new feature-learning frameworks based on compressive sampling and subspace learning techniques for machine condition monitoring. The chapter starts with a concise exposition of the basic theory of compressive sampling and shows how to perform compressive sampling for sparse frequency representations and sparse time-frequency representations. Then, the chapter presents an overview of compressive sampling in machine condition monitoring. The second part of the chapter describes three frameworks based on compressive sampling and presents different method implementation based on these frameworks. In the third part, two case studies and applications of these methods to different classes of machine health conditions are considered.

Chapter 17 presents an original framework combining compressive sampling and a deep neural network based on a sparse autoencoder. Overcomplete features with a different number of hidden layers in the deep neural network are considered in the application of this method to different classes of machine health conditions, using the same two case studies as Chapter 16.

Chapter 18 provides conclusions and recommendations for the application of the different methods studied in this book. These will benefit practitioners and researchers involved in the field of vibration-based machine condition monitoring.

This book is up-to-date and covers many techniques used for machine condition monitoring, including recently developed methods. In addition, this book will provide new methods, including machine learning and compressive sampling, which cover various topics of current research interest. Additional to the material provided in the book, publicly accessible software for most of the introduced techniques in this book and links to publicly available vibration datasets are provided in the appendix.

A work of this magnitude will unfortunately contain errors and omissions. We would like to take this opportunity to apologise unreservedly for all such indiscretions in advance. We would welcome comments and corrections; please send them by email to a.k.nandi@ieee.org or by any other means.

February 2019
London, UK

Hosameldin Ahmed and Asoke K. Nandi

About the Authors

Hosameldin Ahmed received the degree of B.Sc. (Hons.) in Engineering Technology, specialisation in Electronic Engineering, from the Faculty of Science and Technology University of Gezira, Sudan, in 1999 and M.Sc. degree in Computer Engineering and Networks from the University of Gezira, Sudan, in 2010. He has recently received the Ph.D. degree in Electronic and Computer Engineering at Brunel University London, UK. Since 2014, he has been working with his supervisor, Professor Asoke. K. Nandi, in the area of machine condition monitoring. Their collaboration has made several contributions to the advancement of vibration based machine condition monitoring using compressive sampling and modern machine learning algorithms. His work has been published in high-quality journals and international conferences. His research interests lie in the areas of signal processing, compressive sampling, and machine learning with application to vibration-based machine condition monitoring.

Asoke K. Nandi received the degree of Ph.D. in Physics from the University of Cambridge (Trinity College), Cambridge, UK. He has held academic positions in several universities, including Oxford, Imperial College London, Strathclyde, and Liverpool as well as the Finland Distinguished Professorship in Jyvaskyla (Finland). In 2013, he moved to Brunel University London, to become the Chair and Head of Electronic and Computer Engineering. Professor Nandi is a Distinguished Visiting Professor at Tongji University (China) and an Adjunct Professor at the University of Calgary (Canada).

In 1983, Professor Nandi jointly discovered the three fundamental particles known as W^+, W^-, and Z^0, providing the evidence for the unification of the electromagnetic and weak forces, for which the Nobel Committee for Physics in 1984 awarded the prize to his two team leaders for their decisive contributions. His current research interests lie in the areas of signal processing and machine learning, with applications to communications, gene expression data, functional magnetic resonance data, machine condition monitoring, and biomedical data. He has made many fundamental theoretical and algorithmic contributions to many aspects of signal processing and machine learning. He has much expertise in 'Big Data', dealing with heterogeneous data, and extracting information from multiple datasets obtained in different laboratories and at different times. Professor Nandi has authored over 590 technical publications, including 240 journal papers, as well as 4 books: *Automatic Modulation Classification: Principles, Algorithms and Applications* (Wiley, 2015), *Integrative Cluster Analysis in Bioinformatics* (Wiley, 2015), *Blind Estimation Using Higher-Order Statistics* (Springer, 1999), and *Automatic Modulation Recognition of Communications Signals* (Springer, 1996). The h-index of his publications is 73 (Google Scholar) and the ERDOS number is 2.

Professor Nandi is a Fellow of the Royal Academy of Engineering (UK) and of seven other institutions. Among the many awards he has received are the Institute of Electrical and Electronics Engineers (USA) Heinrich Hertz Award in 2012; the Glory of Bengal Award for his outstanding achievements in scientific research in 2010; award from the Society for Machinery Failure Prevention Technology, a division of the Vibration Institute (USA) in 2000; the Water Arbitration Prize of the Institution of Mechanical Engineers (UK) in 1999; and the Mountbatten Premium of the Institution of Electrical Engineers (UK) in 1998. Professor Nandi is an IEEE Distinguished Lecturer (2018–2019).

List of Abbreviations

AANN	Auto-associative neural network
ACO	Ant colony optimisation
ADC	Analog to digital converter
AE	Acoustic emission
AFSA	Artificial fish swarm algorithm
AI	Artificial intelligence
AIC	Akaike's information criterion
AID	Automatic interaction detector
AM	Amplitude modulation
ANC	Adaptive noise cancellation
ANFIS	Adaptive neuro-fuzzy inference system
ANN	Artificial neural network
ANNC	Adaptive nearest neighbour classifier
AR	Autoregressive
ARIMA	Autoregressive integrated moving average
ARMA	Autoregressive moving average
ART2	Adaptive resonance theory-2
AUC	Area under a ROC curve
BFDF	Bearing fundamental defect frequency
BPFI	Bearing pass frequency of inner race
BPFO	Bearing pass frequency of outer race
BPNN	Backpropagation neural network
BS	Binary search
BSF	Ball spin frequency
BSS	Blind source separation
CAE	Contractive autoencoder
CART	Classification and regression tree
CBLSTM	Convolutional bi-directional long short-term memory
CBM	Condition-based maintenance
CBR	Case-based reasoning
CCA	Canonical correlation analysis
CDF	Characteristics defect frequency
CF	Crest factor
CFT	Continuous Fourier transform
CHAID	Chi-square automatic integration detector

Chi-2	Chi-squared
cICA	Constraint-independent component analysis
CLF	Clearance factor
CM	Condition monitoring
CMF	Combined mode function
CMFE	Composite multiscale fuzzy entropy
CNN	Convolutional neural network
CoSaMP	Compressive sampling matching pursuit
CS-Chi-2	Compressive sampling and Chi-square feature selection algorithm
CS-CMDS	Compressive sampling and classical multidimensional scaling
CS-CPDC	Compressive sampling and correlated principal and discriminant components
CS-FR	Compressive sampling and feature ranking
CS-FS	Compressive sampling and Fisher score
CS-GSN	Compressive sampling and GMST, SPE, and neighbourhood component analysis
CS-KLDA	Compressive sampling and kernel linear discriminant analysis algorithm
CS-KPCA	Compressive sampling and kernel principal component analysis method
CS-LDA	Compressive sampling and linear discriminant analysis method
CS-LS	Compressive sampling and Laplacian score
CS-LSL	Compressive sampling and linear subspace learning
CS-NLSL	Compressive sampling and nonlinear subspace learning
CS-PCA	Compressive sampling and principal component analysis
CS-PCC	Compressive sampling and Pearson correlation coefficients
CS-Relief-F	Compressive sampling and Relief-F algorithm
CS-SAE-DNN	Compressive sampling and sparse autoencoder-based deep neural network
CS-SPE	Compressive sampling and stochastic proximity embedding
CVM	Cross-validation method
CWT	Continuous wavelet transform
DAG	Direct acyclic graph
DBN	Deep belief network
DDMA	Discrete diffusion maps analysis
DFA	Detrended-fluctuation analysis
DFT	Discrete Fourier transform
DIFS	Difference signal
DM	Diffusion map
DNN	Deep neural network
DPCA	Dynamic principal component analysis
DRFF	Deep random forest fusion
DT	Decision tree
DTCWPT	Dual-tree complex wavelet packet transform
DWT	Discrete wavelet transform
EBP	Error backpropagation
EDAE	Ensemble deep autoencoder

EEMD	Ensemble empirical mode decomposition
ELM	Extreme learning machine
ELU	Exponential linear unit
EMA	Exponential moving average
EMD	Empirical mode decomposition
ENT	Entropy
EPGS	Electrical power generation and storage
EPSO	Enhanced particle swarm optimisation
ESVM	Ensemble support vector machine
FC-WTA	Fully connected winner-take-all autoencoder
FDA	Fisher discriminant analysis
FDK	Frequency domain kurtosis
FFNN	Feedforward neural network
FFT	Fast Fourier transform
FHMM	Factorial hidden Markov model
FIR	Finite impulse response
FKNN	Fuzzy k-nearest neighbour
FM	Frequency modulation
FMM	Fuzzy min-max
FR	Feature ranking
Fs	Sampling frequency
FS	Fisher score
FSVM	Fuzzy support vector machine
FTF	Fundamental train frequency
GA	Genetic algorithm
GMM	Gaussian mixture model
GMST	Geodesic minimal spanning tree
GP	Genetic programming
GR	Gain ratio
GRU	Gated recurrent unit
HE	Hierarchical entropy
HFD	Higher-frequency domain
HHT	Hilbert-Huang transform
HIST	Histogram
HLLE	Hessian-based local linear embedding
HMM	Hidden Markov model
HOC	Higher-order cumulant
HOM	Higher-order moment
HOS	Higher-order statistics
HT	Hilbert transform
ICA	Independent component analysis
ICDSVM	Inter-cluster distance support vector machine
ID3	Iterative Dichotomiser 3
I-ESLLE	Incremental enhanced supervised locally linear embedding
IF	Impulse factor
IG	Information gain
IGA	Immune genetic algorithm

IIR	Infinite impulse response
IMF	Intrinsic mode function
IMFE	Improved multiscale fuzzy entropy
IMPE	Improved multiscale permutation entropy
ISBM	Improved slope-based method
ISOMAP	Isometric feature mapping
KA	Kernel Adatron
KCCA	Kernel canonical correlation analysis
KFCM	Kernel fuzzy c-means
KICA	Kernel independent component analysis
K-L	Kullback–Leibler divergence
KLDA	Kernel linear discriminant analysis
KNN	Kohonen neural network
k-NN	k-nearest neighbours
KPCA	Kernel principal component analysis
KURT	Kurtosis
LB	Lower bound
LCN	Local connection network
LDA	Linear discriminant analysis
LE	Laplacian eigenmap
Lh	Likelihood
LLE	Local linear embedding
LMD	Local mean decomposition
LOOCV	Leave-one-out cross-validation
LPP	Locality preserving projection
LR	Logistic regression
LRC	Logistic regression classifier
LS	Laplacian score
LSL	Linear subspace learning
LSSVM	Least-square support vector machine
LTSA	Local tangent space alignment
LTSM	Long short-term memory
MA	Moving average
MCCV	Monte Carlo cross-validation
MCE	Minimum classification error
MCM	Machine condition monitoring
MDS	Multidimensional scaling
MED	Minimum entropy deconvolution
MEISVM	Multivariable ensemble-based incremental support vector machine
MF	Margin factor
MFB	Modulated filter-bank structure
MFD	Multi-scale fractal dimension
MFE	Multi-scale fuzzy entropy
MHD	Multilayer hybrid denoising
MI	Mutual information
MLP	Multilayer perceptron
MLR	Multinomial logistic regression

MLRC	Multinomial logistic regression classifier
MMV	Multiple measurement vectors
MRA	Multiresolution analysis
MRF	Markov random field
MSE	Multiscale entropy
MSE	Mean square error
MVU	Maximum variance unfolding
NCA	Neighbourhood component analysis
NILES	Nonlinear estimation by iterative least square
NIPALS	Nonlinear iterative partial least squares
NLSL	Nonlinear subspace learning
NN	Neural network
Nnl	Normal negative log-likelihood
NNR	Nearest neighbour rule
NSAE	Normalised sparse autoencoder
NRS	Neighbourhood rough set
O&M	Operation and maintenance
OLS	Ordinary least squares
OMP	Orthogonal matching pursuit
ONPE	Orthogonal preserving embedding
ORDWT	Overcomplete rational dilation discrete wavelet transform
ORT	Orthogonal criterion
OSFCM	Optimal supervised fuzzy C-means clustering
OSLLTSA	Orthogonal supervised local tangent space alignment analysis
PCA	Principal component analysis
PCC	Pearson correlation coefficient
PCHI	Piecewise cubic Hermite interpolation
PDF	Probability density function
PF	Product function
PHM	Prognostic and health management
PLS	Partial least squares
PLS-PM	Partial least squares path modelling
PLS-R	Partial least squares regression
PNN	Probabilistic neural network
p–p	Peak to peak
PReLU	Parametric rectified linear unit
PSO	Particle swarm optimisation
PSVM	Proximal support vector machine
PWVD	Pseudo Wigner-Ville distribution
QP	Quadratic programming
QPSP-LSSVM	Quantum behaved particle optimisation-least square support vector machine
RBF	Radial basis function
RBM	Restricted Boltzmann machine
RCMFE	Refined composite multi-scale fuzzy entropy
ReLU	Rectified linear unit
RES	Residual signal

RF	Random forest
RFE	Recursive feature elimination
RIP	Restricted isometry property
RL	Reinforcement learning
RMS	Root mean square
RMSE	Root mean square error
RNN	Recurrent neural network
ROC	Receiver operating characteristic
RPM	Revolutions per minute
RSA	Rescaled range analysis
RSGWPT	Redundant second-generation wavelet packet transform
RUL	Remaining useful life
RVM	Relevance vector machine
SAE	Sparse autoencoder
S-ANC	Self-adaptive noise cancellation
SBFS	Sequential backward floating selection
SBS	Sequential backward selection
SCADA	Supervisory control and data acquisition system
SCG	Scale conjugate gradient
SDA	Stacked denoising autoencoder
SDE	Semidefinite embedding
SDOF	Single degree of freedom
SDP	Semidefinite programming
SELTSA	Supervised extended local tangent space alignment
SF	Shape factor
SFFS	Sequential forward floating selection
SFS	Sequential forward selection
SGWD	Second generation wavelet denoising
SIDL	Shift-invariant dictionary learning
SILTSA	Supervised incremental local tangent space alignment
SK	Skewness
SK	Spectral kurtosis
S-LLE	Statistical local linear embedding
SLLTA	Supervised learning local tangent space alignment
SM	Sammon mapping
SMO	Sequential minimal optimisation
SMV	Single measurement vector
SNR	Signal-to-noise ratio
SOM	Self-organising map
SP	Subspace pursuit
SpaEIAD	Sparse extraction of impulse by adaptive dictionary
SPE	Stochastic proximity embedding
SPWVD	Smoothed pseudo Wigner-Ville distribution
SSC	Slope sign change
STD	Standard deviation
STE	Standard error
STFT	Short-time Fourier transform

STGS	Steam turbine-generators
StOMP	Stagewise orthogonal matching pursuit
SU-LSTM	Stacked unidirectional long short-term memory
SVD	Singular value decomposition
SVDD	Support vector domain description
SVM	Support vector machine
SVR	Support vector regression
SWSVM	Shannon wavelet support vector machine
TAWS	Time average wavelet spectrum
TBM	Time-based maintenance
TDIDT	Top-down induction on decision trees
TEO	Teager energy operator
TSA	Time synchronous average
UB	Upper bound
VKF	Vold-Kalman filter
VMD	Variational mode decomposition
VPMCD	Variable predictive model-based class discrimination
VR	Variance
WA	Willison amplitude
WD	Wigner distribution
WFE	Waveform entropy
WKLFDA	Wavelet kernel function and local Fisher discriminant analysis
WL	Wavelength
WPA	Wavelet packet analysis
WPE	Wavelet packet energy
WPT	Wavelet packet transform
WSN	Wireless sensor network
WSVM	Wave support vector machine
WT	Wavelet transform
WTD	Wavelet thresholding denoising
WVD	Wigner-Ville distribution
ZC	Zero crossing

Part I

Introduction

1

Introduction to Machine Condition Monitoring

1.1 Background

The need for an effective condition monitoring (CM) and machinery maintenance program exists wherever complex, expensive machinery is used to deliver critical business functions. For example, manufacturing companies in today's global marketplace use their best endeavours to cut costs and improve product quality to maintain their competitiveness. Rotating machinery is a central part of the manufacturing procedure, and its health and availability have direct effects on production schedules, production quality, and production costs. Unforeseen machine failures may lead to unexpected machine downtime, accidents, and injuries. Recently it has been stated that machine downtime costs UK manufacturers £180bn per year (Ford 2017; Hauschild 2017). Moreover, Mobley stated that based on the specific industry, maintenance costs can represent between 15% and 60% of the cost of goods produced. For instance, in food-related production, average maintenance costs represent approximately 15% of the cost of goods produced, whereas for iron and steel and other heavy industries, maintenance costs represent up to 60% of total production costs (Mobley 2002).

Components including motors, bearings, gearboxes, etc. are engaged to operate effectively to keep the rotating machine in a stable, healthy condition. For that reason, maintenance is performed by repairing, modifying, or replacing these components in order to ensure that machines remain in a healthy condition. Maintenance can be accomplished using two main approaches: corrective and preventive maintenance (Wang et al. 2007). Corrective maintenance is the most basic maintenance technique and is performed after machine failure, which is often very expensive particularly for large-scale applications of rotating machines. Preventive maintenance can be applied to prevent a failure using either time-based maintenance (TBM) or condition-based maintenance (CBM), which can be localised CBM or remote CBM (Higgs et al. 2004; Ahmad and Kamaruddin 2012). TBM uses a calendar schedule that is set in advance to perform maintenance regardless of the health of the machine, which makes this approach expensive in some large and complex machines. In addition, TBM may not prevent machines from failing.

With regard to CBM, it has been reported that 99% of rotating equipment failures are preceded by nonspecific conditions indicating that such a failure is going to happen (Bloch and Geitner 2012). Hence, CBM is regarded as an efficient maintenance approach that can help avoid the unnecessary maintenance tasks of the TBM approach. Numerous studies have shown the economic advantages of CBM in several applications of rotating machines (e.g. McMillan and Ault 2007; Verma et al. 2013; Van Dam and Bond 2015;

Condition Monitoring with Vibration Signals: Compressive Sampling and Learning Algorithms for Rotating Machines, First Edition. Hosameldin Ahmed and Asoke K. Nandi.

Kim et al. 2016). In CBM, decisions about maintenance are made based on the machine's current health , which can be identified through the CM system. Once a fault occurs, an accurate CM technique allows early detection of faults and correct identification of the type of faults. Thus, the more accurate and sensitive the CM system, the more correct the maintenance decision that is made, and the more time available to plan and perform maintenance before machine breakdowns.

Condition monitoring of rotating machine components can minimise the risk of failure by identifying machine health via early fault detection. The main aim of condition monitoring is to avoid catastrophic machine failures that may cause secondary damage, machine downtime, potentially safety incidents, lost production, and higher costs associated with repairs. The CM techniques in rotating machinery encompass the practice of monitoring measurable data (e.g. vibration, acoustic, etc.), which can be used individually or in combination to identify changes in machine condition. This allows the CBM program to be arranged, or other actions to be taken to prevent machine breakdowns (Jardine et al. 2006). Based on the types of sensor data acquired from rotating machines, CM techniques can be grouped into the following: vibration monitoring, acoustic emission (AE) monitoring, electric current monitoring, temperature monitoring, chemical monitoring, and laser monitoring. Of these techniques, vibration-based condition monitoring has been widely studied and has become a well-accepted technique for planned maintenance management (Lacey 2008; Randall 2011). In the real world, different fault conditions generate different patterns of vibration spectrums. Thus, vibration analysis in principle allows us to examine the inner parts of the machine and analyse the health of the operating machine without physically opening it (Nandi et al. 2013). Moreover, various characteristic features can be observed from vibration signals that make this one of the best selections for machine CM.

This chapter describes maintenance approaches for rotating machines failures and applications of machine condition monitoring (MCM). It also provides a description of various CM techniques used for rotating machines.

1.2 Maintenance Approaches for Rotating Machines Failures

As briefly just described, maintenance can be accomplished using two main types: corrective and preventive maintenance. In this section, we will discuss these two types of maintenance in detail.

1.2.1 Corrective Maintenance

Corrective maintenance, also called run-to-failure, is the most basic maintenance technique, performed after a machine breakdown. In this way, when failure happens, it can be catastrophic and result in a long downtime. The procedure of corrective maintenance involves actions, activities, or tasks that are undertaken to restore the machine from breakdown status. This often can be performed by repairing, modifying, or replacing the components that are responsible for the overall system failure, e.g. bearings replacement, gear replacement, etc. Companies that use run-to-failure maintenance, i.e. corrective maintenance, do not need to perform maintenance or spend any money on maintenance while waiting for a machine to fail to operate. However, this type of maintenance is very expensive, particularly for large-scale applications of rotating machines.

1.2.2 Preventive Maintenance

Preventive maintenance is an alternative approach to corrective maintenance. The basic idea of preventive maintenance is to prevent machine breakdowns. This approach consists of actions, activities, or tasks that can be applied to prevent machine failure. Preventive maintenance can be accomplished using either TBM or CBM. These are described as follows.

1.2.2.1 Time-Based Maintenance (TBM)

TBM, also called periodic-based maintenance, uses a calendar schedule that is set in advance and performs maintenance regardless of the health of the machine, which makes this approach expensive for some large, complex machines. In addition, TBM may not prevent machine failures. In TBM, maintenance decisions are determined based on failure-time analysis. In fact, TBM assumes that the failure characteristics of equipment are predictable based on failure-rate trends, which can be grouped into three stages: burn-in, useful life, and wear-out (Ahmad and Kamaruddin 2012).

1.2.2.2 Condition-Based Maintenance (CBM)

As described by Higgs et al. (2004), a CBM system is able to understand and makes decisions without human involvement. CBM is an efficient maintenance approach that can help avoid the unnecessary maintenance tasks of the TBM approach. As described earlier, decisions regarding maintenance are made based on the machine's current health, which can be identified through the CM system. There are two types of CBM systems: localised CBM and remote CBM. Localised CBM is an independent predictive maintenance practice, which is likely to be done by a maintenance engineer or operator. The procedure of localised CBM starts by acquiring and recording CBM data from a component of interest at periodic intervals in order to recognise its current status and then decide whether that status is satisfactory. On the other hand, a remote CBM system can be independent or networked to another business system. Remote CBM monitors the condition of a component at remote areas through wireless sensors networks. Various studies have shown the economic advantages of CBM in several applications of rotating machine (for example, McMillan and Ault 2007; Verma et al. 2013; Van Dam and Bond 2015; Kim et al. 2016).

1.3 Applications of MCM

Having discussed the role of CM in CBM, in this section we present the use of CM in various rotating machine applications including a wind turbine, oil and gas, aerospace and defense, automotive, marine, and locomotive.

1.3.1 Wind Turbines

Wind turbines have become one of the fastest growing renewable energy sources. A major issue of a wind turbine is the relatively high cost of operation and maintenance (O&M). Wind turbines are usually hard-to-access structures, as they are located at remote onshore and offshore places with richer wind sources (Lu et al. 2009; Yang et al. 2013). Hence, they are subject to tough environmental conditions over their lifetime. Accordingly, major failures in wind turbines are expensive to repair and cause loss of revenue due to long downtimes (Bangalore and Patriksson 2018).

Compared to conventional power-generation techniques, wind turbines do not have resource energy costs. However, their O&M costs are high. Hence, CM of wind turbine is utilised as a tool for reducing O&M costs and improving wind farm electricity production by the wind power industry (Zhao et al. 2018). A considerable amount of literature has been published on wind turbine CM. Several systematic reviews of wind turbine CM have been undertaken (Lu et al. 2009; Hameed et al. 2009; Tchakoua et al. 2014; Yang et al. 2014; Qiao and Lu, 2015 2015a,b; Salameh et al. 2018. Gonzalez et al. 2019).

1.3.2   Oil and Gas

The oil and gas industry comprises procedures for exploration, extraction, refining, transporting, and marketing petroleum products, e.g. fuel, oil, and gasoline. Onshore and offshore oil and gas projects are capital-intensive investments with the possibility for serious financial and environmental consequences when a catastrophic failure happen (reza Akhondi et al. 2010; Telford et al. 2011). In their study of multimodel-based process CM of offshore oil and gas production process, Natarajan and Srinivasan stated that offshore oil and gas production platforms are uniquely hazardous, in that the operating personnel have to work in a perilous environment surrounded by extremely flammable hydrocarbons. Therefore, a fault in a piece of equipment may possibly spread to others, resulting in leaks, fires, and explosions, and causing loss of life, capital, and production downtime (Natarajan and Srinivasan 2010). Hence, a technique to monitor equipment in oil and gas production platforms is needed to prevent such failures in these platforms.

Condition monitoring plays an important role in monitoring the condition of equipment while operating and is utilised to predict failure in the mechanical system using a fault-diagnosis technique. Thus far, several studies have investigated the use of CM in the oil and gas industry (e.g. Thorsen and Dalva 1995; reza Akhondi et al. 2010; Natarajan and Srinivasan 2010; Telford et al. 2011).

1.3.3   Aerospace and Defence Industry

An aircraft is an extremely dynamic system where all flight-critical components within the vehicle are exposed to extreme dynamic loads and continuous vibratory and impulsive loads. Hence, to ensure rotorcraft safety and reliability, maintenance inspections, repairs, and parts replacement must be performed regularly. However, this is an expensive and time-consuming task. Even with these measures, early failure due to fatigue leads to many helicopter accidents, resulting in a fatal accident rate per mile flown for rotorcraft (Samuel and Pines 2005). Accordingly, both operating costs and safety are motivating factors for the development of health-monitoring systems. Hence, advanced fault diagnostics and prognostics for aircraft engines are required. Also, the availability and reliability of military spacecraft and aircraft are of vital importance (Keller et al. 2006).

As far as aerospace health management is concerned, Roemer et al. (2001) developed diagnostic and prognostic techniques for aerospace health management applications. Moreover, a survey of aircraft engine health monitoring systems is presented in (Tumer and Bajwa 1999). In terms of helicopter transmission diagnostics, Samuel and Pines reviewed the state of the art in vibration-based helicopter transmission diagnostics (2005).

1.3.4   Automotive

Modern vehicle systems software and hardware are complex, so their maintenance is challenging. Thus, predictive maintenance has become more important. The literature on automotive CM has highlighted several studies focused on vehicle fault diagnosis. For example, Abbas et al. (2007) presented a method for fault diagnosis and failure prognosis of vehicle electrical power generation and storage (EPGS) that includes a battery, a generator, electrical loads, and a voltage controller. Moreover, Shafi et al. (2018) introduced an approach for fault prediction of four main subsystems of a vehicle: fuel system, ignition system, exhaust system, and cooling system.

1.3.5   Marine Engines

The marine diesel engine is one of the most important sources in marine power systems. Therefore, its health and availability are vital to normal operation and efficacy of marine vessels and ships. Unforeseen failures in marine diesel engines may result in substantial economic loss and severe accidents. The problems relating to marine diesel engines, especially medium- and high-speed engines, are due primarily to their large size, which does not allow the use of trial-and-error techniques (Kouremenos and Hountalas 1997; Li et al. 2012a,b). Therefore, CM of marine diesel engines in a ship is very important to ensure vessel safety.

1.3.6   Locomotives

Railway transportation has played an important role in most countries' economic and social development. Thus, the continuous operation of trains is very important in these countries. Unforeseen failure of train components may result in unexpected breakdowns. For instance, locomotive bearings that often rotate at high speeds when the train is running need to be kept in healthy condition to ensure the safety of the locomotive (Shen et al. 2013). Hence, CM of locomotive bearings is very important to ensure the safety of people on trains.

1.4   Condition Monitoring Techniques

Based on the types of sensor data acquired from rotating machines, MCM techniques can be grouped into the following: vibration monitoring, acoustic emission monitoring, a fusion of vibration and acoustic, electric motor current monitoring, oil analysis, thermography, visual inspection, performance monitoring, and trend monitoring.

1.4.1   Vibration Monitoring

As described earlier, vibration-based bearing CM has been extensively used and has become a well-accepted technique for planned maintenance management (Lacey 2008; Randall 2011). In fact, different fault conditions generate different patterns of vibration spectrums. Thus, vibration analysis in principle allows us to examine the inner parts of the machine and analyse the health of the operating machine without physically opening it (Nandi et al. 2013). In addition, various characteristic features can be observed from vibration signals, which makes this one of the best selections for machine CM. In this book, we will describe various techniques for MCM using vibration signals.

1.4.2   Acoustic Emission

Acoustic emission (AE) is a technology for CM of machines such as gearboxes. Emitted sound waves are caused by faults or discontinuities. In electrical machines, sources of AE include impacting, cyclic fatigue, friction, turbulence, material loss, cavitation, leakage, etc. (Goel et al. 2015). AE is often propagated on the surface of the material as Rayleigh waves; and the displacement of these waves is measured by AE sensors, which are usually piezoelectric crystals. Compared to vibration monitoring, AE monitoring can provide a higher signal-to-noise ratio (SNR) in a high-noise environment. However, it has two main drawbacks: (i) it experiences high system costs, and (ii) it requires specialised expertise to acquire AE (Zhou et al. 2007). The application of AE for bearing CM is studied by Li and Li (1995). The results of this study showed that AE is found to be a better signal than vibrations when transducers have to be placed remotely from the bearing. Also, its application for bearing fault detection in another study verified that AE is more sensitive for detecting incipient faults than vibration (Eftekharnejad et al. 2011). Moreover, Caesarendra used AE for low-speed slew bearing CM (Caesarendra et al. 2016).

1.4.3   Fusion of Vibration and Acoustic

In an attempt to obtain informative features, one may use a fusion of vibration signals and AE. This technique has been used for CM in several studies of MCM (e.g. Loutas et al. 2011; Khazaee et al. 2014; Li et al. 2016).

1.4.4   Motor Current Monitoring

Motor current monitoring also can be used for CBM. No additional sensors are needed to implement current monitoring for electric machines. Here, the basic electrical measures associated with electromechanical plants are readily measured by tapping into the existing voltage and current transformers that are often installed as part of the protection system. The two main advantages of current monitoring are as follows: (i) it provides significant economic benefits, and (ii) an overall MCM package is possible (Zhou et al. 2007). Several studies have used current monitoring (Schoen et al. 1995; Benbouzid et al. 1999; Li and Mechefske 2006; Blodt et al. 2008)

1.4.5   Oil Analysis and Lubrication Monitoring

In electrical and mechanical machines, lubrication oil is used to reduce friction between moving surfaces. The lubrication oil is an important source of information for early diagnosis of machine failures, similar to the role of testing human blood samples in order to detect diseases. Compared to vibration-based machine health monitoring techniques, lubrication oil CM is able to provide warnings about machine failure approximately

10 times earlier (Zhu et al. 2013). Moreover, it has been stated that incorrect lubrication, either over-lubrication or under-lubrication, is one of the main reasons for bearing defects (Harris 2001).

1.4.6 Thermography

As described by Garcia-Ramirez et al. (2014), thermographic analysis has been considered a technique that can be used in fault diagnosis with the advantages of being non-invasive and having a wide range of analysis. This technique can be performed through thermographic images that can be captured using a thermographic camera sensor: an infrared detector that absorbs both the energy emitted by the object and the temperature of the surface to be measured, and converts it into a signal called a thermogram. Each pixel of a thermogram has a specific temperature value, and the image contrast can be derived from the differences in temperature of the object surface. Bagavathiappan et al. (2013) presented a comprehensive review of infrared thermography for CM that focused on the advances of infrared thermography as a non-invasive CM tool for machinery, equipment, and processes. Moreover, the most recent contributions related to the application of infrared thermography to different industrial applications have been reviewed in Osornio-Rios et al. (2018).

1.4.7 Visual Inspection

Given high-resolution cameras and advancements in computer hardware and software, visual inspection is considered an alternative technique for MCM, which can be performed by visually inspecting the surface of the components in a machine. The procedure of automated visual inspection involves image acquisition, preprocessing, feature extraction, and classification. The applications of machine vision extract a feature from 2D digital images of a 3D scene. The main aim of an industrial machine vision system is to replace human inspectors with automated visual inspection procedures. Machine vision systems can be categorised into three groups (Ravikumar et al. 2011):

1. *Measurement systems.* The aim of the measurement system is to find the dimensions of an object through digitisation and manipulation of the image of the object.
2. *Guidance systems.* The guidance system instructs a machine to perform specific actions based on what it sees.
3. *Inspection systems.* The inspection system defines whether an object or a scene matches a predefined description.

Various studies have assessed the efficacy of visual inspection in MCM (e.g. Sun et al. 2009; Ravikumar et al. 2011; Chauhan and Surgenor 2015; Liu et al. 2016; Karakose et al. 2017.).

1.4.8 Performance Monitoring

The basic idea here is that a machine's condition can be identified using the information obtained from performance monitoring. This technique requires two predefined conditions to ensure a successful application: (i) the system should be stable in a normal condition, and its stability is reproduced in the parameters under investigation;

and (ii) measurements are taken either manually or automatically. If these conditions are met, any changes from the normal behaviours of the system indicate abnormality (Rao 1996).

1.4.9 Trend Monitoring

Trend monitoring involves continuous or regular measurements of a parameter, e.g. temperature, noise, electric current, etc. It includes the selection of a suitable and measurable indication of machine or component weakening and the study of the trend in this measurement with a running time to indicate when weakening exceeds a critical rate. For instance, the measured data is recorded and plotted on a graph as a function of time. Then, it is compared with other measured data that represent the normal condition of a machine. Here, the difference between the measured data and the predefined data that represent the normal condition is used to recognise any machine abnormality (Davies 2012).

1.5 Topic Overview and Scope of the Book

MCM is a crucial technique for guaranteeing the efficiency and quality of any production process. When machine failure happens, the correct monitoring data analysis helps the engineers to locate the problem and repair it quickly. In an ideal situation, we could predict machine failure in advance and carry out maintenance before the failure happens, and thus machines would always run in a healthy condition and provide satisfactory work.

Given the importance of MCM in various sensitive applications of rotating machines, such as power generation, oil and gas, aerospace and defence, automotive, marine, etc., CM techniques for rotating machinery encompass the practice of monitoring measurable data (e.g. vibration, acoustic, etc.), which can be used individually or in combination to identify changes in a machine's condition. The increased level of complexity in modern rotating machines requires more effective and efficient CM techniques. For that reason, a growing body of literature has resulted from efforts in research and development by many research groups around the world. These publications have a direct impact on the present and future development of MCM. The nature of the MCM problem requires multiple directions for solutions and motivates continuous contributions from generations of researchers.

First, there are various type of CM techniques. As described earlier, based on the types of sensor data acquired from rotating machines, MCM techniques can be grouped into the following: vibration monitoring, acoustic emission monitoring, a fusion of vibration and acoustic, electric motor current monitoring, oil analysis, thermography, visual inspection, performance monitoring, and trend monitoring.

Second, instead of processing the originally acquired signals, a common approach is to compute certain attributes of the raw signal that can describe the signal in essence. In the machine learning community, these attributes are referred to as features. At times, multiple features are computed to form a feature set. Depending on the number of features in the set, one may need to perform further filtering of the set using a feature-selection algorithm. Various techniques of feature extraction and feature selection can be used in MCM.

Third, the core objective is to categorise the acquired signal into the corresponding machine condition correctly, which is generally a multiclass classification problem. A number of classification algorithms can be used to deal with the classification problem.

In this book, we attempt to bring together many techniques in one place and outline a complete guide, from the basics of rotating machines to the generation of knowledge using vibration signals. We will provide an introduction to rotating machines and the vibration signals they produce, at a level that can be easily understood by readers such as post-graduate students, researchers, and practicing engineers (Chapter 2). The introduction will help those readers become familiar with the basic knowledge needed to appreciate the specific applications of the methods in this book. Based on the stages of the MCM framework and the aim to design effective techniques for fault detection and classification of rotating machines, we will cover feature extraction (Chapters 3–8), feature selection (Chapter 9), and classification methods (Chapters 10–13) as well as their applications to machine vibration datasets. Moreover, this book will describe recent trends of deep learning in the field of MCM and provide an explanation of commonly used techniques and examples of their applications in machine fault diagnosis (Chapter 14). Additionally, to assess the efficiency of the classification algorithms introduced in this book, we will describe different validation techniques that can be used to validate the efficiency of classification algorithms in terms of classification results (Chapter 15).

Furthermore, we will present new methods including machine learning and compressive sampling. These offer significant improvements in accuracy, with reduced computational costs. It is important that these are made available to all researchers as well as practitioners and new people coming into this field to help improve safety, reliability, and performance (Chapters 16 and 17). Finally, we will provide conclusions and recommendations for the application of the different methods studied in this book (Chapter 18).

1.6 Summary

In this chapter, we briefly introduced the commonly used maintenance approaches for rotating machines failures as well as the applications of MCM in various sensitive applications of rotating machines such as power generation, oil and gas, aerospace and defence, automotive, marine, etc. In addition, we provided a description of various CM techniques that can be used for rotating machines, including vibration monitoring, acoustic emission monitoring, a fusion of vibration and acoustic, electric motor current monitoring, oil analysis, thermography, visual inspection, performance monitoring, and trend monitoring.

References

Abbas, M., Ferri, A.A., Orchard, M.E., and Vachtsevanos, G.J. (2007). An intelligent diagnostic/prognostic framework for automotive electrical systems. In: *2007 IEEE Intelligent Vehicles Symposium*, 352–357. IEEE.

Ahmad, R. and Kamaruddin, S. (2012). An overview of time-based and condition-based maintenance in industrial application. *Computers & Industrial Engineering* 63 (1): 135–149.

Bagavathiappan, S., Lahiri, B.B., Saravanan, T. et al. (2013). Infrared thermography for condition monitoring–a review. *Infrared Physics & Technology* 60: 35–55.

Bangalore, P. and Patriksson, M. (2018). Analysis of SCADA data for early fault detection, with application to the maintenance management of wind turbines. *Renewable Energy* 115: 521–532.

Benbouzid, M.E.H., Vieira, M., and Theys, C. (1999). Induction motors' faults detection and localization using stator current advanced signal processing techniques. *IEEE Transactions on Power Electronics* 14 (1): 14–22.

Bloch, H.P. and Geitner, F.K. (2012). *Machinery Failure Analysis and Troubleshooting: Practical Machinery Management for Process Plants*. Butterworth-Heinemann.

Blodt, M., Granjon, P., Raison, B., and Rostaing, G. (2008). Models for bearing damage detection in induction motors using stator current monitoring. *IEEE Transactions on Industrial Electronics* 55 (4): 1813–1822.

Caesarendra, W., Kosasih, B., Tieu, A.K. et al. (2016). Acoustic emission-based condition monitoring methods: review and application for low speed slew bearing. *Mechanical Systems and Signal Processing* 72: 134–159.

Chauhan, V. and Surgenor, B. (2015). A comparative study of machine vision based methods for fault detection in an automated assembly machine. *Procedia Manufacturing* 1: 416–428.

Davies, A.e. (2012). *Handbook of Condition Monitoring: Techniques and Methodology*. Springer Science & Business Media.

Eftekharnejad, B., Carrasco, M.R., Charnley, B., and Mba, D. (2011). The application of spectral kurtosis on acoustic emission and vibrations from a defective bearing. *Mechanical Systems and Signal Processing* 25 (1): 266–284.

Ford, J. (2017). Machine downtime costs UK manufacturers £180bn a year. The Engineer. www.theengineer.co.uk/faulty-machinery-machine-manufacturers.

Garcia-Ramirez, A.G., Morales-Hernandez, L.A., Osornio-Rios, R.A. et al. (2014). Fault detection in induction motors and the impact on the kinematic chain through thermographic analysis. *Electric Power Systems Research* 114: 1–9.

Goel, S., Ghosh, R., Kumar, S. et al. (2015). A methodical review of condition monitoring techniques for electrical equipment. NDT.net. https://www.ndt.net/article/nde-india2014/papers/CP0073_full.pdf.

Gonzalez, E., Stephen, B., Infield, D., and Melero, J.J. (2019). Using high-frequency SCADA data for wind turbine performance monitoring: a sensitivity study. *Renewable Energy* 131: 841–853.

Hameed, Z., Hong, Y.S., Cho, Y.M. et al. (2009). Condition monitoring and fault detection of wind turbines and related algorithms: a review. *Renewable and Sustainable Energy Reviews* 13 (1): 1–39.

Harris, T.A. (2001). *Rolling Bearing Analysis*. Wiley.

Hauschild, M. (2017). Downtime costs UK manufacturers £180bn a year. The Manufacturer. https://www.themanufacturer.com/articles/machine-downtime-costs-uk-manufacturers-180bn-year.

Higgs, P.A., Parkin, R., Jackson, M. et al. (2004). A survey on condition monitoring systems in industry. In: *ASME 7th Biennial Conference on Engineering Systems Design and Analysis*, 163–178. American Society of Mechanical Engineers.

Jardine, A.K., Lin, D., and Banjevic, D. (2006). A review on machinery diagnostics and prognostics implementing condition-based maintenance. *Mechanical Systems and Signal Processing* 20 (7): 1483–1510.

Karakose, M., Yaman, O., Baygin, M. et al. (2017). A new computer vision based method for rail track detection and fault diagnosis in railways. *International Journal of Mechanical Engineering and Robotics Research* 6 (1): 22–17.

Keller, K., Swearingen, K., Sheahan, J. et al. (2006). Aircraft electrical power systems prognostics and health management. In: *2006 IEEE Aerospace Conference*, 12. IEEE.

Khazaee, M., Ahmadi, H., Omid, M. et al. (2014). Classifier fusion of vibration and acoustic signals for fault diagnosis and classification of planetary gears based on Dempster–Shafer evidence theory. *Proceedings of the Institution of Mechanical Engineers, Part E: Journal of Process Mechanical Engineering* 228 (1): 21–32.

Kim, J., Ahn, Y., and Yeo, H. (2016). A comparative study of time-based maintenance and condition-based maintenance for optimal choice of maintenance policy. *Structure and Infrastructure Engineering* 12 (12): 1525–1536.

Kouremenos, D.A. and Hountalas, D.T. (1997). Diagnosis and condition monitoring of medium-speed marine diesel engines. *Tribotest* 4 (1): 63–91.

Lacey, S.J. (2008). An overview of bearing vibration analysis. *Maintenance & Asset Management* 23 (6): 32–42.

Li, C.J. and Li, S.Y. (1995). Acoustic emission analysis for bearing condition monitoring. *Wear* 185 (1–2): 67–74.

Li, W. and Mechefske, C.K. (2006). Detection of induction motor faults: a comparison of stator current, vibration and acoustic methods. *Journal of Vibration and Control* 12 (2): 165–188.

Li, Z., Yan, X., Yuan, C., and Peng, Z. (2012a). Intelligent fault diagnosis method for marine diesel engines using instantaneous angular speed. *Journal of Mechanical Science and Technology* 26 (8): 2413–2423.

Li, Z., Yan, X., Guo, Z. et al. (2012b). A new intelligent fusion method of multi-dimensional sensors and its application to tribo-system fault diagnosis of marine diesel engines. *Tribology Letters* 47 (1): 1–15.

Li, C., Sanchez, R.V., Zurita, G. et al. (2016). Gearbox fault diagnosis based on deep random forest fusion of acoustic and vibratory signals. *Mechanical Systems and Signal Processing* 76: 283–293.

Liu, L., Zhou, F., and He, Y. (2016). Vision-based fault inspection of small mechanical components for train safety. *IET Intelligent Transport Systems* 10 (2): 130–139.

Loutas, T.H., Roulias, D., Pauly, E., and Kostopoulos, V. (2011). The combined use of vibration, acoustic emission and oil debris on-line monitoring towards a more effective condition monitoring of rotating machinery. *Mechanical Systems and Signal Processing* 25 (4): 1339–1352.

Lu, B., Li, Y., Wu, X., and Yang, Z. (2009). A review of recent advances in wind turbine condition monitoring and fault diagnosis. In: *2009 IEEE Power Electronics and Machines in Wind Applications, 2009. PEMWA*, 1–7. IEEE.

McMillan, D. and Ault, G.W. (2007). Quantification of condition monitoring benefit for offshore wind turbines. *Wind Engineering* 31 (4): 267–285.

Mobley, R.K. (2002). *An Introduction to Predictive Maintenance*. Elsevier.

Nandi, A.K., Liu, C., and Wong, M.D. (2013). Intelligent vibration signal processing for condition monitoring. In: *Proceedings of the International Conference Surveillance*, vol. 7, 1–15. https://surveillance7.sciencesconf.org/conference/surveillance7/P1_Intelligent_Vibration_Signal_Processing_for_Condition_Monitoring_FT.pdf.

Natarajan, S. and Srinivasan, R. (2010). Multi-model based process condition monitoring of offshore oil and gas production process. *Chemical Engineering Research and Design* 88 (5–6): 572–591.

Osornio-Rios, R.A.A., Antonino-Daviu, J.A., and de Jesus Romero-Troncoso, R. (2018). Recent industrial applications of infrared thermography: a review. *IEEE Transactions on Industrial Informatics* 15 (2): 615–625.

Qiao, W. and Lu, D. (2015a). A survey on wind turbine condition monitoring and fault diagnosis—part I: components and subsystems. *IEEE Transactions on Industrial Electronics* 62 (10): 6536–6545.

Qiao, W. and Lu, D. (2015b). A survey on wind turbine condition monitoring and fault diagnosis—part II: signals and signal processing methods. *IEEE Transactions on Industrial Electronics* 62 (10): 6546–6557.

Randall, R.B. (2011). *Vibration-Based Condition Monitoring: Industrial, Aerospace and Automotive Applications*. Wiley.

Rao, B.K.N. (1996). *Handbook of Condition Monitoring*. Elsevier.

Ravikumar, S., Ramachandran, K.I., and Sugumaran, V. (2011). Machine learning approach for automated visual inspection of machine components. *Expert Systems with Applications* 38 (4): 3260–3266.

reza Akhondi, M., Talevski, A., Carlsen, S., and Petersen, S. (2010). Applications of wireless sensor networks in the oil, gas and resources industries. In: *2010 24th IEEE International Conference on Advanced Information Networking and Applications (AINA)*, 941–948. IEEE.

Roemer, M.J., Kacprzynski, G.J., Nwadiogbu, E.O., and Bloor, G. (2001). *Development of Diagnostic and Prognostic Technologies for Aerospace Health Management Applications*. Rochester, NY: Impact Technologies LLC.

Salameh, J.P., Cauet, S., Etien, E. et al. (2018). Gearbox condition monitoring in wind turbines: a review. *Mechanical Systems and Signal Processing* 111: 251–264.

Samuel, P.D. and Pines, D.J. (2005). A review of vibration-based techniques for helicopter transmission diagnostics. *Journal of Sound and Vibration* 282 (1–2): 475–508.

Schoen, R.R., Habetler, T.G., Kamran, F., and Bartfield, R.G. (1995). Motor bearing damage detection using stator current monitoring. *IEEE Transactions on Industry Applications* 31 (6): 1274–1279.

Shafi, U., Safi, A., Shahid, A.R. et al. (2018). Vehicle remote health monitoring and prognostic maintenance system. *Journal of Advanced Transportation* 2018: 1–10.

Shen, C., Liu, F., Wang, D. et al. (2013). A doppler transient model based on the Laplace wavelet and spectrum correlation assessment for locomotive bearing fault diagnosis. *Sensors* 13 (11): 15726–15746.

Sun, H.X., Zhang, Y.H., and Luo, F.L. (2009). Visual inspection of surface crack on labyrinth disc in aeroengine. *Optics and Precision Engineering* 17: 1187–1195.

Tchakoua, P., Wamkeue, R., Ouhrouche, M. et al. (2014). Wind turbine condition monitoring: state-of-the-art review, new trends, and future challenges. *Energies* 7 (4): 2595–2630.

Telford, S., Mazhar, M.I., and Howard, I. (2011). Condition based maintenance (CBM) in the oil and gas industry: an overview of methods and techniques. In: *Proceedings of the 2011 International Conference on Industrial Engineering and Operations Management*. Kuala Lumpur, Malaysia: IEOM Research Solutions Pty Ltd.

Thorsen, O.V. and Dalva, M. (1995). A survey of faults on induction motors in offshore oil industry, petrochemical industry, gas terminals, and oil refineries. *IEEE Transactions on Industry Applications* 31 (5): 1186–1196.

Tumer, I. and Bajwa, A. (1999). A survey of aircraft engine health monitoring systems. In: *35th Joint Propulsion Conference and Exhibit*, 2528. American Institute of Aeronautics and Astronautics.

Van Dam, J. and Bond, L.J. (2015). Economics of online structural health monitoring of wind turbines: cost benefit analysis. *AIP Conference Proceedings* 1650 (1): 899–908.

Verma, N.K., Khatravath, S., and Salour, A. (2013). Cost benefit analysis for condition-based maintenance. In: *2013 IEEE Conference Prognostics and Health Management (PHM)*, 1–6.

Wang, L., Chu, J., and Wu, J. (2007). Selection of optimum maintenance strategies based on a fuzzy analytic hierarchy process. *International Journal of Production Economics* 107 (1): 151–163.

Yang, W., Court, R., and Jiang, J. (2013). Wind turbine condition monitoring by the approach of SCADA data analysis. *Renewable Energy* 53: 365–376.

Yang, W., Tavner, P.J., Crabtree, C.J. et al. (2014). Wind turbine condition monitoring: technical and commercial challenges. *Wind Energy* 17 (5): 673–693.

Zhao, J., Deng, W., Yin, Z. et al. (2018). A portable wind turbine condition monitoring system and its field applications. *Clean Energy* 2 (1): 58–71.

Zhou, W., Habetler, T.G., and Harley, R.G. (2007). Bearing condition monitoring methods for electric machines: a general review. In: *2007 IEEE International Symposium on Diagnostics for Electric Machines, Power Electronics and Drives, 2007. SDEMPED*, 3–6. IEEE.

Zhu, J., Yoon, J.M., He, D. et al. (2013). Lubrication oil condition monitoring and remaining useful life prediction with particle filtering. *International Journal of Prognostics and Health Management* 4: 124–138.

2

Principles of Rotating Machine Vibration Signals

2.1 Introduction

As described in Chapter 1, condition-based maintenance (CBM) of a rotating machine is regarded as an efficient maintenance approach in which the decision regarding maintenance is made based on the current health condition of the machine. The current health condition can be identified through condition monitoring (CM) systems, including the practice of monitoring measurable data, e.g. vibrations, acoustics, motor current, etc. Of these types of data, vibration-based CM has been widely studied and has become a well-accepted technique for planned maintenance management. In practice, different fault conditions generate different patterns of vibration spectrums. Thus, vibration analysis in principle allows us to examine the inner parts of the machine and analyse the health conditions of the operating machine without physically opening it. Moreover, various characteristic features can be observed from vibration signals, which make vibration analysis one of the best selections for machine CM.

This chapter presents the principles of rotating machine vibration and acquisition techniques. The first part of this chapter is a presentation of vibration basics, vibration signals produced by rotating machines, and types of vibration signals. The second part is concerned with vibration data acquisition techniques and highlights the advantages and limitations of vibration signals.

2.2 Machine Vibration Principles

Vibration is the repetitive, periodic, or oscillatory response of a mechanical system (De Silva 2006). Typically, all rotating machines produce vibrations due to dynamic forces, i.e. repeating forces, which mostly come from imbalanced, misaligned, worn, or improperly driven rotating parts. The rate of the vibration cycles is called *frequency*. Based on the repetitive motions of a mechanical system and the rate of the vibration cycles, we define two terms as follows:

1. *Oscillations.* Repetitive motions that are clean and regular and happen at low frequencies.
2. *Vibrations.* Any repetitive motions, even at high frequencies, with low amplitudes and irregular and random behaviour.

Condition Monitoring with Vibration Signals: Compressive Sampling and Learning Algorithms for Rotating Machines, First Edition. Hosameldin Ahmed and Asoke K. Nandi.
© 2020 John Wiley & Sons Ltd. Published 2020 by John Wiley & Sons Ltd.

The vibrations of linear systems can be categorised into four main groups (Norton and Karczub 2003):

1. *Free vibrations.* Happens when a system vibrates in the absence of any externally applied forces: i.e. the system vibrates under the action of internal forces. A finite system undergoing free vibrations will vibrate in one or more of a series of specific patterns, and each of these specific patterns vibrates at a constant frequency called a *natural frequency.*
2. *Forced vibrations.* Occur under the excitation of external forces, which can be categorised as being (i) harmonic, (ii) periodic, (iii) nonperiodic (pulse or transient), or (iv) random (stochastic). These type of vibrations happen at excitation frequencies that are often independent of the natural frequencies of the system.
3. *Damped vibrations.* With time, the amplitude keeps decreasing.
4. *Undamped vibrations.* The response of a vibration system without a damping component.

The phenomenon of resonance is encountered when a natural frequency of the system occurs simultaneously with one of the exciting frequencies.

Every real structure has an infinite number of natural frequencies, but many machinery vibration problems involve just one of these frequencies. Therefore, the single degree of freedom (SDOF) model with one natural frequency can be useful for analysing vibrations from machines. A vibrating system can be described using three main elements: (i) mass, which is the motion of an unbending body that stores kinetic energy, (ii) stiffness, which is the deflection of a flexible component that stores potential energy, and (iii) damping, which absorbs vibration energy (Vance et al. 2010). Figure 2.1 presents the SDOF model as described in (Vance et al. 2010, Brandt 2011), where k, c, and m are the stiffness, damping, and mass, respectively.

The undamped natural frequency can be defined using Eq. (2.1),

$$w_n = \sqrt{\frac{k}{m}} \ \text{rad/sec} \tag{2.1}$$

where w_n can be represented as *cycles/sec*, i.e. hertz (Hz), using Eq. (2.2)

$$f_n = \frac{w_n}{2\pi} \tag{2.2}$$

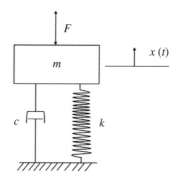

Figure 2.1 Single degree of freedom vibration model (SDOF) (Vance et al. 2010).

which can be defined as revolutions per minute (rpm) as follows,

$$N = 60f_n \tag{2.3}$$

where N is the number of revolutions per minute. Nevertheless, the real-life system is often damped, and the equation of motion of such a damped system can be given as follows (Mohanty 2014):

$$m\frac{d^2x}{dt^2} + c\frac{dx}{dt} + kx = 0 \tag{2.4}$$

Here, the damping factor ξ can be described as follows:

$$\xi = \frac{c}{2\sqrt{km}} \tag{2.5}$$

The response to such a damped oscillator can be described using Eq. (2.6),

$$x(t) = A^{-\xi wt} \sin(w_d t + \phi) \tag{2.6}$$

where $w_d = w_n\sqrt{1 - \xi^2}$.

Equation (2.4) represents the linear motion of a system, which can be used to describe the motion of machines (in particular, gearboxes) and consequently describe the vibrations. Another way to describe the dynamics of machines is by using torsional systems, which is known as *torsional vibrations*. The equation of motion representing the free vibration response of a SDOF-damped torsional vibration system can be represented using Eq. (2.7),

$$m\frac{d^2\theta}{dt^2} + c_t\frac{d\theta}{dt} + k_t\theta = 0 \tag{2.7}$$

where c_t is the torsional viscous damping coefficient, k_t is the torsional stiffness, θ is the rotational displacement, $\frac{d\theta}{dt}$ is rotational velocity, and $\frac{d^2\theta}{dt^2}$ is the rotational acceleration.

Machine vibration is the back-and-forth motion of a machine or machine components. The vibration of the rotating machine can be due to repeating forces, looseness, or resonances. Repeating forces can be due to the rotation of imbalanced, misaligned, worn, or improperly driven machine components. The equation of motion of a damped SDOF system subjected to external force $F(t)$ on the mass (see Figure 2.1), can be given by Eq. (2.8):

$$m\frac{d^2x}{dt^2} + c\frac{dx}{dt} + kx = F(t) \tag{2.8}$$

The dynamic forces or torques that excite the machine system can be represented using harmonic functions such that

$$F(t) = F_0 \cos w_f t \tag{2.9}$$

where w_f is the excitation frequency, also called the *input frequency* or *forcing frequency*. The rotational speed of a rotating machine corresponds to w_f. The response of such a damped harmonic oscillator to the harmonic force in Eq. (2.9) can be described using Eq. (2.10):

$$x(t) = A^{-iwt} \sin(w_d t + \theta) + A_0 \cos(w_f t - \phi) \tag{2.10}$$

Here, $A^{-iwt} \sin(w_d t + \theta)$ represents the transient response, $A_0 \cos(w_f t - \phi)$ represents the steady-state response, and ϕ can be computed using Eq. (2.11),

$$\phi = tan^{-1} \frac{2\xi r}{1 - r^2} \qquad (2.11)$$

where $r = \frac{w_f}{w_n}$ is the frequency ration. At resonance r = 1, the excitation frequency w_f is equal to the natural frequency w_n, as mentioned earlier.

A vibration analysis system often comprises four main parts: signal transducer, analyser, analysis software, and a computer for data analysis and storage (Scheffer and Girdhar 2004). Generally, machine vibration signals are obtained from several sources in a rotating machine, and they are often described by two main numerical descriptors: (i) the amplitude of the vibration, which is its magnitude. It usually describes the severity of the vibration and consequently the severity of the fault. The larger the amplitude, the larger the fault severity in the machine. In addition, (ii) the frequency of the vibration is the number of times one vibration cycle happens in one second. It usually indicates the type of fault.

2.3 Sources of Rotating Machines Vibration Signals

The most common sources of vibrations in machinery are related to the inertia of moving parts. Some parts have a reciprocating motion, accelerating back and forth. The forces are usually periodic and therefore produce periodic displacements observed as vibrations. Even without reciprocating parts, most machines have rotating shafts and wheels that cannot be perfectly balanced, so according to Newton's laws, there must be a rotating force vector at bearing supports of each rotor to produce the centripetal acceleration of the mass centre. Most of these force vectors are rotating and therefore produce a rotating displacement vector that can be observed as an orbit if two orthogonal vibration transducers are employed (Vance et al. 2010).

Typical machinery consists of a driver (the main mover), e.g. electric motor; and driven equipment, such as pumps, compressors, mixers, agitators, fans, blowers, etc. Sometimes, when the driven equipment has to be driven at speeds different than the speed of the main mover, a gearbox or belt drive is utilised (Scheffer and Girdhar 2004). Each of the rotating parts in the machine consists of other simple components: for instance, stators, rotors, bearings, seals, couplings, gears, etc. These components are engaged to operate effectively to maintain the rotating machine's stable, healthy condition. For that reason, maintenance is performed in order to ensure that the machine remains in a healthy condition by repairing, modifying, or replacing these components. When faults develop in one or more components, they result in high vibration levels, i.e. high levels of vibration amplitude. Examples of faults that result in high levels of vibration amplitude in rotating machines include:

(1) Unbalanced rotating components
(2) Misalignment of bearings
(3) Damaged bearings and gears
(4) Bad drive belts and chains
(5) Looseness
(6) Resonance
(7) Bent shaft

The following subsections describe some of the widely investigated sources of rotating machinery vibration signals.

2.3.1 Rotor Mass Unbalance

Rotor balancing is one of the most important and frequently addressed daily operations in achieving smooth-running rotating machinery. If a rotor mass axis does not coincide with its axis of rotation, we often say a rotor mass is *unbalanced* (Crocker 2007). Unbalance is the most common source of vibration in machines with rotating parts (MacCamhaoil 2012). For example, in a motor in normal, healthy condition, the rotor is often centrally aligned with the stator, and the axis of rotation of the rotor is the same as the geometrical axis of the stator. Consequently, the air gap between the outer surface of the rotor and the inner surface of the stator will be identical. On the other hand, if the rotor is not centrally aligned with the stator, this will result in a non-identical air gap that is also called *air-gap eccentricity*.

Mass-unbalanced rotors can be categorised into three main types: (i) static mass unbalanced, which is eccentricity of the rotor's centre of gravity caused by a point mass at a certain radius from the centre of rotation; (ii) couple unbalanced, caused by two equal masses placed symmetrically about the centre of gravity but positioned 180° from each other, such that when the rotor rotates, the two masses cause a shift in the inertia axis that consequently results in unbalance; and (iii) dynamic unbalanced, which is a combination of static and couple unbalance (MacCamhaoil 2012; Karmakar et al. 2016).

2.3.2 Misalignment

As described by Piotrowski (2006), shaft misalignment happens when the centrelines of rotation of two or more machinery shafts are not in line with each other. More precisely, *shaft misalignment* is the deviation of the relative shaft position from the collinear axis of rotation measured at the flexing points, also called *flexing planes*, in the coupling when equipment is running under normal operating conditions. On the other hand, the main goal of shaft alignment is to position the machinery casings such that all of these deviations are below certain acceptable values, which are related to the rotating machines' alignment tolerances. There are three main factors that can affect the alignment of rotating machinery: (i) the speed of the drive(s), (ii) the maximum deviation at flexing points or points of power transmission and power reception, and (iii) the distance between flexing points or points of power transmission.

Misalignment is often measured at the flexing points in the coupling where the coupling causes the misalignment condition. In cases where only one flexing point exists and there is an offset between the shafts or a combination of an angle and an offset, very high radial forces will be transferred through the coupling into the bearings of the two machines. In cases where more than two flexing planes exist, there will be a large amount of uncontrolled motion between the two connected shafts, which often results in very high levels of vibration in the rotating machines. Figure 2.2 shows illustrations of offset, angular, and combination shaft misalignment (Sofronas 2012).

2.3.3 Cracked Shafts

A shaft in a rotating machine is a mechanical part for transmitting torque and rotation to other components that are not connected directly to a drive train. Shafts are among the

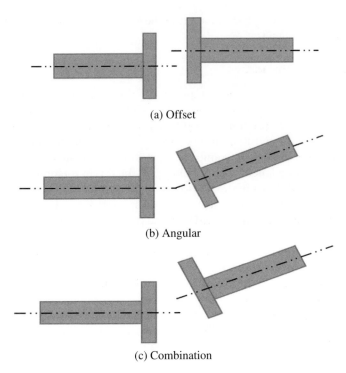

(a) Offset

(b) Angular

(c) Combination

Figure 2.2 Illustrations of offset, angular, and combination shaft misalignment (Sofronas 2012). (See insert for a colour representation of this figure.)

most critical components of high-performance rotating machines such as high-speed compressors, steam and gas turbines, generators, etc., which are subjected to the most arduous working conditions (Sabnavis et al. 2004). Due to the speedily changing nature of bending stresses, the presence of many stress raisers, and likely design or manufacturing errors, shafts are subjected to fatigue cracks. *Shaft cracks* can be defined as any accidental gaps in the shaft material that happen due to various mechanisms, e.g. high and low cycle fatigue, or stress corrosion. Figure 2.3 shows an illustration of a fatigue crack nucleated at the top of the surface and propagated along a 45° helix (Stephens et al. 2000).

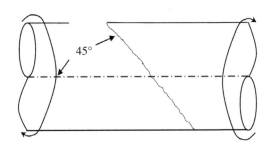

45°

Figure 2.3 Illustrations of fatigue failure of a torsion shaft (Stephens et al. 2000).

Bloch and Geitner (2012) described several causes of shaft failures as follows:

1. Shafts in petrochemical plant machinery operate under various conditions, e.g. corrosive environments, and temperatures that vary from extremely low, as in cold ethylene vapour and liquid service, to extremely high, as in gas turbines.
2. Shafts are subjected to one or more of different types of loads, including tension, compression, bending, and torsion.
3. Shafts are often exposed to high vibratory stresses.
4. Shaft metal fatigue is the most common cause of shaft failures that start at the most vulnerable point in a dynamically stressed area.

In addition, Sabnavis et al. (2004) described a typical chronology of events that lead to total failure by cracking in a ductile material as follows:

1. *Crack initiation*. Small discontinuities start in the shaft material, which can be caused by mechanical stress raisers (e.g. sharp keyways, abrupt cross-sectional changes, grooves, etc.) or factors (e.g. metallurgical factors such as forging flaws).
2. *Crack propagation*. The discontinuity increases in size as a result of cyclic stresses induced in the component.
3. *Failure*. This happens when the shaft material that has not been affected by the crack cannot withstand the applied loads.

2.3.4 Rolling Element Bearings

Considering their role in maintaining motion between static and moving parts in rotating machinery, rolling bearings are critical components of the whole system of rotating machines. In practice, rolling bearing failures may lead to major failures in machines. It is stated that approximately 40–90% of rotating machine failures are related to bearing faults (Immovilli et al. 2010) based on machine size. Thus, in most production processes in industry, roller bearings need to be kept in a healthy condition to guarantee continuity of production. Therefore, it is very important to monitor roller bearings to avoid machine breakdowns. There are two types of bearings: (i) plain (sliding) bearings that maintain motion through sliding contact, and (ii) rolling element bearings that maintain motion through rolling contact (Collins et al. 2010). The latter are widely used in most applications of rotating machinery and can be categorised into two main groups: ball bearings (spherical rolling elements) and roller bearings (nominally cylindrical rolling elements). As shown in Figure 2.4, a roller bearings consists of several components: (i) the inner race, in which the shaft drives; (ii) the outer race, normally positioned in a hole or housing; (iii) the rolling elements, normally placed between the inner race and the outer race; and (iv) the cage, also called the retainer, which can be made of plastic or metal and is used to keep the rolling elements separated equally.

Bearings defects can occur for many reasons, including (i) fatigue, which happens when there is too much load on the bearing; (ii) incorrect lubrication, either over-lubrication or under-lubrication; (iii) contamination and corrosion; and (iv) incorrect bearing installation (Harris 2001; Nandi et al. 2005). Faults in rolling bearings generate a series of impulses that repeat periodically at a rate called the bearing fundamental defect frequency (BFDF), which relies on the shaft speed, the geometry of the bearing (Figure 2.5), and the site of the faults. Based on the damaged part, BFDFs can be categorised into four types: bearing pass frequency of the outer race (BPFO), bearing

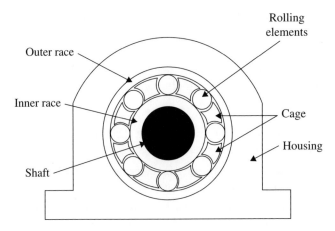

Figure 2.4 A typical roller bearing. (Ahmed and Nandi 2018).

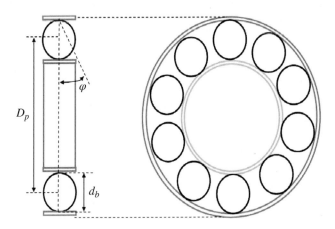

Figure 2.5 Rolling element bearing geometry. (See insert for a colour representation of this figure.)

pass frequency of the inner race (BPFI), ball spin frequency (BSF), and fundamental train frequency (FTF), which relate to the fault at the outer race, the inner race, the rolling element, and the cage, respectively (McFadden and Smith 1985; Rai and Upadhyay 2016). The BFDFs can be computed using the following formulas:

$$BPFO = \frac{N_b S_{sh}}{2}\left(1 - \frac{d_b}{D_p}cos\varphi\right) \tag{2.12}$$

$$BPFI = \frac{N_b S_{sh}}{2}\left(1 + \frac{d_b}{D_p}cos\varphi\right) \tag{2.13}$$

$$BSF = \frac{D_p}{2d_b}\left(1 - \left(\frac{d_b}{D_p}cos\varphi\right)^2\right) \tag{2.14}$$

$$FTF = \frac{S_{sh}}{2}\left(1 - \frac{d_b}{D_p}cos\varphi\right) \tag{2.15}$$

Here N_b is the number of rolling elements, S_{sh} is the shaft speed, d_b is the rolling element diameter, D_p is the pitch diameter, and φ is the angle of the load from the radial plane. Based on these BFDFs, the frequency of the collected bearing vibration signal indicates the source of the fault, and the amplitude indicates the fault severity.

2.3.5 Gears

In view of their role in power transmission and changing rotation speed, gears are critical components of the whole system of rotating machines. The power transmission from a source, e.g. an engine or motor, through a machine to an output actuation is one of the most common machine tasks (Shigley 2011). This can be done by means of transmitting power through the rotary motion of a shaft that is supported by bearings. The forces transmitted between meshing gears provide torsional moments to shafts for motion and power transmission. Gears, belt pulleys, or chain teeth are combined to provide speed changes between shafts. The speed is changed and transmitted to the output shaft by the gear transmission inside a gearbox.

There are four principal types of gears: spur, helical, bevel, and worm gears. These types of gears can be described briefly as follows:

1. *Spur gears.* Have teeth parallel to the axis of rotation and are utilised to transmit motion from one shaft to another parallel shaft.
2. *Helical gears.* Have teeth inclined to the axis of rotation. Similar to spur gears, helical gears can be used to transmit motion between parallel shafts. Additionally, sometimes they can be used to transmit motion between non-parallel shafts.
3. *Bevel gears.* Have teeth shaped on conical surfaces and are used for transmitting motion between intersecting shafts.
4. *Worm gears.* Are mostly used when the speed ratios of the two shafts are high.

As described in Randall (2011), typically the situation is not so perfect, as the teeth deform under load, providing a *meshing error* or *transmitting error*. Moreover, there are geometric deviations from the tooth ideal profiles, both intentional (typically due to *tip relief*) and unintentional. The transmitting error varies with tooth deflection that varies with load. Hence, the amplitude of the produced vibration at the tooth-meshing frequency differs directly with the load variation in service and can be regarded as an amplitude modulation effect. There are various types of gear failures, which Fakhfakh et al. categorised into three main groups:

(1) Manufacturing defects (tooth profile error, the eccentricity of wheels, etc.)
(2) Installation defects (parallelism, etc.)
(3) Defects occurring during transmission (e.g. tooth wear, cracks, etc.) (Fakhfakh et al. 2005)

2.4 Types of Vibration Signals

As described previously, when faults develop in machine components, they result in high vibration levels. These specific vibration signals can often be separated from others, in addition, to differentiate faulty and normal conditions. Vibration signals produced by machines can be categorised into two main groups: (i) stationary and (ii) nonstationary

(Randall 2011). The following subsections present a brief description of these two types of signals.

2.4.1 Stationary

Vibration signals are categorised as stationary signals when their statistical properties do not vary with time. There are two main types of stationary signals: (i) deterministic, which include periodic and quasi-periodic, and (ii) random. Deterministic signals, i.e. periodic and quasi-periodic, are made up of discrete sinusoidal components that can only be seen in the frequency domain, which will be described in Chapter 4. Random signals can only be described using their statistical properties, e.g. mean value, mean square value, etc., as will be described in detail in Chapter 3.

2.4.2 Nonstationary

Nonstationary signals have frequency content that varies with time. It can be grouped into two main classes: (i) continuous, which can be divided into continuously varying and cyclostationary; and (ii) transit. These types of signals are often analysed using time-frequency domain analysis, which will be described in Chapter 5.

2.5 Vibration Signal Acquisition

Vibration signal acquisition includes the measurement of three main quantities: (i) displacement, (ii) velocity, and (iii) acceleration. The quantity to be measured depends on the size of the machine to be monitored and the frequency range to be used. Usually, when the vibration frequency increases, it is expected that displacement levels will decrease and acceleration levels will increase (Tavner et al. 2008). It is generally accepted that between 10 Hz (600 rpm) and 1000 Hz (60 krpm), velocity provides a good indication of the severity of vibrations; above 1000 Hz (60 krpm), acceleration is the only good quantity to use (Scheffer and Girdhar 2004).

2.5.1 Displacement Transducers

A displacement transducer measures the relative distance between two surfaces using low-frequency responses. There are two types of noncontacting displacement transducers (Adams 2009):

(1) *Capacitance type.* Works on the principle of measuring the electrical capacitance of the gap between the transducer tip and the target whose position is measured.
(2) *Inductance type.* Has been proven to be the optimum rotor-to-stator position measurement method.

2.5.2 Velocity Transducers

A velocity transducer is composed of a mass suspended in very soft springs and surrounded by an electrical coil. Its springs are configured to generate a very low natural frequency. It is often used to measure vibrations in bearing housings or machinery casings. It is effective in the low- to mid-frequency range (10 Hz to around 1500 Hz). In a velocity transducer, the vibration of the electrical coil firmly attached to the housing causes the magnetic flux lines to induce a voltage in the coil proportional to the velocity of housthe ing vibration (Adams 2009).

2.5.3 Accelerometers

Accelerometers are transducers that produce a signal related to acceleration, which is the second derivative of the displacement. Accelerometers operate in a wide frequency range up to tens of kHz. Usually, accelerometers are lightweight, ranging from 0.4 to 50 g (Nandi et al. 2013). An accelerometer produces an electrical output that is directly proportional to the acceleration in the sensing axis. The most widely used types of accelerometer in machine CM is the piezoelectric accelerometers.

The basic construction of an accelerometer comprises a spring-mounted mass in contact with a piezoelectric element. The mass causes a dynamic force, which is proportional to the acceleration level of the vibration, to the piezoelectric element. There are two main types of accelerometers: (i) compression type, where a compressive force is exerted on the piezoelectric element, often used for measuring high shock levels; and (ii) shear type, where a shear force is exerted on the piezoelectric element, often used for general-purpose applications.

There are five common ways to mount accelerometers on a vibration structure (Norton and Karczub 2003):

(1) Through a connecting threaded stud
(2) Through a cementing stud
(3) Through a thin layer of wax
(4) Through a magnet
(5) Through a hand-held probe

2.6 Advantages and Limitations of Vibration Signal Monitoring

Vibration signal processing has some obvious advantages, which can be summarised as follows:

1. Vibration sensors are non-intrusive and at times noncontact. As such, we can perform diagnostics in a nondestructive manner.
2. As almost 80% of common rotating machine problems are due to misalignment and unbalance, vibration analysis represents an effective tool to monitor machine conditions.
3. Trending vibration levels can be used to identify improper practices, such as using equipment beyond its design specifications (higher temperatures, speeds, or loads). These trends can also be utilised for comparing the performance of similar machines from different manufacturers (Scheffer and Girdhar 2004).
4. Vibration signals can be obtained online and in situ. This is a desired feature for production lines. The trending capability also provides a means for predictive maintenance of the machinery. As such, unnecessary downtime for preventive maintenance can be minimised.

5. Vibration sensors are inexpensive and widely available. Modern mobile smart devices typically are equipped with one triaxial accelerometer.
6. The technologies to acquire and convert the analog outputs from the sensors are affordable nowadays.
7. Techniques for diagnosing a wide range of problems exist in the literature, including cracks (e.g. Sekhar and Prabhu 1998; Loutas et al. 2009), worn or faulty bearings (e.g. Jack et al. 1999; Guo et al. 2005; Rojas and Nandi 2005; Ahmed et al. 2018; Ahmed and Nandi 2019, 2018), shaft unbalance (e.g. McCormick and Nandi 1997; Soua et al. 2013), faulty gearboxes (e.g. He et al. 2007; Bartelmus and Zimroz 2009; Ottewill and Orkisz 2013), etc.

Although techniques based on vibration signals are generally versatile, their usefulness is limited by several factors. As mentioned earlier, different vibration sensors have different operating characteristics, and attention should be paid when selecting a suitable sensor. This requires engineers to understand the physical characteristics of vibration sources. Also, useful characteristics of the vibration signal can easily be masked by inappropriate mounting of sensors. The adhesion mechanism may also damp the high-frequency component if it is not done correctly. Vibration sensors also come with a wide variety of dynamic ranges and sensitivities. A careful perusal of the datasheet is always recommended when choosing a suitable sensor (Nandi et al. 2013).

2.7 Summary

In this chapter, we briefly introduced the principles of rotating machine vibration as well as vibration acquisition techniques. The first part of the chapter presented vibration basics, vibration signals produced by rotating machines, and types of vibration signals. The second part described vibration data–acquisition techniques and highlighted the advantages of vibration analysis for condition monitoring of rotating machines, and the care that must be exercised to benefit from this.

References

Adams, M.L. (2009). *Rotating Machinery Vibration: From Analysis to Troubleshooting*. CRC Press.

Ahmed, H. and Nandi, A. (2019). Three-stage hybrid fault diagnosis for rolling bearings with compressively-sampled data and subspace learning techniques. *IEEE Transactions on Industrial Electronics* 66 (7): 5516–5524.

Ahmed, H. and Nandi, A.K. (2018). Compressive sampling and feature ranking framework for bearing fault classification with vibration signals. *IEEE Access* 6: 44731–44746.

Ahmed, H.O.A., Wong, M.L.D., and Nandi, A.K. (2018). Intelligent condition monitoring method for bearing faults from highly compressed measurements using sparse over-complete features. *Mechanical Systems and Signal Processing* 99: 459–477.

Bartelmus, W. and Zimroz, R. (2009). Vibration condition monitoring of planetary gearbox under varying external load. *Mechanical Systems and Signal Processing* 23 (1): 246–257.

Bloch, H.P. and Geitner, F.K. (2012). *Machinery Failure Analysis and Troubleshooting: Practical Machinery Management for Process Plants*. Butterworth-Heinemann.

Brandt, A. (2011). *Noise and Vibration Analysis: Signal Analysis and Experimental Procedures*. Wiley.

Collins, J.A., Busby, H.R., and Staab, G.H. (2010). *Mechanical Design of Machine Elements and Machines: A Failure Prevention Perspective*. Wiley.

Crocker, M.J.e. (2007). *Handbook of Noise and Vibration Control*. Wiley.

De Silva, C.W. (2006). *Vibration: Fundamentals and Practice*. CRC press.

Fakhfakh, T., Chaari, F., and Haddar, M. (2005). Numerical and experimental analysis of a gear system with teeth defects. *The International Journal of Advanced Manufacturing Technology* 25 (5–6): 542–550.

Guo, H., Jack, L.B., and Nandi, A.K. (2005). Feature generation using genetic programming with application to fault classification. *IEEE Transactions on Systems, Man, and Cybernetics, Part B (Cybernetics)* 35 (1): 89–99.

Harris, T.A. (2001). *Rolling Bearing Analysis*. Wiley.

He, Q., Kong, F., and Yan, R. (2007). Subspace-based gearbox condition monitoring by kernel principal component analysis. *Mechanical Systems and Signal Processing* 21 (4): 1755–1772.

Immovilli, F., Bellini, A., Rubini, R., and Tassoni, C. (2010). Diagnosis of bearing faults in induction machines by vibration or current signals: a critical comparison. *IEEE Transactions on Industry Applications* 46 (4): 1350–1359.

Jack, L.B., Nandi, A.K., and McCormick, A.C. (1999). Diagnosis of rolling element bearing faults using radial basis function networks. *Applied Signal Processing* 6 (1): 25–32.

Karmakar, S., Chattopadhyay, S., Mitra, M., and Sengupta, S. (2016). *Induction Motor Fault Diagnosis*. Singapore: Publisher Springer.

Loutas, T.H., Sotiriades, G., Kalaitzoglou, I., and Kostopoulos, V. (2009). Condition monitoring of a single-stage gearbox with artificially induced gear cracks utilizing on-line vibration and acoustic emission measurements. *Applied Acoustics* 70 (9): 1148–1159.

MacCamhaoil, M. (2012). Static and dynamic balancing of rigid rotors. Bruel & Kjaer application notes. https://www.bksv.com/doc/BO0276.pdf.

McCormick, A.C. and Nandi, A.K. (1997). Real-time classification of rotating shaft loading conditions using artificial neural networks. *IEEE Transactions on Neural Networks* 8 (3): 748–757.

McFadden, P.D. and Smith, J.D. (1985). The vibration produced by multiple point defects in a rolling element bearing. *Journal of Sound and Vibration* 98 (2): 263–273.

Mohanty, A.R. (2014). *Machinery Condition Monitoring: Principles and Practices*. CRC Press.

Nandi, S., Toliyat, H.A., and Li, X. (2005). Condition monitoring and fault diagnosis of electrical motors—a review. *IEEE Transactions on Energy Conversion* 20 (4): 719–729.

Nandi, A.K., Liu, C., and Wong, M.D. (2013). Intelligent vibration signal processing for condition monitoring. In: *Proceedings of the International Conference Surveillance*, vol. 7, 1–15. https://surveillance7.sciencesconf.org/resource/page/id/20.

Norton, M.P. and Karczub, D.G. (2003). *Fundamentals of Noise and Vibration Analysis for Engineers*. Cambridge university press.

Ottewill, J.R. and Orkisz, M. (2013). Condition monitoring of gearboxes using synchronously averaged electric motor signals. *Mechanical Systems and Signal Processing* 38 (2): 482–498.

Piotrowski, J. (2006). *Shaft Alignment Handbook*. CRC Press.

Rai, A. and Upadhyay, S.H. (2016). A review on signal processing techniques utilized in the fault diagnosis of rolling element bearings. *Tribology International* 96: 289–306.

Randall, R.B. (2011). *Vibration-Based Condition Monitoring: Industrial, Aerospace and Automotive Applications*. Wiley.

Rojas, A. and Nandi, A.K. (2005). Detection and classification of rolling-element bearing faults using support vector machines. In: *2005 IEEE Workshop on Machine Learning for Signal Processing*, 153–158. IEEE.

Sabnavis, G., Kirk, R.G., Kasarda, M., and Quinn, D. (2004). Cracked shaft detection and diagnostics: a literature review. *Shock and Vibration Digest* 36 (4): 287.

Scheffer, C. and Girdhar, P. (2004). *Practical Machinery Vibration Analysis and Predictive Maintenance*. Elsevier.

Sekhar, A.S. and Prabhu, B.S. (1998). Condition monitoring of cracked rotors through transient response. *Mechanism and Machine Theory* 33 (8): 1167–1175.

Shigley, J.E. (2011). *Shigley's Mechanical Engineering Design*. Tata McGraw-Hill Education.

Sofronas, A. (2012). *Case Histories in Vibration Analysis and Metal Fatigue for the Practicing Engineer*. Wiley.

Soua, S., Van Lieshout, P., Perera, A. et al. (2013). Determination of the combined vibrational and acoustic emission signature of a wind turbine gearbox and generator shaft in service as a pre-requisite for effective condition monitoring. *Renewable Energy* 51: 175–181.

Stephens, R.I., Fatemi, A., Stephens, R.R., and Fuchs, H.O. (2000). *Metal Fatigue in Engineering*. Wiley.

Tavner, P., Ran, L., Penman, J., and Sedding, H. (2008). *Condition Monitoring of Rotating Electrical Machines*, vol. 56. IET.

Vance, J.M., Zeidan, F.Y., and Murphy, B.G. (2010). *Machinery Vibration and Rotordynamics*. Wiley.

Part II

Vibration Signal Analysis Techniques

3

Time Domain Analysis

3.1 Introduction

Vibration signals collected from a rotating machine using vibration transducers are often in the time domain. They are a collection of time-indexed data points collected over historical time, representing acceleration, velocity, or proximity based on the type of transducer used to collect the signals. In practice, the vibration signals usually include a large collection of responses from several sources in the rotating machine and some background noise. This makes it challenging to directly use the acquired vibration signals for fault diagnosis of the machine, either by manual inspection or automatic monitoring. As an alternative to processing the vibration signals, the common approach is to compute certain attributes of the raw signal that can describe the signal in essence. In the machine learning community, these attributes are also called *characteristics*, *signatures*, or *features*.

For integrative fault diagnosis, several types of methods need to be adopted in a cascade of steps starting from raw vibration datasets and ending at final mature sets of results. These include vibration analysis techniques that have the ability to obtain useful information about a machine's condition from the raw vibration datasets, which can be successfully used for fault diagnosis. We focus on such methods, such as time domain methods, frequency domain methods, and time-frequency domain methods, in this book.

The manual inspection of vibration signals as part of time domain fault diagnosis may be divided into two main types: (i) visual inspection and (ii) feature-based inspection.

3.1.1 Visual Inspection

In this type of inspection, a machine's condition can be assessed by comparing a measured vibration signal to a previously measured vibration signal from a machine in a normal condition, i.e. measured from a new or healthy machine. In this case, both signals should be measured on the same frequency range. Vibration measurements that are higher than normal indicate that the machine is in a fault condition, which causes the machine to produce more vibration. For instance, Figure 3.1 a shows a typical time domain vibration signal for brand-new roller bearings, and Figure 3.1b represents an inner race (IR) fault condition for roller bearings. Obviously, in the case shown in this figure, the time waveform in Figure 3.1b shows spikes with a high level of amplitude in some locations of the vibration signal, while other parts of the vibration remain in the

Condition Monitoring with Vibration Signals: Compressive Sampling and Learning Algorithms for Rotating Machines, First Edition. Hosameldin Ahmed and Asoke K. Nandi.
© 2020 John Wiley & Sons Ltd. Published 2020 by John Wiley & Sons Ltd.

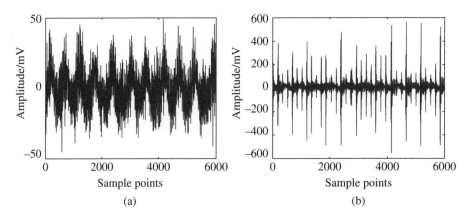

Figure 3.1 Time domain vibration signal of a roller bearing: (a) brand-new condition; (b) inner race fault condition.

lower amplitude level compared to the normal condition vibration signal in Figure 3.1a. This tells us that the machine is in an abnormal condition. This technique is a simple and cost-effective method of condition monitoring, which uses an oscilloscope to view the vibration signal or computer-based aids to collect data and record or display information. Readers who are interested in more details of visual inspection systems are referred to Davies (1998).

Nevertheless, this type of inspection is not dependable in the field for monitoring the condition of rotating machines, for the following four reasons: (i) not all time waveform signals from rotating machines provide clear visual differences (Guo et al. 2005). For example, Figure 3.2 presents two typical vibration signals from roller bearing in a worn but undamaged condition (Figure 3.2a) and an outer race (OR) fault condition (Figure 3.2b); in this case, it is difficult to depend on visual inspection to analyse the time waveform characteristics and identify whether a machine is in a fault condition. (ii) In practice, we deal with a large collection of vibration signals that usually contain

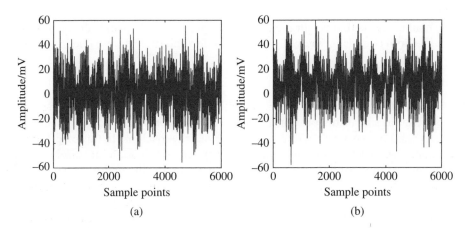

Figure 3.2 Time domain vibration signal of a roller bearing: (a) worn but undamaged condition; (b) outer race fault condition.

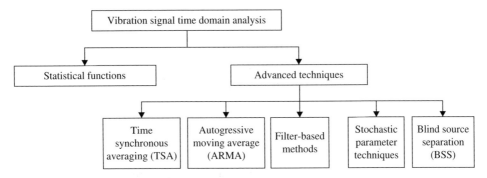

Figure 3.3 Vibration signal time domain analysis techniques.

some background noise. (iii) We sometimes deal with low-amplitude signals measured in noisy backgrounds. And (iv) the demand for early detection of faults makes the manual inspection of all collected signals impractical.

3.1.2 Features-Based Inspection

In this type of inspection, the machine's condition can be evaluated by computing certain features of the raw vibration signal that can describe the signal in essence. These features can be used to identify the difference between two vibration signals. On the other hand, automatic monitoring of a rotating machine uses machine learning classifiers to classify the signal with its correct condition type using either the raw vibration signal or computed features of the raw vibration signal in the time domain.

For a complete view of the field, this chapter introduces vibration signal processing in the time domain by giving an explanation of statistical functions and other advanced techniques that can be used to extract features from time-indexed raw vibration datasets, which sufficiently represent machine health. This will position time domain–based features in their place within the wider context of machine fault diagnosis. The other two types of vibration signal analysis, i.e. frequency domain and time-frequency domain analysis, will be covered in detail in Chapters 4 and 5, respectively.

Various time domain–based techniques are used for vibration signal analysis. They are summarised in Figure 3.3.

3.2 Statistical Functions

As mentioned in Chapter 2, acquired vibration signals are usually obtained from several sources in a rotating machine with random behaviour. Due to their randomness characteristics, these vibration signals cannot be defined by a direct mathematical formula and can only be analysed using statistical techniques with respect to time. Therefore, it is not surprising to find that earlier works in this area focus on time domain descriptive statistics-based features that can be used for either manual inspection or automatic monitoring. Numerous types of statistical functions have been heavily used to extract features from vibration signals in the time domain based on signal amplitude. The following subsections discuss those statistical functions in more detail.

3.2.1 Peak Amplitude

The peak amplitude, x_p, is the maximum positive amplitude of the vibration signal; it can also be defined as half the difference between the maximum and minimum vibration amplitude, i.e. the maximum positive peak amplitude and the maximum negative peak amplitude. This can be mathematically given by Eq. (3.1):

$$x_p = \frac{1}{2}[x_{max}(t) - x_{min}(t)] \qquad (3.1)$$

3.2.2 Mean Amplitude

The mean amplitude, $\bar{x}$, is the average of the vibration signal over a sampled interval, which can be computed using Eq. (3.2),

$$\bar{x} = \frac{1}{T} \int x(t)\, dt \qquad (3.2)$$

where T is the sampled signal duration and $x(t)$ is the vibration signal. For a discrete sampled signal, Eq. (3.2) can be rewritten as:

$$\bar{x} = \frac{1}{N} \sum_{i=1}^{N} x_i \qquad (3.3)$$

where N is the number of sampled points and x_i is an element of signal x.

3.2.3 Root Mean Square Amplitude

The root mean square (RMS) amplitude, x_{RMS}, is the variance of the vibration signal magnitude. The mathematical expression of x_{RMS} is shown in Eq. (3.4),

$$x_{RMS} = \sqrt{\frac{1}{T} \int |x(t)|^2 dt} \qquad (3.4)$$

where T is the sampled signal duration and $x(t)$ is the vibration signal. The RMS amplitude is resilient to spurious peaks in the steady-state operating condition. If the vibration signal is in discrete form, Eq. (3.4) can be rewritten as follows:

$$x_{RMS} = \sqrt{\frac{1}{N} \sum_{i=1}^{N} |x_i|^2} \qquad (3.5)$$

3.2.4 Peak-to-Peak Amplitude

The peak-to-peak amplitude, also called the range, x_{p-p}, is the range of the vibration signal, $x_{max}(t) - x_{min}(t)$, which denotes the difference between the maximum positive peak amplitude and the maximum negative peak amplitude.

3.2.5 Crest Factor (CF)

The crest factor (CF), x_{CF}, is defined as the ratio of the peak amplitude value, x_p, and the RMS amplitude, x_{RMS}, of the vibration signal. This can be computed using as follows:

$$x_{CF} = \frac{x_p}{x_{RMS}} \qquad (3.6)$$

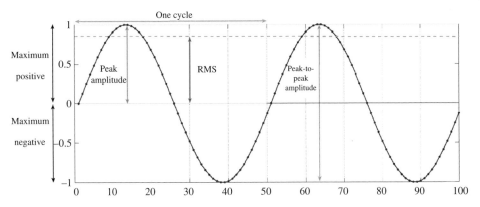

Figure 3.4 A pure sine wave with amplitude of 1 and 100 sample points. (See insert for a colour representation of this figure.)

In a time series vibration signal, peaks will result in an increase of the CF. Thus, the CF is useful in detecting the early stages of a fault condition and is used in online monitoring. It is often utilised as a measure of the impulsive nature of a vibration signal that will give basic information about how much change is occurring in a normal-condition vibration waveform. For instance, in a fixed period of a pure sine wave (x) (Figure 3.4), with 100 samples, maximum positive amplitude of 1, and maximum negative amplitude of -1, the x_{RMS} value is equal to 0.707 and x_{CF} is 1.414. Hence, a signal with a value of x_{CF} higher than 1.414 indicates an abnormal state in the signal.

3.2.6 Variance and Standard Deviation

The variance, σ_x^2, defines deviations of the vibration signal energy from the mean value, which can be mathematically given as follows:

$$\sigma_x^2 = \frac{\sum (x_i - \bar{x})^2}{N - 1} \tag{3.7}$$

The square root of the variance, i.e. σ_x, is called the standard deviation of the signal x, and is expressed as:

$$\sigma_x = \sqrt{\frac{\sum (x_i - \bar{x})^2}{N - 1}} \tag{3.8}$$

Here, x_i represents an element of x, $\bar{x}$ is the mean of x, and N is the number of sampled points.

3.2.7 Standard Error

The standard error of a predicted y for an individual x in the regression, y_{STE}, can be expressed in the following equation:

$$y_{STE} = \sqrt{\frac{1}{N-1} \left[\sum (y - \bar{y})^2 - \frac{\left[\sum (x - \bar{x})(y - \bar{y}) \right]^2}{\sum (x - \bar{x})^2} \right]} \tag{3.9}$$

Here N is the sample size, and $\bar{x}$ and $\bar{y}$ are the sample means.

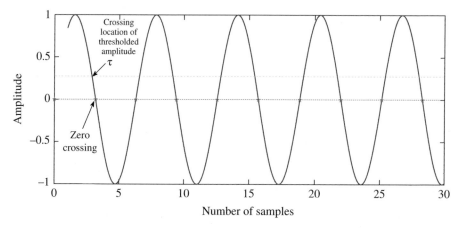

Figure 3.5 Crossing locations of amplitude equal to zero and a τ threshold amplitude. (See insert for a colour representation of this figure.)

3.2.8 Zero Crossing

The digitised vibration signal has a portion above zero and a portion below zero; therefore, when the signal crosses the x-axis, the amplitude value is equal to zero. This location where the signal crosses the x-axis is called a zero crossing (ZC). Hence, the zero crossing, x_{ZC}, can be identified as the number of times the signal crosses the x-axis if it satisfies the following criteria:

$$x_i > 0 \text{ and } x_{i-1} < 0$$
$$x_i < 0 \text{ and } x_{i-1} > 0 \tag{3.10}$$

where x_i is the current signal value and x_{i-1} is the previous signal value. To avoid background noise contained in the vibration signal, a threshold τ may be used instead of the amplitude value of zero, i.e. instead of counting the number of crossings when the amplitude value equals zero, the number of crossings may be counted for the amplitude of value τ (Figure 3.5) such that

$$|x_i - x_{i-1}| \geq \tau \tag{3.11}$$

In ZC-based feature extraction, the density of the time intervals between successive ZCs, and the excess threshold measurement, are the two commonly used measurements for representing the information contained in the ZC features. William and Hoffman (2011) empirically demonstrated that ZC features extracted from time domain vibrations using the duration between successive ZCs are useful in early detection and identification of bearing faults. The ZC technique is useful in finding ZC locations and counts the number of cycles that occur in the time interval to obtain the frequency estimation (f) as follows (Zhen et al. 2013),

$$f = \frac{Fs}{number\ of\ cycles} \tag{3.12}$$

where Fs is sampling frequency.

3.2.9   Wavelength

The wavelength, x_{WL}, is a measure of distance between two consecutive positive peaks or negative peaks of the vibration signal that can be computed using Eq. (3.13),

$$x_{WL} = \sum_{i=1}^{N} |x_i - x_{i-1}| \tag{3.13}$$

Also, the relation between signal wave frequency, wave speed, and x_{WL} can be expressed mathematically using Eq. (3.14),

$$x_{WL} = \frac{Wave\ speed}{f} \tag{3.14}$$

Hence, the value of x_{WL} decreases as the frequency of the vibration signal increases.

3.2.10   Willison Amplitude

The Willison amplitude, x_{WA}, is defined as the number of times the difference between a vibration signal amplitude amongst two adjacent samples exceeds a predefined threshold τ. This can be computed using the following equation:

$$x_{WA} = \sum_{i=1}^{N} f(|x_i - x_{i+1}|)$$

$$s.t \quad f(x) = \begin{cases} 1 & if\ x \geq \tau \\ 0 & otherwise \end{cases} \tag{3.15}$$

3.2.11   Slope Sign Change

The slope sign change, x_{ssc}, is defined as the number of times the slope of the vibration signal changes sign. Similar to x_{ZC}, x_{ssc} needs to introduce a threshold τ to reduce the background noise made at slope sign changes. Given three continuous data points x_{i-1}, x_i, and x_{i+1}, x_{ssc} can be computed using Eq. (3.16),

$$x_{ssc} = \sum_{i=1}^{N} [g((x_i - x_{i-1})(x_i - x_{i+1}))]$$

$$s.t \quad g(x) = \begin{cases} 1 & if\ x \geq \tau \\ 0 & otherwise \end{cases} \tag{3.16}$$

3.2.12   Impulse Factor

The impulse factor, x_{IF}, is defined as the ratio of the peak value to the average of the absolute value of the vibration signal and can be expressed as:

$$x_{IF} = \frac{x_{peak}}{\frac{1}{N} \sum_{i=1}^{N} |x_i|} \tag{3.17}$$

The impulse factor is useful in measuring the impact of a fault generated in the vibration signal.

3.2.13 Margin Factor

The margin factor, x_{MF}, can be calculated using the following equation:

$$x_{MF} = \frac{x_{peak}}{\left(\frac{1}{N}\sum_{i=1}^{N}\sqrt{|x_i|}\right)^2}$$

(3.18)

The margin factor value changes significantly with changes in the peak value, which makes it very sensitive to impulse faults in particular.

3.2.14 Shape Factor

The shape factor, x_{SF}, is defined as the ratio of the RMS value to the average of the absolute value of the vibration signal and can be expressed as:

$$x_{SF} = \frac{x_{RMS}}{\frac{1}{N}\sum_{i=1}^{N}|x_i|}$$

(3.19)

The shape factor is useful in measuring the change resulting in the vibration signal due to unbalance and misalignment defects.

3.2.15 Clearance Factor

The clearance factor, x_{CLF}, is defined as the ratio of the maximum value of the input vibration signal to the mean square root of the absolute value of the input vibration signal and can be expressed as:

$$x_{CLF} = \frac{x_{max}}{\left(\frac{1}{N}\sum_{i=1}^{N}\sqrt{|x_i|}\right)^2}$$

(3.20)

3.2.16 Skewness

The skewness, also called the third normalised central statistical moment, x_{SK}, is a measure of the asymmetrical behaviour of the vibration signal through its probability density function (PDF): i.e. it measures whether the vibration signal is skewed to the left or right side of the distribution of the normal state of the vibration signal. For a signal with N sample points, x_{SK} can be presented by Eq. (3.21),

$$x_{SK} = \frac{\sum_{i=1}^{N}(x_i - \bar{x})^3}{N\,\sigma_x^3}$$

(3.21)

The value of x_{SK} for a normal condition is zero.

3.2.17 Kurtosis

The kurtosis, also called the fourth normalised central statistical moment, x_{KURT}, is a measure of the peak value of the input vibration signal through its PDF: i.e. it measures whether the peak is higher or lower than the peak of the distribution corresponding to a normal condition of the vibration signal. For a signal with N sample points, x_{KURT} can be formulated as shown in Eq. (3.22),

$$x_{KURT} = \frac{\sum_{i=1}^{N}(x_i - \bar{x})^4}{N\,\sigma_x^4} \tag{3.22}$$

Other higher moments from the fifth (HOM5) to the ninth (HOM9) central statistical moments can be calculated by raising the power expression in Eq. (3.21) correspondingly. These can be represented by Eqs. (3.23)–(3.27):

$$HOM5 = \frac{\sum_{i=1}^{N}(x_i - \bar{x})^5}{N\,\sigma_x^5} \tag{3.23}$$

$$HOM6 = \frac{\sum_{i=1}^{N}(x_i - \bar{x})^6}{N\,\sigma_x^6} \tag{3.24}$$

$$HOM7 = \frac{\sum_{i=1}^{N}(x_i - \bar{x})^7}{N\,\sigma_x^7} \tag{3.25}$$

$$HOM8 = \frac{\sum_{i=1}^{N}(x_i - \bar{x})^8}{N\,\sigma_x^8} \tag{3.26}$$

$$HOM9 = \frac{\sum_{i=1}^{N}(x_i - \bar{x})^9}{N\,\sigma_x^9} \tag{3.27}$$

3.2.18 Higher-Order Cumulants (HOCs)

Higher-order cumulants (HOCs) are closely related to the higher-order moments of the signal. The moments of the signal are related to its probability density function (p.d.f) through the moment-generating function. The HOCs of the signal are obtained by evaluating the derivatives of the logarithm of moment-generation function. Let m_n represent the nth moment of a signal x, where the first- and second-order moments represent the mean and variance, respectively. The first four cumulants can be written in terms of moments using Eqs. (3.28)–(3.31):

$$CU_1 = m_1 \tag{3.28}$$

$$CU_2 = m_2 - m_1^2 \tag{3.29}$$

$$CU_3 = m_3 - 3m_2 m_1 + 2m_1^3 \tag{3.30}$$

$$CU_4 = m_4 - 3m_2^2 - 4m_3 m_1 + 12m_2 m_1^2 - 6m_1^4 \tag{3.31}$$

If the vibration signal is normalised to zero mean and unit variance, then the computation simplifies, leaving only the non-m_1 terms. Hence, among other type of normalisation techniques, it is customary to normalise the vibration signal to zero mean and unit variance. Here, the third- and fourth-order cumulants are commonly referred to as *skewness* and *kurtosis*. Moreover, where there is more than one signal, it is possible to compute the cross cumulants. Thus far, a number of studies have used moments and cumulant features in machine fault diagnosis (e.g. McCormick and Nandi 1996, 1997, 1998; Jack and Nandi 2000; Guo et al. 2005; Zhang et al. 2005; Rojas and Nandi 2006; Saxena and Saad 2006. Tian et al. 2015).

3.2.19 Histograms

A histogram can be assumed to be a discrete PDF of the vibration signal. Two types of features can be obtained from the histogram: the lower bound (LB) and upper bound (UB). These can be expressed using the following equations:

$$LB = x_{min} - 0.5 \left(\frac{x_{max} - x_{min}}{N - 1} \right) \tag{3.32}$$

$$UB = x_{max} - 0.5 \left(\frac{x_{max} - x_{min}}{N - 1} \right) \tag{3.33}$$

3.2.20 Normal/Weibull Negative Log-Likelihood Value

The negative log-likelihood of the time domain vibration signal (x) can be expressed using the following equation:

$$-logL = - \sum_{i=1}^{N} log[f(a, b \backslash x_i)] \tag{3.34}$$

where $f(a, b \backslash x_i)$ is the PDF of the vibration signal. The normal negative log-likelihood (Nnl) and the Weibull negative log-likelihood (Wnl) can be used as features of the time domain vibration signals. The Nnl and Wnl can be computed using Eq. (3.34), and their PDFs can be expressed using Eqs. (3.35) and (3.36):

$$Normal\ PDF = \frac{1}{\sigma \sqrt{2\pi}} exp^{-\left(x_i - \frac{\mu}{2\sigma^2} \right)^2} \tag{3.35}$$

$$Weibull\ PDF = \frac{b}{a} \left(\frac{x_i}{a} \right)^{b-1} exp^{-\left(\frac{x_i}{a} \right)} \tag{3.36}$$

where μ is the signal mean and σ is the standard deviation.

3.2.21 Entropy

The entropy, x_{ENT}, is a measure of uncertainty of the probability distribution of the vibration signal. For a vibration signal with N sample points, the entropy can be represented as shown in Eq. (3.37),

$$x_{ENT} = \sum_{i=1}^{N} p_{x_i} log\ p_{x_i} \tag{3.37}$$

where p_{x_i} are the probabilities computed from the distribution of x.

A considerable amount of literature has been published on vibration monitoring using statistical time domain techniques to preprocess vibration signals as input features, individually or in combination with other techniques. These studies are summarised in Table 3.1. As can be seen from this table, all the listed studies used more than one time domain statistical techniques to extract features from the raw vibration data. Each study used at least six techniques, and some used more than 10 techniques to extract features from the raw vibration data. Of these techniques, kurtosis is used in all the mentioned studies. Moreover, skewness, shape factor, impulse factor, variance, CF, peak-to-peak, RMS, and mean are among the most-used techniques in these studies.

Table 3.1 Summary of the time domain statistical features that have been used in different studies of monitoring a machine's condition.

Studies	Peak	Mean	RMS	Max	Min	Sum	P-P	CF	VR	STD	ZC	WL	WA	SSC	IF	MF	SF	CLF	SK	KURT	HIST	NnL	WnL	STE	LF	HO5 to HO9
McCormick and Nandi 1996.	✓								✓																	
McCormick and Nandi 1997.		✓						✓											✓	✓						
McCormick and Nandi 1998.		✓						✓											✓	✓						
Jack and Nandi 2000.	✓							✓											✓	✓						
Samanta et al. 2003.			✓																✓	✓						✓
Sun et al. 2004.		✓	✓					✓							✓				✓	✓						
Guo et al. 2005.		✓	✓					✓	✓										✓	✓						
Zhang et al. 2005	✓								✓				✓													
Rojas and Nandi 2006.	✓							✓											✓	✓						
Saxena and Saad 2006.	✓							✓	✓										✓	✓						
Yang et al. 2007		✓	✓			✓	✓	✓	✓						✓		✓		✓	✓						
Sugumaran and Ramachandran 2007.					✓	✓	✓												✓	✓						
Sassi et al. 2007.	✓	✓	✓					✓		✓					✓		✓		✓	✓						
Sreejith et al. 2008.	✓	✓	✓		✓	✓	✓	✓	✓	✓					✓	✓	✓	✓	✓	✓		✓	✓			
Chebil et al. 2011.	✓	✓	✓		✓	✓	✓	✓	✓	✓					✓	✓	✓	✓	✓	✓		✓	✓			
Kankar et al. 2011.	✓	✓	✓		✓	✓	✓	✓		✓					✓		✓		✓	✓						
Saimurugan et al. 2011					✓	✓	✓	✓	✓	✓					✓		✓		✓	✓						
Yiakopoulos et al. 2011		✓	✓		✓	✓	✓	✓	✓	✓					✓		✓		✓	✓	✓			✓		
Sugumaran and Ramachandran 2011					✓	✓	✓												✓	✓	✓					
Prieto et al. 2013.	✓	✓	✓			✓	✓	✓	✓	✓					✓		✓		✓	✓						✓
Lakshmi et al. 2014.	✓	✓	✓					✓											✓	✓						✓
Ali et al. 2015.	✓	✓	✓					✓	✓						✓	✓	✓		✓	✓						
Rauber et al. 2015	✓	✓	✓		✓	✓	✓	✓	✓	✓		✓	✓	✓	✓	✓	✓		✓	✓						
Nayana and Geethanjali 2017		✓										✓	✓	✓	✓		✓		✓	✓						
Tahir et al. 2017	✓	✓	✓				✓	✓									✓		✓	✓						

VR, variance; STD, standard deviation; HIST, histogram; STE, standard error; HO5–HO9, fifth higher-order moment to ninth higher-order moment.

3.3 Time Synchronous Averaging

3.3.1 TSA Signals

The time synchronous average (TSA), x_{TSA}, can be defined as a periodicity feature of the vibration signal. It extracts periodic waveforms from noisy vibration data, as first introduced by Braun (1975), and it is still of interest as part of research on monitoring the condition of rotating machines, especially gearboxes (Wegerich 2004; Combet and Gelman 2007; Bechhoefer and Kingsley 2009; Ha et al. 2016). Also, it has been used to diagnose bearing faults (McFadden and Toozhy 2000; Christian et al. 2007; Ahamed et al. 2014). This technique can be performed by averaging the time domain vibration signal in synchronisation with the sampling frequency or sampling time used to acquire vibration signals from the rotating machine of interest. The mathematical expression is shown in Eq. (3.38),

$$x_{TSA} = \frac{1}{N} \sum_{n=0}^{N-1} x(t + nT) \tag{3.38}$$

where T is the period of averaging and N is the number of sample points.

This is considered one of the most useful techniques for vibration signal analysis; it removes any periodic events that are not synchronous with the specific sampling frequency or sampling time of a vibration signal. This also may allow time domain vibration signals that are concealed in noise to be viewed. For example, Figure 3.6a presents a typical vibration signal of a roller bearing with an IR fault condition, and Figure 3.6b shows its corresponding 12 kHz synchronised signal in red and time-synchronous averaging signal in blue.

3.3.2 Residual Signal (RES)

The residual signal (RES), x_{RES}, is defined as the signal that results from subtracting the TSA from the synchronised vibration signal. Zakrajsek et al. (1993) developed two vibration-based analysis method based on the RES – NA4 and NA4* – which are successfully used to monitor the condition of gearboxes (McClintic et al. 2000; Sait and Sharaf-Eldeen 2011).

3.3.2.1 NA4

The NA4 of the RES, $x_{RES-NA4}$, is defined as the ratio of the fourth statistical moment of the RES to the square of the current time-averaged variance of the RES. The mathematical expression is shown in Eq. (3.39):

$$x_{RES-NA4} = \frac{N \sum_{i=1}^{N} (r_i - \bar{r})^4}{\left\{ \frac{1}{M} \sum_{j=1}^{M} \left[\sum_{i=1}^{N} (r_{ij} - r_j^2) \right] \right\}^2} \tag{3.39}$$

Here r is the RES, $\bar{r}$ is the mean of the RES, N is the total number of sample points, i is the data point number in the time series, j is the time record number in the run ensemble, and M is the number of records in the current time.

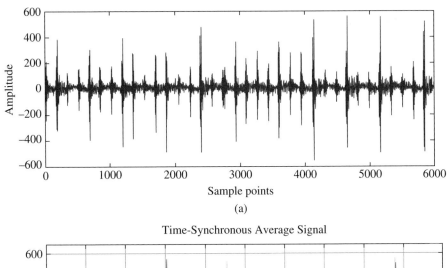

Sample points

(a)

Time-Synchronous Average Signal

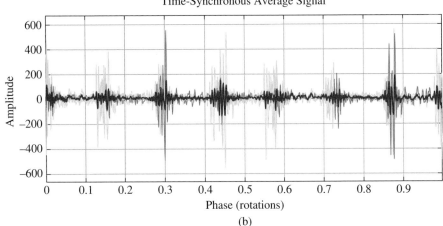

Phase (rotations)

(b)

Figure 3.6 (a) Time-domain vibration signal of a roller bearing with an inner race fault condition; (b) its time-synchronous average signal. (See insert for a colour representation of this figure.)

3.3.2.2 NA4*

NA4* is an improved version of NA4. The NA4* of the RES, $x_{RES-NA4^*}$, is defined as the ratio of the fourth statistical moment of the RES to the squared variance of the RES for a gear in healthy condition. This can be expressed using the following equation:

$$x_{RES-NA4^*} = \frac{N \sum_{i=1}^{N} (r_i - \overline{r})^4}{(\sigma_{r-h}^2)^2} \tag{3.40}$$

where r is the RES, $\overline{r}$ is the mean of the RES, N is the total number of sample points, $r - h$ is the RES for a gearbox in healthy condition, and (σ_{r-h}^2) is the variance of the RES for a gearbox in healthy condition.

3.3.3 Difference Signal (DIFS)

The difference signal (DIFS), x_{DIFS}, can be defined as the signal that results from removing the regular meshing, i.e. the shaft frequency and harmonics, primary meshing frequency, and harmonics, along with their first-order sidebands, from the TSA signal x_{TSA} of the vibration signal. Several vibration analysis techniques based on the DIFS have been proposed for gear-fault detection: FM4, M6A, and M8A (Stewart 1977; Martin 1989; Zakrajsek et al. 1993). These are detailed in the following subsections.

3.3.3.1 FM4

The FM4 of the input vibration signal, $x_{DIFS-FM4}$, can be determined by first computing the DIFS d and then computing its normalised kurtosis. This can be expressed using the following equation,

$$x_{RES-FM4} = \frac{N\sum_{i=1}^{N}(d_i - \bar{d})^4}{\left[\sum_{i=1}^{N}(d_i - \bar{d})^2\right]^2} \tag{3.41}$$

where $\bar{d}$ is the mean of d, and N is the total number of sample points of the input vibration signal.

3.3.3.2 M6A

The M6A of the DIFS d, $x_{DIFS-MA6}$, can be computed by using the sixth moment normalised by the variance to the third power, as shown in the following equation:

$$x_{RES-M6A} = \frac{N^2\sum_{i=1}^{N}(d_i - \bar{d})^6}{\left[\sum_{i=1}^{N}(d_i - \bar{d})^2\right]^3} \tag{3.42}$$

The M6A was suggested as an indicator of surface damage on the components of rotating machines.

3.3.3.3 M8A

The M8A of the DIFS d, $x_{DIFS-M8A}$, can be computed by using the eighth moment normalised by the variance to the fourth power as shown in the following equation:

$$x_{RES-M8A} = \frac{N^3\sum_{i=1}^{N}(d_i - \bar{d})^8}{\left[\sum_{i=1}^{N}(d_i - \bar{d})^2\right]^4} \tag{3.43}$$

The M8A was suggested as an indicator of surface damage on the components of rotating machines.

3.4 Time Series Regressive Models

Model-based techniques for vibration monitoring can provide a means of detecting machine faults even if data are only available from the machine in its normal condition (McCormick et al. 1998). In regressive model-based vibration monitoring, the autoregressive (AR), autoregressive moving average (ARMA – also known as the mixture of AR and the moving average (MA) – and autoregressive integrated moving average

(ARIMA) methods have been the most-used techniques. The subsequent subsections discuss those types of models in more detail.

3.4.1 AR Model

The autoregressive model (AR), AR (p), is basically a linear regression analysis of the current signal values, i.e. the estimated signal values, of the vibration time series against previous values of the time series, i.e. the values of the measured time series signal. This can be expressed mathematically as in Eq. (3.44),

$$x_t = a_1 x_{t-1} + a_2 x_{t-2} + \dots + a_p x_{t-p} + \mu_t = \mu_t + \sum_{i=1}^{p} a_i x_{t-i} \tag{3.44}$$

where x_t is the stationary signal, a_1-a_p are the model parameters, μ_t is white noise (also called random shock or innovation), and p is the model order. The parameter estimation of AR modelling can be accomplished using the covariance method, modified covariance method, or Yule-Walker method. Of these methods, the Yule-Walker method is commonly used to estimate the parameters of the AR model for a given vibration signal, and the model order can be chosen using Akaike's information criterion (AIC) (Garga et al. 1997; McCormick et al. 1998; Endo and Randall 2007; Ayaz 2014). Figures 3.7a–d show examples of the original autoregressive signal and the linear predictor-based estimated vibration signal for a brand-new bearing (see Figure 3.1a) using different values of p.

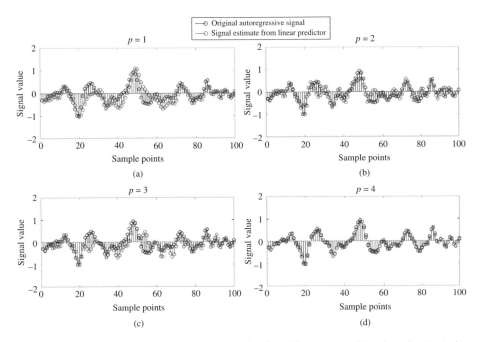

Figure 3.7 Examples of the original autoregressive signal and the linear predictor-based estimated signal of a vibration signal from a brand-new bearing using different values of p. (See insert for a colour representation of this figure.)

3.4.2 MA Model

The MA model, MA (q), is basically a linear regression analysis where the current signal series is modelled based on the weighted sum of values of the time series. This can be expressed as in Eq. (3.45),

$$x_t = b_1\mu_{t-1} + b_2\mu_{t-2} + \ldots + b_q\mu_{t-q} + \mu_t = \mu_t + \sum_{i=1}^{q} b_i\,\mu_{t-i} \qquad (3.45)$$

where b_1–b_q are the model parameters, μ is white noise, and q is the model order.

3.4.3 ARMA Model

The ARMA model, ARMA (p, q), is a combination of AR (p) and MA (q) to achieve better flexibility in fitting actual time series (Box et al. 2015). This can be expressed mathematically as in Eq. (3.46),

$$x_t = a_1 x_{t-1} + \ldots + a_p x_{t-p} + \mu_t + b_1\mu_{t-1} + \ldots + b_q\mu_{t-q}$$

$$= \mu_t + \sum_{i=1}^{p} a_i x_{t-i} + \sum_{i=1}^{q} b_i\mu_{t-i} \qquad (3.46)$$

where a_1–a_p and b_1–b_q are the model parameters, μ is white noise, and p and q are the model orders for AR and MA, respectively.

3.4.4 ARIMA Model

The AR and ARMA models just discussed can be used for stationary time series vibration data. Box et al. (2015) introduced the idea of using the ARMA model for applications of non-stationary time series by applying the differencing technique. This is done by computing the difference between consecutive observations on the non-stationary time series to produce the stationary time series. This developed model is normally called ARIMA (p, D, q), which is a combination of AR (P), integration (I), and MA (q), where p and q are the model orders for AR and MA, respectively, and D is the number of differencing operators. This can be represented as:

$$\Delta^D x_t = a_1 \Delta^D x_{t-1} + \ldots + a_p \Delta^D x_{t-p} + \mu_t + b_1\mu_{t-1} + \ldots + b_q\mu_{t-q} \qquad (3.47)$$

where Δ^D is the difference, a_1–a_p and b_1–b_q are the model parameters, and μ is white noise.

There are also many other types of regressive models. Readers who are interested in more details of the algorithms introduced here and other types of algorithms are referred to (Palit and Popovic 2006; Box et al. 2015).

Numerous studies have used AR models to diagnose bearing faults. For instance, Baillie and Mathew compared three techniques of AR modelling for their performance and reliability under conditions of several vibration signal lengths of induced faults in rolling element (RE) bearings. It has been shown that AR modelling has the ability to diagnosis faults from a limited amount of vibration data, and consequently it might be ideal for classifying health conditions in slow or varying-speed machinery (Baillie and Mathew 1996). In a study by McCormick and colleagues (McCormick et al. 1998), the application of periodic time-variant AR models to the problems of fault detection and diagnosis in machinery has been demonstrated. In this study, a comparison was made between time-invariant and time-varying systems, and it was found that the time-varying model performed better, detecting a cage (CA) fault that the time-invariant model failed to detect. In addition, Junsheng et al. (2006) proposed a fault feature extraction method based on empirical mode decomposition (EMD) technique and AR model for roller bearings fault diagnosis. Poulimenos and Fassois (2006) presented a survey and comparisons of seven time-dependent ARMA-based methods for non-stationary random vibration modelling and analysis. Furthermore, Li et al. (2007) proposed a higher-order-statistics-based ARMA model for detecting bearing faults that has the ability to eliminate the effects of noise and obtain clearer information. Pham and Yang (2010) introduced a hybrid model of ARMA and generalised autoregressive conditional heteroscedasticity (GARCH) to estimate and forecast the machine state based on vibration signals. Moreover, Ayaz (2014) investigated the effectiveness of AR modelling with orders in the range of 1–200 for extracting bearing fault characteristics from vibration signals. Lu et al. (2018) presented a prognostic algorithm for the degradation of bearings using the variable forgetting factor recursive least-square (VFF-RLS) combined with an ARMA model.

3.5 Filter-Based Methods

The acquired vibration signal often contains background noise and interference with unknown frequencies. Many researchers have utilised filter-based methods to remove noise and isolate signals from raw signals. These include demodulation, the Prony model, and adaptive noise cancellation (ANC). The subsequent subsections discuss those methods in more detail.

3.5.1 Demodulation

As illustrated in the previous chapter, characteristic of vibration frequencies can be computed using standard formulas, e.g. the bearing fundamental defect frequency (BFDF). However, the acquired vibration signals may contain amplitude, phase, or frequency modulation (FM) caused by faults in the rotating machine or by noise. This often makes fault diagnosis difficult. Modulation happens when the amplitude or frequency of a signal wave, also called the carrier signal, is changed according to the intensity of the signal. For example, Figure 3.8 shows the modulation of a sine wave: Figure 3.8a shows a sine wave signal in blue with 20 Hz frequency and amplitude of value 8 that represents the carrier (the original signal), and another sine wave in red with 4 Hz frequency and amplitude of value 2 (the modulating signal) that typically has a lower amplitude and frequency than the carrier signal; Figure 3.8b shows the corresponding modulated signal with respect to amplitude modulation (AM), where the amplitude of the sine wave changes periodically; and Figure 3.8c shows the corresponding modulated signal with respect to frequency modulation (FM), where the frequency of the sine wave changes periodically.

The amplitude and frequency modulation of rotating machinery vibration signals, which are often caused by fault conditions in the machinery, are discussed in several studies. For instance, by way of illustration, Nakhaeinejad and Ganeriwala (2009) show

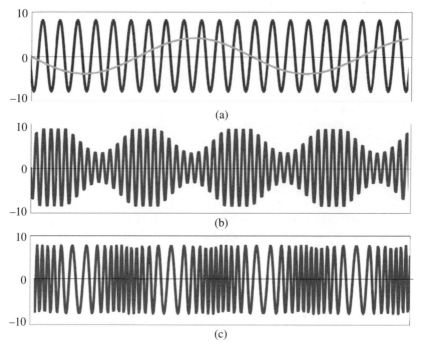

Figure 3.8 Example of sine wave amplitude and frequency modulation. (See insert for a colour representation of this figure.)

that several parallel and angular misalignments can generate low-frequency modulations in rotating machinery vibration signals. According to (Randall 2011), torque fluctuations result in phase and frequency modulation of the rotor speed in synchronous motors and induction motors, respectively. Another example of phase and frequency modulation is shaft torsional vibration. Also, an example of AM is the amplitude of the resulting vibration of tooth-meshing frequency in gears, which varies with load fluctuations in service. The AM within a single frequency of a signal often results in pairs of sidebands in the vibration spectrum, which are located around each modulated frequency component. Some types of rotating machine faults, e.g. gear faults, can be diagnosed by detailed analysis of these sidebands, which we will detail in the next chapter. Moreover, the characteristic vibration signatures of rolling bearings are often generated in the form of modulation of the BFDF (Lacey 2008). Figure 3.9 depicts some typical time series plots of real vibration data for roller bearings in six conditions. These include two normal conditions, brand-new (NO) and worn but undamaged (NW); and four fault conditions, inner race (IR) fault, outer race (OR) fault, rolling element (RE) fault, and cage (CA) fault. Depending on the fault conditions, the defects modulate the vibration signals with their own patterns. The IR and OR fault conditions have a fairly periodic signal; the RE fault may or may not be periodic, depending upon several factors including the level of damage to the rolling element, the load of the bearing, and the track that the ball describes within the raceway itself. The CA fault generates a random distortion, which also depends on the degree of damage and the bearing load.

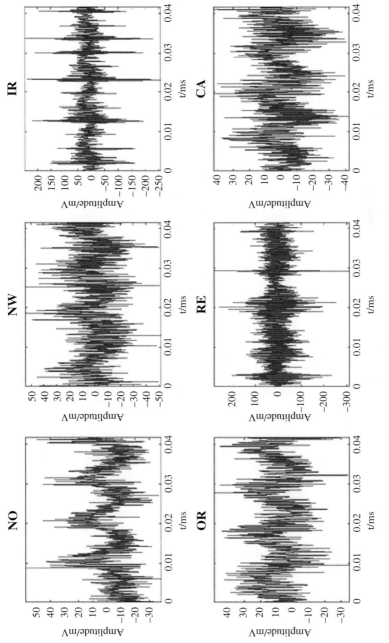

Figure 3.9 Typical time domain vibration signals for six different bearing health conditions. Source: (Ahmed and Nandi 2018).

The demodulation process is the inverse of the modulation process and can be amplitude demodulation or phase demodulation. That amplitude demodulation process (also called high-frequency resonance, resonance demodulation, or envelop analysis) separates low-level frequencies from high-frequency background noise (Singh and Vishwakarma 2015). The demodulation process consists of three steps: (i) the raw vibration signal is band-pass filtered; (ii) it is rectified or enveloped by folding the bottom part of the time waveform over the top part, usually using the Hilbert-Huang transform (HHT); and finally (iii) it is transformed utilising a fast Fourier transform (FFT). Many researchers have utilised envelop analysis to diagnose bearing faults (Toersen 1998; Randall et al. 2000; Konstantin-Hansen 2003; Patidar and Soni 2013). More details regarding this process are provided in the next chapter.

3.5.2 Prony Model

Similar to the AR, ARMA, and ARIMA models, the Prony model attempts to fit a model to the sampled time domain signal. The Prony analysis estimates the model parameters by fitting the summation of the damped sinusoidal components to an equally spaced sample. It computes modal information such as amplitude, damping, frequency, and phase shift, which can be utilised for fault diagnosis or to recover the original signal. Given a time series signal $x(t)$ with N samples, its Prony model can be computed using the following equation:

$$\widehat{x}(t) = \sum_{i=1}^{L} A_i e^{-\sigma_i t} \cos(w_i t + \phi_i), \tag{3.48}$$

where L is the total number of damped exponential components, $\widehat{x}(t)$ is the estimation of the observed signal $x(t)$, A_i is the amplitude, $w_i = 2\pi f_i t$ is the angular frequency, σ_i is the damping coefficients, and ϕ_i is the phase shift of the ith sample (Tawfik and Morcos 2001). Equation (3.48) can be rewritten as Eq. (3.49),

$$\widehat{x}(t) = \sum_{i=1}^{n} H_i Q_i^k, \tag{3.49}$$

where H_i and Q_i can be computed using Eqs. (3.50) and (3.51), respectively, $1 < n > N$, and $k = 0, 2, \ldots . N - 1$:

$$H_i = \frac{A_i}{2} e^{j\phi_i} \tag{3.50}$$

$$Q_i = e^{(\sigma_i + j2\pi f_i)T} \tag{3.51}$$

where T is the sampled signal duration. The output of Eq. (3.49), which fulfils its own characteristic equation, is assumed to satisfy Eq. (3.52):

$$\widehat{x}_k = a_1 \widehat{x}_{k-1} + a_1 \widehat{x}_{k-1} + \ldots + a_n \widehat{x}_{k-n} \tag{3.52}$$

the characteristic equation can be represented by Eq. (3.53):

$$p(Q) = Q^n - (a_1 Q^{n-1} + a_2 Q^{n-2} + \ldots + a_n) \tag{3.53}$$

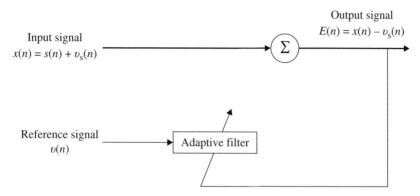

Figure 3.10 Diagram of a typical ANC process to separate two signals (Randall 2011).

This method formed the central focus of a study by Chen and Mechefske (2002) in which the authors found that the Prony model-based method can be an efficient technique for diagnosing machine faults using short-duration transient vibration signals.

3.5.3 Adaptive Noise Cancellation (ANC)

An adaptive filter is utilised to model the relationship between two signals in an iterative manner. ANC is a technique to remove the background noise contained in the time waveform. This technique utilises a primary input signal $x(n)$, which contains background noise $v_s(n)$ and a source signal $s(n)$, e.g. a vibration signal from one component of a rotating machine and an unwanted signal $v(n)$ that is also called a reference signal, e.g. a signal from another component of the machine. To obtain the estimated signal, the reference signal is adaptively filtered and subtracted from the primary input signal (Widrow et al. 1975). This reference signal is often acquired from one or more sensors located at points in the noise field where the signal of interest is weak or undetectable (Benesty and Huang 2013). A diagram of a typical ANC process to separate two signals is depicted in Figure 3.10, where the input signal $x(n)$ is the sum of the noise $v_s(n)$ and the source signal $s(n)$. To eliminate the noise contained in the input signal, the reference signal $v(n)$ is passed through an adaptive filter to generate an estimated reference signal $\hat{v}(n)$ that is subtracted from the input signal $x(n)$ to obtain the error signal $E(n)$. The system output $E(n)$ is fed back to the adaptive filter, where the filter parameters are adjusted to minimise the output $E(n)$.

ANC has been used by many studies to diagnose roller bearing faults (Chaturvedi and Thomas 1982; Li and Fu 1990; Lu et al., 2009; Elasha et al. 2016). However, the problem with ANC in machine fault diagnosis is that it is not always easy to identify the reference signal $v(n)$ that is correlated with the noise $v_s(n)$. To overcome this problem, a further development of ANC was formulated using a delayed version of the input signal (Widrow et al. 1975; Ho and Randall 2000; Ruiz-Cárcel et al. 2015). This version was called self-adaptive noise cancellation (S-ANC), and a typical S-ANC process to separate two signals is depicted in Figure 3.11.

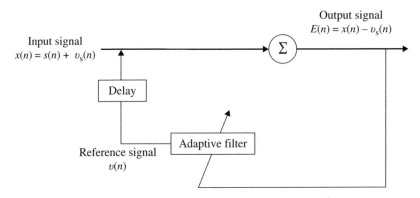

Figure 3.11 Diagram of a typical S-ANC process to separate two signals (Randall 2011).

3.6 Stochastic Parameter Techniques

Stochastic parameters such as chaos, correlation dimensions, and thresholding methods, i.e. soft threshold and hard threshold, are considered effective techniques to analyse time series vibration signals. For instance, Jiang and Shao proposed a fault diagnosis method for rolling bearings using chaos theory (Jiang and Shao 2014). Logan and Mathew (1996) introduced the correlation dimension technique to diagnose vibration faults in rolling bearings. Yang et al. (2007) applied the capacity dimension, information dimension, and correlation dimension to classify fault conditions of rolling bearings. Moreover, the application of thresholding methods for rolling bearings is studied in (Lin and Qu 2000).

3.7 Blind Source Separation (BSS)

Vibration signals acquired from rotating machines are often a mixture of vibrations from several sources in the machine and some background noise. This often makes the signal-to-noise ratio (SNR) very poor, and, consequently, feature extraction becomes difficult. This often happens in industrial environments. Blind source separation (BSS) is a signal-processing method that recovers the unobserved signals from a set of observations of numerous signal combinations (Gelle et al. 2003); this technique is useful in cases where there is a lack of knowledge about the different combinations of signals received by each sensor. The standard BSS model assumes the presence of L statistically independent signals $X(n) = [x_1(n), \dots, x_L(n)]$, the observations of a mixture of L signals $Y(n) = [y_1(n), \dots, y_L(n)]$, and additive noise $\aleph(n) = [n_1(n), \dots, n_L(n)]$ such that

$$Y(n) = f(X(n), X(n-1), \dots, X(0)) + \aleph(n) \tag{3.54}$$

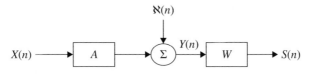

Figure 3.12 BSS model.

where f is a nonlinear and time-dependent function and $\aleph(n)$ is additive noise. In the case of a noise-free, time-independent mixing process, we can rewrite Eq. (3.54) as (3.55),

$$Y(n) = A * X(n) + \aleph(n) \tag{3.55}$$

where A is an unknown mixing matrix. The general model of BSS is shown Figure 3.12.
The BSS model estimates a separating matrix W such that,

$$S(n) = W * Y(n) = W * A * X(n) \tag{3.56}$$

More details of the mathematical formulation of BSS can be found in (Yu et al. 2013). Many researchers have considered BSS-based techniques for machine fault diagnosis. For example, Gelle et al. (2003) examined the recovery of vibration information acquired from a single rotating machine working in a noisy environment by separating the sensor signal from the influence of another working machine. Thus, they demonstrated that BSS can be employed as a pre-processing step for diagnosing faults in rotating machines using vibration signals. A BSS-based technique that involves signal separation in a context of spatially correlated noise is proposed in (Serviere and Fabry 2005). Chen et al. (2012) presented a fault-diagnosis method for rolling bearings based on BSS to separate signals collected from the rolling bearings and the gearbox.

3.8 Summary

This chapter has provided a review of monitoring the condition of vibration-based machines, and covered several vibration analysis techniques that have been used for machine fault diagnosis. Numerous techniques have been proposed using features extracted in the time domain from raw vibration signals using statistical parameters, e.g. peak-to-peak value, RMS, crest factor, skewness, and kurtosis; all of the studies presented in this chapter used multiple domain statistical techniques to extract features from the raw vibration data. Each study used at least six techniques, and some used more than 10 techniques to extract features from the raw vibration data. Of these techniques, kurtosis is used in all the mentioned studies. Moreover, skewness, shape factor, impulse factor, variance, CF, peak-to-peak, RMS, and mean are among the most-used techniques in these studies. In addition, this chapter has introduced other advanced techniques such as TSA, which is considered one of the most useful techniques for vibration signal analysis and removes any periodic events that are not

Table 3.2 Summary of some of the techniques introduced and their publically accessible software.

Algorithm name	Platform	Package	Function
Max	MATLAB	Signal processing toolbox; measurement and feature extraction; descriptive statistics	max
Min			min
Mean			mean
Peak-to-peak			Peak2peak
RMS			rms
STD			std
VAR			var
Crest factor			Peak2rms
Zero crossing	MATLAB	Bruecker (2016)	crossing
Skewness	MATLAB	Statistics and machine learning toolbox	skewness
Kurtosis			kurtosis
TSA	MATLAB	Signal-processing toolbox	tsa
AR-covariance			arcov
AR-Yule-Walker			aryule
Prony			prony
ARIMA	MATLAB	Econometric toolbox	arima
ANC	MATLAB	Clemens (2016)	Adapt-filt-tworef
S-ANC	MATLAB	NJJ (2018)	sanc
BSS	R	JADE and BSSasymp (Miettinen et al. 2017)	BSS
BSS	MATLAB	Gang (2015)	YGBSS

synchronous with a specific sampling frequency or the sampling time of a vibration signal; AR and ARIMA, which provide a means of detecting machine faults even if data are only available from the machine in its normal condition; and filter-based methods, which have been used by many researchers to remove noise and isolate signals from raw signals. Most of the techniques introduced and their publicly accessible software are summarised in Table 3.2.

References

Ahamed, N., Pandya, Y., and Parey, A. (2014). Spur gear tooth root crack detection using time synchronous averaging under fluctuating speed. *Measurement* 52: 1–11.

Ahmed, H. and Nandi, A.K. (2018). Compressive sampling and feature ranking framework for bearing fault classification with vibration signals. *IEEE Access* 6: 44731–44746.

Ali, J.B., Fnaiech, N., Saidi, L. et al. (2015). Application of empirical mode decomposition and artificial neural network for automatic bearing fault diagnosis based on vibration signals. *Applied Acoustics* 89: 16–27.

Ayaz, E. (2014). 1315. Autoregressive modeling approach of vibration data for bearing fault diagnosis in electric motors. *Journal of Vibroengineering*, 16 (5), pp. 2130–2138.

Baillie, D.C. and Mathew, J. (1996). A comparison of autoregressive modeling techniques for fault diagnosis of rolling element bearings. *Mechanical Systems and Signal Processing* 10 (1): 1–17.

Bechhoefer, E. and Kingsley, M. (2009). A review of time synchronous average algorithms. In: *Annual Conference of the Prognostics and Health Management Society*, 24–33. San Diego, CA: http://ftp.phmsociety.org/sites/phmsociety.org/files/phm_submission/2009/phmc_09_5.pdf.

Benesty, J. and Huang, Y. (eds.) (2013). *Adaptive signal processing: applications to real-world problems*. Springer Science & Business Media.

Box, G.E., Jenkins, G.M., Reinsel, G.C., and Ljung, G.M. (2015). *Time Series Analysis: Forecasting and Control*. Wiley.

Braun, S. (1975). The extraction of periodic waveforms by time domain averaging. *Acta Acustica united with Acustica* 32 (2): 69–77.

Bruecker, S. (2016). Crossing. Mathworks File Exchange Center. https://www.mathworks.com/matlabcentral/fileexchange/2432-crossing.

Chaturvedi, G.K. and Thomas, D.W. (1982). Bearing fault detection using adaptive noise cancelling. *Journal of Mechanical Design* 104 (2): 280–289.

Chebil, J., Hrairi, M., and Abushikhah, N. (2011). Signal analysis of vibration measurements for condition monitoring of bearings. *Australian Journal of Basic and Applied Sciences* 5 (1): 70–78.

Chen, C.Z., Meng, Q., Zhou, H., and Zhang, Y. (2012). Rolling bearing fault diagnosis based on blind source separation. *Applied Mechanics and Materials* 217: 2546–2549. Trans Tech Publications.

Chen, Z. and Mechefske, C.K. (2002). Diagnosis of machinery fault status using transient vibration signal parameters. *Modal Analysis* 8 (3): 321–335.

Christian, K., Mureithi, N., Lakis, A., and Thomas, M. (2007). On the use of time synchronous averaging, independent component analysis and support vector machines for bearing fault diagnosis. In: *Proceedings of the 1st International Conference on Industrial Risk Engineering, Montreal, December 2007*, 610–624. https://pdfs.semanticscholar.org/87fb/93bae0ae23afe8f39205d34b98f24660c97e.pdf.

Clemens, R. (2016). Noise canceling adaptive filter. Mathworks File Exchange Center. https://www.mathworks.com/matlabcentral/fileexchange/10447-noise-canceling-adaptive-filter.

Combet, F. and Gelman, L. (2007). An automated methodology for performing time synchronous averaging of a gearbox signal without speed sensor. *Mechanical Systems and Signal Processing* 21 (6): 2590–2606.

Davies, A. (1998). Visual inspection systems. In: *Handbook of Condition Monitoring*, 57–77. Dordrecht: Springer.

Elasha, F., Mba, D., and Ruiz-Carcel, C. (2016). A comparative study of adaptive filters in detecting a naturally degraded bearing within a gearbox. *Case Studies in Mechanical Systems and Signal Processing* 3: 1–8.

Endo, H. and Randall, R.B. (2007). Enhancement of autoregressive model based gear tooth fault detection technique by the use of minimum entropy deconvolution filter. *Mechanical Systems and Signal Processing* 21 (2): 906–919.

Gang, Y. (2015). A novel BSS. Mathworks File Exchange Center. https://www.mathworks
.com/matlabcentral/fileexchange/50867-a-novel-bss.

Garga, A.K., Elverson, B.T., and Lang, D.C. (1997). AR modeling with dimension reduction
for machinery fault classification. In: *Critical Link: Diagnosis to Prognosis*, 299–308.
Haymarket: http://php.scripts.psu.edu/staff/k/p/kpm128/pubs/ResDifAnalysisFinal
.PDF.

Gelle, G., Colas, M., and Servière, C. (2003). Blind source separation: a new pre-processing
tool for rotating machines monitoring? *IEEE Transactions on Instrumentation and
Measurement* 52 (3): 790–795.

Guo, H., Jack, L.B., and Nandi, A.K. (2005). Feature generation using genetic programming
with application to fault classification. *IEEE Transactions on Systems, Man, and
Cybernetics, Part B (Cybernetics)* 35 (1): 89–99.

Ha, J.M., Youn, B.D., Oh, H. et al. (2016). Autocorrelation-based time synchronous
averaging for condition monitoring of planetary gearboxes in wind turbines. *Mechanical
Systems and Signal Processing* 70: 161–175.

Ho, D. and Randall, R.B. (2000). Optimisation of bearing diagnostic techniques using
simulated and actual bearing fault signals. *Mechanical Systems and Signal Processing* 14
(5): 763–788.

Jack, L.B. and Nandi, A.K. (2000). Genetic algorithms for feature selection in machine
condition monitoring with vibration signals. *IEE Proceedings-Vision, Image and Signal
Processing* 147 (3): 205–212.

Jiang, Y.L. and Shao, Y.X. (2014). Fault diagnosis of rolling bearing based on fuzzy neural
network and chaos theory. *Vibroengineering PROCEDIA* 4: 211–216.

Junsheng, C., Dejie, Y., and Yu, Y. (2006). A fault diagnosis approach for roller bearings
based on EMD method and AR model. *Mechanical Systems and Signal Processing* 20 (2):
350–362.

Kankar, P.K., Sharma, S.C., and Harsha, S.P. (2011). Fault diagnosis of ball bearings using
machine learning methods. *Expert Systems with Applications* 38 (3): 1876–1886.

Konstantin-Hansen, H. (2003). Envelope analysis for diagnostics of local faults in rolling
element bearings. *Bruel & Kjaer*: 1–8.

Lacey, S.J. (2008). An overview of bearing vibration analysis. *Maintenance & Asset
Management* 23 (6): 32–42.

Lakshmi Pratyusha, P., Shanmukha Priya, V., and Naidu, V.P.S. (2014). Bearing health
condition monitoring: time domain analysis. *International Journal of Advanced Research
in Electrical, Electronics and Instrumentation Engineering*: 75–82.

Li, F.C., Ye, L., Zhang, G.C., and Meng, G. (2007). Bearing fault detection using
higher-order statistics based ARMA model. *Key Engineering Materials* 347: 271–276.
Trans Tech Publications.

Li, Z. and Fu, Y. (1990). Adaptive noise cancelling technique and bearing fault diagnosis.
Journal of Aerospace Power 5: 199–203.

Lin, J. and Qu, L. (2000). Feature extraction based on Morlet wavelet and its application for
mechanical fault diagnosis. *Journal of Sound and Vibration* 234 (1): 135–148.

Logan, D. and Mathew, J. (1996). Using the correlation dimension for vibration fault
diagnosis of rolling element bearings—I. Basic concepts. *Mechanical Systems and Signal
Processing* 10 (3): 241–250.

Lu, B., Nowak, M., Grubic, S., and Habetler, T.G. (2009). An adaptive noise-cancellation
method for detecting generalized roughness bearing faults under dynamic load

conditions. In: *Proceedings of the Energy Conversion Congress and Exposition,* 1091–1097. IEEE.

Lu, Y., Li, Q., Pan, Z., and Liang, S.Y. (2018). Prognosis of bearing degradation using gradient variable forgetting factor RLS combined with time series model. *IEEE Access* 6: 10986–10995.

Martin, H.R. (1989). Statistical moment analysis as a means of surface damage detection. In: *Proceedings of the 7th International Modal Analysis Conference,* 1016–1021. Las Vegas, USA: Society for Experimental Mechanics.

McClintic, K., Lebold, M., Maynard, K. et al. (2000). Residual and difference feature analysis with transitional gearbox data. In: *54th Meeting of the Society for Machinery Failure Prevention Technology,* 1–4. Virginia Beach, VA.

McCormick, A.C. and Nandi, A.K. (1996). A comparison of artificial neural networks and other statistical methods for rotating machine condition classification. In: *Colloquium Digest-IEE,* 2–2. IEE Institution of Electrical Engineers.

McCormick, A.C. and Nandi, A.K. (1997). Neural network autoregressive modeling of vibrations for condition monitoring of rotating shafts. In: *International Conference on Neural Networks, 1997,* vol. 4, 2214–2218. IEEE.

McCormick, A.C. and Nandi, A.K. (1998). Cyclostationarity in rotating machine vibrations. *Mechanical systems and signal processing* 12 (2): 225–242.

McCormick, A.C., Nandi, A.K., and Jack, L.B. (1998). Application of periodic time-varying autoregressive models to the detection of bearing faults. *Proceedings of the Institution of Mechanical Engineers, Part C: Journal of Mechanical Engineering Science* 212 (6): 417–428.

McFadden, P.D. and Toozhy, M.M. (2000). Application of synchronous averaging to vibration monitoring of rolling element bearings. *Mechanical Systems and Signal Processing* 14 (6): 891–906.

Miettinen, J., Nordhausen, K., and Taskinen, S. (2017). Blind source separation based on joint diagonalization in R: the packages JADE and BSSasymp. *Journal of Statistical Software* vol 76: 1–31.

Nakhaeinejad, M. and Ganeriwala, S. (2009). Observations on dynamic responses of misalignments. Technological notes. SpectraQuest Inc.

Nayana, B.R. and Geethanjali, P. (2017). Analysis of statistical time-domain features effectiveness in identification of bearing faults from vibration signal. *IEEE Sensors Journal* 17 (17): 5618–5625.

NJJ1. (2018). Self adaptive noise cancellation. Mathworks File Exchange Center. https://www.mathworks.com/matlabcentral/fileexchange/65840-sanc-x-l-mu-delta.

Palit, A.K. and Popovic, D. (2006). *Computational Intelligence in Time Series Forecasting: Theory and Engineering Applications.* Springer Science & Business Media.

Patidar, S. and Soni, P.K. (2013). An overview on vibration analysis techniques for the diagnosis of rolling element bearing faults. *International Journal of Engineering Trends and Technology (IJETT)* 4 (5): 1804–1809.

Pham, H.T. and Yang, B.S. (2010). Estimation and forecasting of machine health condition using ARMA/GARCH model. *Mechanical Systems and Signal Processing* 24 (2): 546–558.

Poulimenos, A.G. and Fassois, S.D. (2006). Parametric time-domain methods for non-stationary random vibration modelling and analysis—a critical survey and comparison. *Mechanical Systems and Signal Processing* 20 (4): 763–816.

Prieto, M.D., Cirrincione, G., Espinosa, A.G. et al. (2013). Bearing fault detection by a novel condition-monitoring scheme based on statistical-time features and neural networks. *IEEE Transactions on Industrial Electronics* 60 (8): 3398–3407.

Randall, R.B. (2011). *Vibration-Based Condition Monitoring: Industrial, Aerospace and Automotive Applications*. Wiley.

Randall, R.B., Antoni, J., and Chobsaard, S. (2000). A comparison of cyclostationary and envelope analysis in the diagnostics of rolling element bearings. In: *Proceedings of the 2000 IEEE International Conference on Acoustics, Speech, and Signal Processing*, vol. 6, 3882–3885. IEEE.

Rauber, T.W., de Assis Boldt, F., and Varejão, F.M. (2015). Heterogeneous feature models and feature selection applied to bearing fault diagnosis. *IEEE Transactions on Industrial Electronics* 62 (1): 637–646.

Rojas, A. and Nandi, A.K. (2006). Practical scheme for fast detection and classification of rolling-element bearing faults using support vector machines. *Mechanical Systems and Signal Processing* 20 (7): 1523–1536.

Ruiz-Cárcel, C., Hernani-Ros, E., Chandra, P. et al. (2015). Application of linear prediction, self-adaptive noise cancellation, and spectral kurtosis in identifying natural damage of rolling element bearing in a gearbox. In: *Proceedings of the 7th World Congress on Engineering Asset Management (WCEAM 2012)*, 505–513. Cham: Springer.

Saimurugan, M., Ramachandran, K.I., Sugumaran, V., and Sakthivel, N.R. (2011). Multi component fault diagnosis of rotational mechanical system based on decision tree and support vector machine. *Expert Systems with Applications* 38 (4): 3819–3826.

Sait, A.S. and Sharaf-Eldeen, Y.I. (2011). A review of gearbox condition monitoring based on vibration analysis techniques diagnostics and prognostics. In: *Rotating Machinery, Structural Health Monitoring, Shock and Vibration, Volume 5*, 307–324. New York, NY: Springer.

Samanta, B., Al-Balushi, K.R., and Al-Araimi, S.A. (2003). Artificial neural networks and support vector machines with genetic algorithm for bearing fault detection. *Engineering Applications of Artificial Intelligence* 16 (7–8): 657–665.

Sassi, S., Badri, B., and Thomas, M. (2007). A numerical model to predict damaged bearing vibrations. *Journal of Vibration and Control* 13 (11): 1603–1628.

Saxena, A. and Saad, A. (2006). Genetic algorithms for artificial neural net-based condition monitoring system design for rotating mechanical systems. In: *Applied Soft Computing Technologies: The Challenge of Complexity*, 135–149. Berlin, Heidelberg: Springer.

Serviere, C. and Fabry, P. (2005). Principal component analysis and blind source separation of modulated sources for electro-mechanical systems diagnostic. *Mechanical Systems and Signal Processing* 19 (6): 1293–1311.

Singh, S. and Vishwakarma, D.M. (2015). A review of vibration analysis techniques for rotating machines. *International Journal of Engineering Research & Technology* 4 (3): 2278–0181.

Sreejith, B., Verma, A.K., and Srividya, A. (2008). Fault diagnosis of rolling element bearing using time-domain features and neural networks. In: *IEEE Region 10 and the Third International Conference on Industrial and Information Systems, 2008. ICIIS 2008*, 1–6. IEEE.

Stewart, R.M. (1977). *Some Useful Data Analysis Techniques for Gearbox Diagnostics*. University of Southampton.

Sugumaran, V. and Ramachandran, K.I. (2007). Automatic rule learning using decision tree for fuzzy classifier in fault diagnosis of roller bearing. *Mechanical Systems and Signal Processing* 21 (5): 2237–2247.

Sugumaran, V. and Ramachandran, K.I. (2011). Effect of number of features on classification of roller bearing faults using SVM and PSVM. *Expert Systems with Applications* 38 (4): 4088–4096.

Sun, Q., Chen, P., Zhang, D., and Xi, F. (2004). Pattern recognition for automatic machinery fault diagnosis. *Journal of Vibration and Acoustics* 126 (2): 307–316.

Tahir, M.M., Khan, A.Q., Iqbal, N. et al. (2017). Enhancing fault classification accuracy of ball bearing using central tendency based time domain features. *IEEE Access* 5: 72–83.

Tawfik, M.M. and Morcos, M.M. (2001). ANN-based techniques for estimating fault location on transmission lines using Prony method. *IEEE Transactions on Power Delivery* 16 (2): 219–224.

Tian, X., Cai, L., and Chen, S. (2015). Noise-resistant joint diagonalization independent component analysis based process fault detection. *Neurocomputing* 149: 652–666.

Toersen, H. (1998). Application of an envelope technique in the detection of ball bearing defects in a laboratory experiment. *Tribotest* 4 (3): 297–308.

Wegerich, S.W. (2004). Similarity based modeling of time synchronous averaged vibration signals for machinery health monitoring. In: *2004 IEEE Aerospace Conference Proceedings, 2004*, vol. 6, 3654–3662. IEEE.

Widrow, B., Glover, J.R., McCool, J.M. et al. (1975). Adaptive noise cancelling: principles and applications. *Proceedings of the IEEE* 63 (12): 1692–1716.

William, P.E. and Hoffman, M.W. (2011). Identification of bearing faults using time domain zero-crossings. *Mechanical Systems and Signal Processing* 25 (8): 3078–3088.

Yang, J., Zhang, Y., and Zhu, Y. (2007). Intelligent fault diagnosis of rolling element bearing based on SVMs and fractal dimension. *Mechanical Systems and Signal Processing* 21 (5): 2012–2024.

Yiakopoulos, C.T., Gryllias, K.C., and Antoniadis, I.A. (2011). Rolling element bearing fault detection in industrial environments based on a K-means clustering approach. *Expert Systems with Applications* 38 (3): 2888–2911.

Yu, X., Hu, D., and Xu, J. (2013). *Blind Source Separation: Theory and Applications*. Wiley.

Zakrajsek, J.J., Townsend, D.P., and Decker, H.J. (1993). An analysis of gear fault detection methods as applied to pitting fatigue failure data. No. NASA-E-7470. National Aeronautics and Space Administration.

Zhang, L., Jack, L.B., and Nandi, A.K. (2005). Fault detection using genetic programming. *Mechanical Systems and Signal Processing* 19 (2): 271–289.

Zhen, D., Wang, T., Gu, F., and Ball, A.D. (2013). Fault diagnosis of motor drives using stator current signal analysis based on dynamic time warping. *Mechanical Systems and Signal Processing* 34 (1–2): 191–202.

4

Frequency Domain Analysis

4.1 Introduction

As described in Chapter 3, various time domain–based techniques are used for vibration signal analysis, such as mean, peak, kurtosis, etc.; these deal with the sum of included sine waves in the raw vibration measurements. In the frequency domain, each sine wave will be presented as a spectral component (see Figure 4.1). Frequency analysis, also called spectral analysis, is one of the most commonly used vibration analysis techniques for monitoring the condition of machines. In fact, frequency domain analysis techniques have the ability to divulge information based on frequency characteristics that are not easily observed in the time domain. The measured time domain vibration signals are often generated by several components of a rotating machine, e.g. bearings, shaft, and fan, where a single motion of a component produces a sine wave with a single frequency and amplitude and other components add further frequencies. In other words, each component of a rotating machine produces a single frequency. However, we do not often see these produced frequencies individually in the measured signal: we see a summation of the signals that the sensor measured. The spectrum of the frequency components generated from the time domain waveforms makes it easier to see each source of vibration (see Figure 4.1).

As mentioned in Chapter 2, rotating machine components generate various types of vibration signals in the time domain: random signals and periodic signals, for both healthy and faulty conditions. The frequency representations of these signals in the frequency domain are often performed using Fourier analysis, which is generally classified into three types: Fourier series, continuous Fourier transform (CFT), and discrete Fourier transform (DFT). The basic idea of Fourier analysis was discovered in the nineteenth century by the French mathematician J. Fourier, who showed that any periodic function can be expressed as a summation of complex exponential functions. Of these types, DFT is an important tool in the frequency analysis of discrete time signals (Van Loan 1992; Diniz et al. 2010).

In this chapter, we will focus on basic concepts and methodologies of Fourier analysis that are commonly used to transform time domain vibration signals to the frequency domain. In addition, this chapter gives an explanation of different techniques that can be used to extract various frequency spectrum features that can more efficiently represent a machine's health.

Condition Monitoring with Vibration Signals: Compressive Sampling and Learning Algorithms for Rotating Machines, First Edition. Hosameldin Ahmed and Asoke K. Nandi.
© 2020 John Wiley & Sons Ltd. Published 2020 by John Wiley & Sons Ltd.

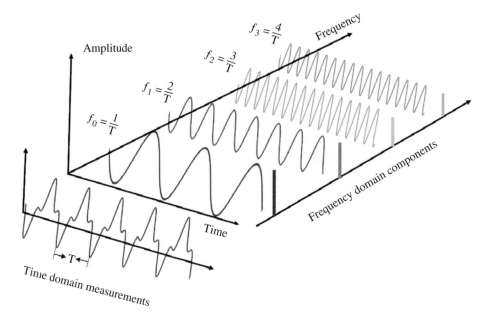

Figure 4.1 Time domain measurements vs. frequency domain measurements. Source: (Brandt 2011). (See insert for a colour representation of this figure.)

4.2 Fourier Analysis

4.2.1 Fourier Series

Fourier analysis, also called harmonic analysis, of a periodic signal $x(t)$, is the decomposition of the series into a summation of sinusoidal components, where each sinusoid has a specific amplitude and phase. A periodic signal contains frequencies that are integer harmonics of the fundamental frequency $f_0 = 1/T$, where T is the period of the signal. Thus, the only frequencies present in the signal are $\frac{1}{T}, \frac{2}{T}, \ldots, \frac{n}{T}$ (see Figure 4.1). The Fourier series of a periodic signal $x(t)$ can be represented as a linear combination of all sine and cosine functions that have the same period T. This can be mathematically expressed using the following equation:

$$x(t) = \frac{1}{2}a_0 + \sum_{k=1}^{\infty} a_k \, cos(kw_0t) + \sum_{k=1}^{\infty} b_k \, sin(kw_0t) \tag{4.1}$$

where a_k and b_k are the coefficients, which can be calculated using Eqs. (4.2) and (4.3),

$$a_k = \frac{2}{T} \int_{-\frac{T}{2}}^{\frac{T}{2}} x(t) \, cos(kw_0t) \, dt, \quad k = 0, 1, 2, \ldots \tag{4.2}$$

$$b_k = \frac{2}{T} \int_{-\frac{T}{2}}^{\frac{T}{2}} x(t) \, sin(kw_0t) \, dt, \quad k = 0, 1, 2, \ldots \tag{4.3}$$

and

$$a_0 = \frac{1}{T} \int_0^T x(t)dt \qquad (4.4)$$

Here, $w_0 = \frac{2\pi}{T} \, rad/s$ is the fundamental angular frequency of a period T.

As vibration signals typically exhibit the form of a high-frequency sinusoidal signal modulated by multiple lower frequencies, and sometimes periodic signals, sines are important in vibration signal analysis. According to the theory of Fourier series, all periodic signals are composed of a sum of sines. For example, for a series of sinusoids with the frequencies $f_0, f_1, f_2,$ and f_3, where $f_1, f_2,$ and f_3 are multiples of f_0 – i.e. $f_1 = 2f_0 = \frac{2}{T}$, $f_2 = 3f_0 = \frac{3}{T}$, and $f_3 = 4f_0 = \frac{4}{T}$, as illustrated in Figure 4.1 – the periodic signal $x(t)$ that composed this series of sinusoids can be mathematically expressed using Eq. (4.5):

$$x(t) = \frac{1}{2}a_0 + a_1 \cos t\left(\frac{2\pi}{T}t\right) + a_2 \cos t\left(\frac{4\pi}{T}t\right) + a_3 \cos t\left(\frac{6\pi}{T}t\right) + a_4 \cos t\left(\frac{8\pi}{T}t\right)$$

$$+ b_1 \sin\left(\frac{2\pi}{T}t\right) + b_2 \sin\left(\frac{4\pi}{T}t\right) + b_3 \sin\left(\frac{6\pi}{T}t\right) + b_4 \sin\left(\frac{8\pi}{T}t\right) \qquad (4.5)$$

The coefficients a_1 and b_1 represent w_0, i.e. the fundamental frequency (the first harmonic component); a_2 and b_2 represent the second harmonic component ($2w_0$); a_3 and b_3 represent the third harmonic component ($3w_0$); and a_4 and b_4 represent the fourth harmonic component ($4w_0$).

The Fourier series of a signal can also be expressed in a complex form. Hence, instead of performing two integrations to calculate the coefficients using Eqs. (4.2) and (4.3) in the trigonometric form of the series, the Fourier series can be represented in exponential form by using Euler's formula $e^{jw_0t} = \cos(w_0t) + j \sin(w_0t)$, where $j = \sqrt{-1}$ and the sin and cos can be mathematically expressed using Eqs. (4.6) and (4.7):

$$\cos(w_0t) = \frac{e^{jw_0t} + e^{-jw_0t}}{2} \qquad (4.6)$$

$$\sin(w_0t) = \frac{e^{jw_0t} - e^{-jw_0t}}{j2} \qquad (4.7)$$

Also, to make the physical interpretation of Eq. (4.1) simple, instead of using two sinusoids at each frequency, one may rewrite Eq. (4.1) using one sinusoid with phase angle $\emptyset_k$ at each frequency, such that,

$$x(t) = \frac{1}{2}a_0 + \sum_{k=1}^{\infty} c_k \cos(kw_0t + \emptyset_k) \qquad (4.8)$$

where

$$c_k = \sqrt{a_k^2 + b_k^2} \qquad (4.9)$$

$$\emptyset_k = \tan^{-1}\left(\frac{b_k}{a_k}\right) \qquad (4.10)$$

By using the exponential form described in (4.6), Eq. (4.8) can be rewritten as the following equation,

$$x(t) = \frac{1}{2}a_0 + \sum_{k=1}^{\infty} \frac{c_k}{2}(e^{j(w_k t + \emptyset_k)} + e^{-j(w_k t + \emptyset_k)}) \tag{4.11}$$

where $w_k = kw_0$.

In the analysis of vibrations in rotating machines, the Fourier series is mainly used for periodic signals that are often produced by a machine rotating at a constant speed. However, real-world vibration signals are not always periodic. These can be represented in an integral form, also called the Fourier integral. The Fourier integral of a function $f(t)$ can be expressed using the following equation,

$$f(t) = \frac{1}{\pi} \int_0^{\infty} [A(w) \cos(wt) + B(w) \sin(wt)] \, dw \tag{4.12}$$

where

$$A(w) = \int_{-\infty}^{\infty} f(t) \cos(wt) \, dt \tag{4.13}$$

$$B(w) = \int_{-\infty}^{\infty} f(t) \sin(wt) \, dt \tag{4.14}$$

4.2.2 Discrete Fourier Transform

The Fourier transform (FT) of a signal $x(t)$ can be mathematically given by Eq. (4.15),

$$x(w) = \int_{-\infty}^{\infty} x(t)e^{-jwt} \, dt \tag{4.15}$$

where $w = 2\pi/T$ represents the frequency. Moreover, the corresponding inverse relationship can be calculated using Eq. (4.16):

$$x(t) = \frac{1}{2\pi} \int_{-\infty}^{\infty} x(jw)e^{jwt} dw \tag{4.16}$$

Here, the factor 2π is used to allow the recovery of the original signal $x(t)$ from successive use of the two transforms. When the signal is discrete, one may use the DFT. In fact, discrete methods are often required for computerised implementation and analysis. The DFT can be expressed mathematically using the following equation:

$$X_{DFT}(k) = \sum_{n=0}^{N-1} x(n)e^{-j2\pi nk/N}, \; k = 0, 1, \dots, N-1 \tag{4.17}$$

Or, more efficiently,

$$X_{DFT}(k) = \sum_{n=0}^{N-1} x(n)W_N^{nk}, \; k = 0, 1, \dots, N-1 \tag{4.18}$$

where

$$W_N = e^{-\frac{j2\pi}{N}} = \cos\left(\frac{2\pi}{N}\right) - j\sin\left(\frac{2\pi}{N}\right) \tag{4.19}$$

Here, k is the index, and N is the length of the signal. The inverse of that DFT that transforms $X_{DFT}(k)$ back into $x(n)$ can be expressed using the following equation:

$$x(n) = \frac{1}{N}\sum_{n=0}^{N-1} X_{DFT}(k)W_N^{-nk}, \; n = 0, 1, \ldots, N-1 \tag{4.20}$$

To compute the DFT of a signal of length N, one needs N^2 complex multiplications; this limits its practical use for signals with a large number of samples (Diniz et al. 2010).

4.2.3 Fast Fourier Transform (FFT)

The fast Fourier transform (FFT) is an efficient algorithm that computes the DFT and its inverse for a stationary time series signal with a significant reduction in complexity. In fact, the FFT computes the DFT of a signal of length N using $N\,log_2 N$ complex multiplications instead of N^2 complex multiplications. The FFT technique was first proposed by Cooley and Tukey (1965) based upon sparse factorization of the DFT (Van Loan 1992). Briefly, we present the simplified FFT form as follows.

Suppose we have a discrete time series signal x with length N sample points, where N is chosen to be a power of 2, i.e. $N = 2^m$. The length N is separated into two parts: x_{even} ($x_0, x_2, x_4, \ldots$), composed of even-numbered points; and x_{odd} ($x_1, x_3, x_5, \ldots$), composed of odd-numbered points. Each has half of the total sampled points ($N/2$). Based on this, Eq. (4.18) can be rewritten as Eq. (4.21):

$$X_{DFT}(k) = \sum_{n-even=0}^{N-2} x(n)W_N^{nk} + \sum_{n-odd=0}^{N-1} x(n)W_N^{nk} \tag{4.21}$$

By replacing n with $2m$ in the even-indexed summation and replacing n with $2m+1$ in the odd-indexed summation, where $m = 0, 1, 2, \ldots, \frac{N}{2} - 1$, Eq. (4.21) can be rewritten as

$$X_{DFT}(k) = \sum_{m=0}^{\frac{N}{2}-1} x(2m)W_N^{2mk} + \sum_{m=0}^{\frac{N}{2}-1} x(2m+1)W_N^{(2m+1)k} \tag{4.22}$$

$$= \sum_{m=0}^{\frac{N}{2}-1} x(2m)(W_N^2)^{mk} + \sum_{m=0}^{\frac{N}{2}-1} x(2m+1)(W_N^2)^{mk} W_N^k \tag{4.23}$$

Here, W_N^2 can be simplified such that

$$W_N^2 = W_{N/2} \tag{4.24}$$

Then, (Eq. 4.22) can be rewritten as Eq. (4.25):

$$X_{DFT}(k) = \sum_{m=0}^{\frac{N}{2}-1} x(2m)W_{N/2})^{mk} + \sum_{m=0}^{\frac{N}{2}-1} x(2m+1)(W_{N/2})^{mk} W_N^k \tag{4.25}$$

Based on x_{n-even} and x_{n-odd}, the DFT of signal $x\,(n)$ generally can be given by the following equation:

$$X_{DFT}(k) = DFT_{N/2}\{x_{even}(m), k\} + W_N^k . DFT_{\frac{N}{2}}\{x_{odd}(m), k\}, \; k = 0, 1, \ldots, N-1 \tag{4.26}$$

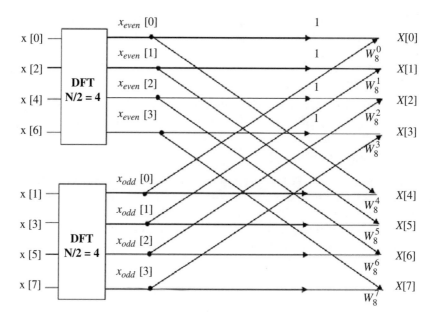

Figure 4.2 Illustration of the computation of eight points of the DFT using the FFT. Source: (Kuo et al. 2006).

For illustrative purposes, Figure 4.2 shows the computation of DFT with $N = 8$ by separating the length into two parts, i.e. each part with $N/2$ points. As examples, the computation of $x[0]$ and $x[1]$ can be expressed using the following equations:

$$X[0] = x_{even}[0] + x_{odd}[0]W_8^0 \tag{4.27}$$

$$X[1] = x_{even}[1] + x_{odd}[1]W_8^1 \tag{4.28}$$

Moreover, Figure 4.3 depicts six types of vibration signals for healthy and faulty conditions from Figure 3.9, with 1200 points and their corresponding absolute values of Fourier coefficients, using the implementation of FFT from the *fft* function in the Fourier Analysis and Filtering Toolbox in MATLAB.

Another significant aspect of FFT is that it retains phase information for the signal, which makes the inverse transformation possible. The inverse FFT can be performed using Eq. (4.20). Using MATLAB, we can easily obtain the inverse FFT by implementing the *ifft* function in the Fourier Analysis and Filtering Toolbox in MATLAB. An additional analysis technique related to the inverse DFT is cepstrum analysis, which is defined as the inverse DFT of the log magnitude of the DFT of a signal. Thus, the cepstrum of a discrete signal $x[n]$, $x_{cepstrum}[n]$ can expressed using the following equation,

$$x_{cepstrum}[n] = \mathcal{F}^{-1}\{log|\mathcal{F}\{x[n]\}|\} \tag{4.29}$$

where $\mathcal{F}$ is the DFT and $\mathcal{F}^{-1}$ is the inverse DFT.

Readers who are interested in more details of the mathematical formulation of FFT for DFT calculation are referred to (Cochran et al. 1967; Van Loan 1992; Diniz et al. 2010).

Moving on now to consider the application of FFT in diagnosing machine faults, early studies can be found in (Harris 1978; Renwick 1984; Renwick and Babson 1985;

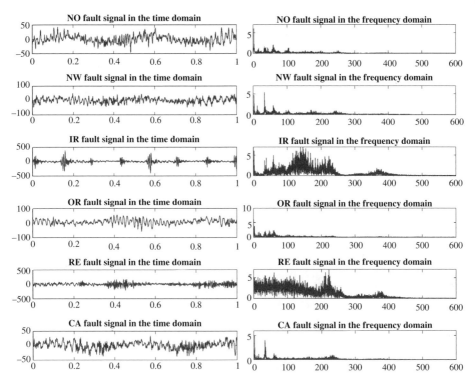

Figure 4.3 Six types of vibration signals of healthy and faulty conditions from Figure 3.9, with 1200 points and their corresponding absolute values of Fourier coefficients obtained using FFT.

Burgess 1988.). In 1991, Zeng and Wang introduced a framework for machine fault diagnosis that involves data acquisition, data processing, feature extraction, fault clustering, and fault assignments. During data processing, the collected data is first converted into the frequency domain using FFT (Zeng and Wang 1991). McCormick and Nandi described the use of artificial neural networks (ANNs) for classification of rotating shaft conditions using moments of time series as input features, as they can quickly be computed from the measured data. In comparison with frequency domain analysis, they used FFT as a suitable choice for real-time implementation. Since FFT produces many frequency bins, related to the number of samples used to compute it, they stated that there a decision is still required about which frequency components to use as features. Based on the fact that harmonics of the rotating frequency dominate the signal, they selected those bins that corresponded to the first 10 harmonics of the signal. They showed that this choice of features clearly performs less well than using moments. They provided several reasons for that: (i) some of the rig signals have spectral peaks higher than the 10th harmonic (770–1 kHz), and the inclusion of these could possibly improve the results. (ii) It is possible that spectral energy lies at frequencies that are not harmonics of the rotation speed, and therefore a better choice of FFT bins not arbitrarily chosen could possibly give improved results. Finally, (iii) it may be that problems such as the random nature of the signal, spectral leakage, and low spectral resolution caused the estimated features to be less-reliable indicators of condition than might have been expected (McCormick and Nandi 1997). Li et al.

introduced a technique for detection of common motor bearing faults using frequency domain vibration signals and ANN. In this method, the sensors-collected time domain vibration signals are transformed into the frequency domain using the FFT technique. Using the frequency spectrum of the vibration signal to train an ANN, the authors achieved excellent results with minimal data (Li et al. 1998). Betta et al. presented a DSP-based FFT analyser for fault diagnosis of rotating machines using vibration analysis. The results of using this method to estimate the vibration signal in a faulty condition proved the effectiveness of the proposed method in diagnosing machine faults (Betta et al. 2002).

Furthermore, Farokhzad proposed a vibration-based fault-detection technique for centrifugal pumps using FFT and an adaptive-fuzzy inference system. In this technique, the FFT was used to extract features from vibration signals, and then these features were fed into an adaptive neuro-fuzzy inference system as input vectors (Farokhzad 2013). Zhang et al. introduced a method for classification of faults and prediction of degraded components and machines in a manufacturing system using vibration signals. In this method, first, the acquired vibration signals were decomposed into several signals containing one approximation and some details using the wavelet packet transform (WPT). Then FFT was used to transform these signals into the frequency domain. The FFT-based extracted features were used to train an ANN, which can be used to predict degradation and identify a fault of the components and machines. The results demonstrated the effectiveness of this method (Zhang et al. 2013). Talhaoui et al. presented techniques to evaluate faults in the case of broken rotor bars in an induction motor. Processes were applied with closed-loop control. Electrical and mechanical variables are analysed using FFT, discrete wavelet transform (DWT) at startup, and steady state. (Talhaoui et al. 2014). Randall described two main areas of applicability for cepstrum analysis to mechanical problems: (i) detecting, classifying, and perhaps removing families of harmonics and sidebands in vibration and acoustic signals; and (ii) the blind separation of source and transfer function effects (Randall 2017).

Moreover, a methodology for improving the FFT-based spectral estimation for fault diagnosis is presented in (de Jesus Romero-Troncoso 2017). In this method, the FFT is used to extract the characteristic parameters of the fault. Then, multirate signal-processing techniques are introduced to improve the FFT-based methods by reducing spectral leakage with fractional resampling. The experimental results showed improvement in FFT-based methods for detecting faults in an induction motor. Gowid and colleagues compared and evaluated the performance of an FFT-based segmentation, feature-selection, and fault-identification algorithm; and neural network (NN), multilayer perceptron (MLP). The results showed a number of advantages of the FFT-based segmentation, feature-selection, and fault-identification algorithm over NN, including ease of implementation and a reduction of cost and time (Gowid et al. 2017). Ahmed and Nandi utilised FFT as a sparsifying transform to obtain a sparse representation of collected roller bearing vibration signals. The obtained sparse reorientation is then used to produce compressively sampled signals using a compressive sampling (CS) framework. Then, various feature-selection techniques are used to select fewer features from the compressively sampled signal. With these selected features, MLP, ANN, and support vector machines are used to deal with the classification problem. The results showed the effectiveness of this method in classifying the condition of roller bearings (Ahmed and Nandi 2018).

4.3   Envelope Analysis

As described in Chapter 3, demodulation analysis is needed due to the fact that signals from a faulty machine are modulated by fault characteristics and/or background noise. For example, envelope analysis can extract related diagnosis information from the acquired signals. Envelope analysis, also called high-frequency resonance analysis or resonance demodulation, is a signal-processing technique that is considered a powerful and dependable method for fault detection in rolling element bearings. Its typically involves the following three steps: (i) the raw vibration signal is band-pass filtered; (ii) it is then rectified or enveloped by folding the bottom part of the time waveform over the top part, usually using the Hilbert-Huang transform (HHT); and finally (iii) it is transformed utilising FFT. Finally, the envelope spectrum is obtained, showing obvious amplitude peaks that are not easy to see in the FFT domain. For illustration purposes, Figure 4.4 depicts the upper and lower peak envelopes of a vibration signal from a worn but undamaged bearing (NW) from Figure 3.9. Figure 4.5 presents the envelope signal and the corresponding envelope spectrum of a vibration signal from a worn but undamaged bearing (NW) from Figure 3.9.

Moving on now to consider the application of the envelope analysis in machine fault diagnosis, an early study can be found in (McFadden and Smith 1984), where the authors reviewed the vibration monitoring of rolling element bearings by the high-frequency resonance method. They showed that the procedures for obtaining the spectrum of the envelope signal are well established. Furthermore, many researchers have utilised envelope analysis for diagnosing bearing faults. For example, Toersen conducted a study to judge the diagnostic capabilities of an envelope technique, based on enveloping the resonance frequency of a transducer. The results showed that vibration spectrum components of OR faults showed up in the envelope spectra of undamaged bearings; IR and ball faults showed the well-known effects of load modulation (Toersen 1998).

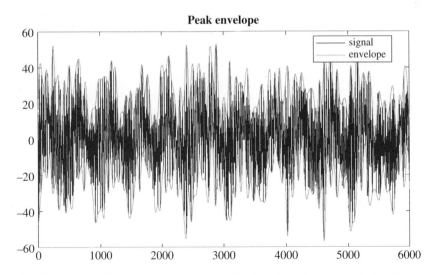

Figure 4.4 The upper and lower peak envelopes of a vibration signal from a worn but undamaged bearing (NW) from Figure 3.9, with length 1200 points. (See insert for a colour representation of this figure.)

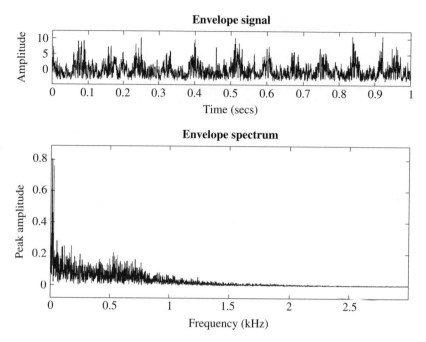

Figure 4.5 The envelope signal and envelope spectrum of a vibration signal from a worn but undamaged bearing (NW) from Figure 3.9.

Randall et al. compared the cyclostationary and envelope analyses in the diagnostics of rolling element bearings. They stated that some machine signals, while being nearly periodic, are not precisely phase-locked to shaft speeds, e.g. impulsive signals in roller bearings that are cyclostationary of the second order. However, they concluded that there does not appear to be much benefit in analysing them using spectral correlation. The results showed that better results can be obtained by analysing the squared envelope of the corresponding analytic signal (Randall et al. 2000). Tsao et al. proposed a method that selects appropriate intrinsic mode functions (IMFs) for envelope analysis. In this method, instead of examining all the resonant frequency bands during the process of detecting bearing faults, as in traditional envelope analysis, this method used empirical mode decomposition to choose an appropriate resonant frequency band for characterising the features of bearing faults using envelope analysis. The experimental results showed the effectiveness of the proposed method (Tsao et al. 2012). Patidar and Soni attempted to summarise the recent research and developments in the vibration analysis techniques for diagnosis of rolling element bearing faults. They found that the time domain techniques can only indicate the faults present in the bearing but can't identify the location. Frequency domain techniques have the ability to identify the location of faults in bearings. Moreover, they showed that envelope analysis is a very useful method to detect incipient failures of rolling element bearings (Patidar and Soni 2013).

4.4 Frequency Spectrum Statistical Features

With the invention of FFT and digital computers, the efficient computation of the signal's power spectrum became feasible. Analogous to the mathematical duals, frequency domain features are popularly used in condition monitoring. Similar to the amplitude features, these frequency domain features provide a quick overview of a machine's condition without specific diagnostic capability (Nandi et al. 2013). The following subsections briefly describe some of these features.

4.4.1 Arithmetic Mean

The arithmetic mean of a frequency spectrum, $\widehat{x}(w)$ (in dBm), can be defined using the following equation,

$$\widehat{x}(w) = 20 \, log \left\{ \frac{\frac{1}{N} \sum_N A_n}{10^{-5}} \right\} \tag{4.30}$$

where A_n denotes the amplitude of the nth component in a total of N components. The feature gives the average amplitude value within a frequency range.

4.4.2 Geometric Mean

The geometric mean, $\widehat{x}_{geo}(w)$, is an alternative to the arithmetic mean, which can be expressed mathematically (in dBm) using the following equation:

$$\widehat{x}_{geo}(w) = \left\{ \sum_N 20 \, log \left(\frac{\frac{A_n}{\sqrt{2}}}{10^{-5}} \right) \right\} \tag{4.31}$$

4.4.3 Matched Filter RMS

The matched filter root mean square (RMS), M_{frms}, can be computed when a reference spectrum is available. The M_{frms} provides the RMS amplitude of the frequency component with reference to the reference spectrum, and can be expressed using the following equation,

$$M_{frms} = 10 log \left\{ \frac{1}{N} \sum_N \left(\frac{A_i}{A_n^{ref}} \right) \right\} \tag{4.32}$$

where A_n^{ref} denotes the nth component of the reference spectrum.

4.4.4 The RMS of Spectral Difference

The RMS of spectral difference, R_d, can be computed using the following equation:

$$R_d = \sqrt{\left\{ \frac{1}{N} \sum_N (P_n - P_n^{ref})^2 \right\}} \tag{4.33}$$

Here, P_n and P_n^{ref} denote the amplitude (in dB) of the incident and reference spectra.

4.4.5 The Sum of Squares Spectral Difference

The sum of squares of spectral difference, s_d, gives greater weight to high-amplitude components and can be calculated using the following equation:

$$S_d = \frac{1}{N} \sum_N \sqrt{(P_n + P_n^{ref})} * |P_n - P_n^{ref}| \tag{4.34}$$

4.4.6 High-Order Spectra Techniques

Higher-order spectra, also called polyspectra, are defined in terms of the higher-order cumulants described in the previous chapter. Specific types of higher-order spectra are the third-order spectrum, also called the bispectrum, which is the Fourier transform of the third-order statistics; and the trispectrum, which is the Fourier transform of the fourth-order statistics of a stationary signal (Nikias and Mendel 1993). It can provide more diagnostic information than the power spectrum if the signals contain non-Gaussian components. The bispectrum and trispectrum can be normalised with respect to the power spectrum to produce bicoherence and tricoherence, which are useful in detecting quadratic phase coupling; but with machine condition signals, a significant component may be periodic, and this would result in a "bed-of-nails" (McCormick and Nandi 1999).

Efficient methods of computing the bispectrum (B_s) and trispectrum (T_s) can be expressed mathematically using the following equations,

$$B_s(f_1, f_2) = E(X(f_1) X(f_2) X^*(f_1 + f_2)) \tag{4.35}$$

$$T_s(f_1, f_2, f_3) = E(X(f_1) X(f_2) X(f_3) X^*(f_1 + f_2 f_3)) \tag{4.36}$$

where $X(f)$ is the Fourier transform of $x[n]$.

Having defined what is meant by bispectrum and trispectrum, we will now move on to discuss their application in diagnosing machine faults. To date, several studies have investigated the application of bispectral and trispectral machine condition diagnosis. For example, McCormick and Nandi investigated the application of two higher-order spectra-based approaches for condition monitoring of mechanical machinery. With the first approach, the entire bispectral or trispectral analysis was used as a set of input features to a classifier, which could then define the machine's condition. Six classification algorithms were used to deal with the classification problem. With the second approach, the high-order spectrum of machine vibration signals was examined at a higher resolution. Experimental results described the effectiveness of these approaches in machine fault diagnosis (McCormick and Nandi 1999). Furthermore, in their study, Yunusa-Kaltungo and Sinha (2014) showed the possibility of enhancing fault diagnosis and maintenance cost-effectiveness through the combination of bispectrum and trispectrum analysis.

Moreover, spectral kurtosis has been investigated in many studies of diagnosing faults in rolling bearings: for example, Vrabie et al. (2004). Tian et al. (2016) proposed a method for fault detection and degradation of bearings in electric motors. This method first extracts fault features using spectral kurtosis and cross-correlation and then combines these features using principal component analysis (PCA) and semisupervised k-nearest neighbour (k-NN). Furthermore, a good review of the application of spectral kurtosis for fault detection, diagnosis, and prognostics of rotating machines can be found in (Wang et al. 2016).

4.5 Summary

This chapter has presented signal processing in the frequency domain, which has the ability to divulge information based on frequency characteristics that are not easy to observe in the time domain. The first part described Fourier analysis, including Fourier series, DFT, and FFT, which are the most commonly used signal transformation techniques and allow one to transform time domain signals to the frequency domain. In addition, this chapter has provided an explanation of different techniques that can be used to extract various frequency spectrum features that can more efficiently represent a machine's health. These include: (i) envelope analysis, also called high-frequency resonance analysis or resonance demodulation, which is a signal-processing technique that is considered a powerful and dependable method for detecting faults in rolling element bearing; and (ii) frequency domain features, which provide a quick overview of a machine's condition without specific diagnostic capability, and which include arithmetic mean, matched filter RMS, the RMS of spectral difference, the sum of squares spectral difference, and high-order spectra techniques. Most of the technique introduced and their publically accessible software are summarised in Table 4.1.

Table 4.1 Summary of the publicly accessible software for some of the techniques introduced in this chapter.

Algorithm name	Platform	Package	Function
Discrete Fourier transform	MATLAB	Communication System Toolbox	fft
Inverse discrete Fourier transform			ifft
Fast Fourier transform		Fourier analysis and filtering	fft
Inverse fast Fourier transform			ifft
Signal envelope	MATLAB	Signal Processing Toolbox	envelope
Envelope spectrum for machinery diagnosis			envspectrum
Average spectrum vs. order for vibration signal			Orderspectrum
Complex cepstral analysis			cceps
Inverse complex cepstrum			Icceps
Real cepstrum and minimum phase reconstruction			rceps

References

Ahmed, H. and Nandi, A.K. (2018). Compressive sampling and feature ranking framework for bearing fault classification with vibration signals. *IEEE Access* 6: 44731–44746.

Betta, G., Liguori, C., Paolillo, A., and Pietrosanto, A. (2002). A DSP-based FFT-analyzer for the fault diagnosis of rotating machine based on vibration analysis. *IEEE Transactions on Instrumentation and Measurement* 51 (6): 1316–1322.

Brandt, A. (2011). *Noise and Vibration Analysis: Signal Analysis and Experimental Procedures*. Wiley.

Burgess, P.F. (1988). Antifriction bearing fault detection using envelope detection. *Transactions of the Institution of Professional Engineers New Zealand: Electrical/Mechanical/Chemical Engineering Section* 15 (2): 77.

Cochran, W.T., Cooley, J.W., Favin, D.L. et al. (1967). What is the fast Fourier transform? *Proceedings of the IEEE* 55 (10): 1664–1674.

Cooley, J.W. and Tukey, J.W. (1965). An algorithm for the machine calculation of complex Fourier series. *Mathematics of Computation* 19 (90): 297–301.

Diniz, P.S., Da Silva, E.A., and Netto, S.L. (2010). *Digital Signal Processing: System Analysis and Design*. Cambridge University Press.

Farokhzad, S. (2013). Vibration based fault detection of centrifugal pump by fast fourier transform and adaptive neuro-fuzzy inference system. *Journal of Mechanical Engineering and Technology* 1 (3): 82–87.

Gowid, S., Dixon, R., and Ghani, S. (2017). Performance comparison between fast Fourier transform-based segmentation, feature selection, and fault identification algorithm and neural network for the condition monitoring of centrifugal equipment. *Journal of Dynamic Systems, Measurement, and Control* 139 (6): 061013.

Harris, F.J. (1978). On the use of windows for harmonic analysis with the discrete Fourier transform. *Proceedings of the IEEE* 66 (1): 51–83.

de Jesus Romero-Troncoso, R. (2017). Multirate signal processing to improve FFT-based analysis for detecting faults in induction motors. *IEEE Transactions on Industrial Informatics* 13 (3): 1291–1300.

Kuo, S.M., Lee, B.H., and Tian, W. (2006). *Real-Time Digital Signal Processing: Implementations and Applications*. Wiley.

Li, B., Goddu, G., and Chow, M.Y. (1998). Detection of common motor bearing faults using frequency-domain vibration signals and a neural network based approach. In: *Proceedings of the 1998 American Control Conference, 1998*, vol. 4, 2032–2036. IEEE.

McCormick, A.C. and Nandi, A.K. (1997). Real-time classification of rotating shaft loading conditions using artificial neural networks. *IEEE Transactions on Neural Networks* 8 (3): 748–757.

McCormick, A.C. and Nandi, A.K. (1999). Bispectral and trispectral features for machine condition diagnosis. *IEE Proceedings-Vision, Image and Signal Processing* 146 (5): 229–234.

McFadden, P.D. and Smith, J.D. (1984). Vibration monitoring of rolling element bearings by the high-frequency resonance technique – a review. *Tribology International* 17 (1): 3–10.

Nandi, A.K., Liu, C., and Wong, M.D. (2013). Intelligent vibration signal processing for condition monitoring. In: *Proceedings of the International Conference Surveillance*, vol. 7, 1–15. https://surveillance7.sciencesconf.org/resource/page/id/20.

Nikias, C.L. and Mendel, J.M. (1993). Signal processing with higher-order spectra. *IEEE Signal Processing Magazine* 10 (3): 10–37.

Patidar, S. and Soni, P.K. (2013). An overview on vibration analysis techniques for the diagnosis of rolling element bearing faults. *International Journal of Engineering Trends and Technology (IJETT)* 4 (5): 1804–1809.

Randall, R.B. (2017). A history of cepstrum analysis and its application to mechanical problems. *Mechanical Systems and Signal Processing* 97: 3–19.

Randall, R.B., Antoni, J., and Chobsaard, S. (2000). A comparison of cyclostationary and envelope analysis in the diagnostics of rolling element bearings. In: *2000 IEEE International Conference on Acoustics, Speech, and Signal Processing, 2000. ICASSP'00. Proceedings*, vol. 6, 3882–3885. IEEE.

Renwick, J.T. (1984). Condition monitoring of machinery using computerized vibration signature analysis. *IEEE Transactions on Industry Applications* (3): 519–527.

Renwick, J.T. and Babson, P.E. (1985). Vibration analysis – a proven technique as a predictive maintenance tool. *IEEE Transactions on Industry Applications* (2): 324–332.

Talhaoui, H., Menacer, A., Kessal, A., and Kechida, R. (2014). Fast Fourier and discrete wavelet transforms applied to sensorless vector control induction motor for rotor bar faults diagnosis. *ISA Transactions* 53 (5): 1639–1649.

Tian, J., Morillo, C., Azarian, M.H., and Pecht, M. (2016). Motor bearing fault detection using spectral kurtosis-based feature extraction coupled with K-nearest neighbor distance analysis. *IEEE Transactions on Industrial Electronics* 63 (3): 1793–1803.

Toersen, H. (1998). Application of an envelope technique in the detection of ball bearing defects in a laboratory experiment. *Tribotest* 4 (3): 297–308.

Tsao, W.C., Li, Y.F., Du Le, D., and Pan, M.C. (2012). An insight concept to select appropriate IMFs for envelope analysis of bearing fault diagnosis. *Measurement* 45 (6): 1489–1498.

Van Loan, C. (1992). *Computational Frameworks for the Fast Fourier Transform*, vol. 10. Siam.

Vrabie, V., Granjon, P., Maroni, C.S. et al. (2004). Application of spectral kurtosis to bearing fault detection in induction motors. 5th International Conference on Acoustical and Vibratory Surveillance Methods and Diagnostic Techniques (Surveillance5).

Wang, Y., Xiang, J., Markert, R., and Liang, M. (2016). Spectral kurtosis for fault detection, diagnosis, and prognostics of rotating machines: a review with applications. *Mechanical Systems and Signal Processing* 66: 679–698.

Yunusa-Kaltungo, A. and Sinha, J.K. (2014). Combined bispectrum and trispectrum for faults diagnosis in rotating machines. *Proceedings of the Institution of Mechanical Engineers, Part O: Journal of Risk and Reliability* 228 (4): 419–428.

Zeng, L. and Wang, H.P. (1991). Machine-fault classification: a fuzzy-set approach. *The International Journal of Advanced Manufacturing Technology* 6 (1): 83–93.

Zhang, Z., Wang, Y., and Wang, K. (2013). Fault diagnosis and prognosis using wavelet packet decomposition, Fourier transform and artificial neural network. *Journal of Intelligent Manufacturing* 24 (6): 1213–1227.

5

Time-Frequency Domain Analysis

5.1 Introduction

As mentioned in the previous chapter, the Fourier transform allows a distinctive transform of a signal from the time domain to the frequency domain. The frequency domain representation delivers the spectral components of the signal, and the time distribution information of the spectral components is often included in the phase characteristic of the Fourier transform. However, it is not easy to use this information about the time distribution in the frequency domain. In addition, the basic assumption when we transform a signal to its frequency domain is that its frequency components do not change over time, i.e. the signal is stationary. Thus, the Fourier transform in the frequency domain does not have the ability to provide a time distribution information of the spectral components. Rotating machines, in general, generate stationary vibration signals. Nevertheless, Brandt stated that most analysis of rotating machines is based on examining the vibrations during a speed sweep, where machines are either speeded up from low to high revolutions per minute (RPM) or slowed down from high to low RPM. This often results in nonstationary signals whose frequency content changes over time (Brandt 2011).

The time-frequency domain has been used for nonstationary waveform signals, which are very common when machinery faults occur. Thus far, several time-frequency analysis techniques have been developed and applied to machinery fault diagnosis: e.g. short-time Fourier transform (STFT), wavelet transform (WT), Hilbert-Huang transform (HHT), empirical mode decomposition (EMD), local mean decomposition (LMD), etc. This chapter introduces signal processing in the time-frequency domain and gives an explanation of several techniques that can be used to examine the time-frequency characteristics of the time-indexed series signal, which can be figured more effectively than Fourier transform and its corresponding frequency spectrum features.

5.2 Short-Time Fourier Transform (STFT)

The STFT (Cohen 1995) is the first modified version of the Fourier transform that allows one to analyse nonstationary signals in the time-frequency domain. It was first introduced by Gabor (1946). The STFT of a time domain vibration signal, $x(t)$, $STFT_{x(t)}(t,f)$, can be expressed using the following equation,

$$STFT_{x(t)}(t, w) = \int_{-\infty}^{+\infty} x(t)w(t - \tau)exp(-jwt)d\tau \tag{5.1}$$

Condition Monitoring with Vibration Signals: Compressive Sampling and Learning Algorithms for Rotating Machines, First Edition. Hosameldin Ahmed and Asoke K. Nandi.

where $w(\tau)$ is a window function and τ is a time variable. For a vibration signal in discrete form, Eq. (5.1) can be rewritten as Eq. (5.2):

$$STFT_{x(n)}(n, w) = \sum_{-\infty}^{+\infty} x(n)w(n - m)exp(-jwn) \tag{5.2}$$

The basic idea of the STFT is that instead of computing the discrete Fourier transform (DFT) of the whole signal, we decompose a signal into shorter segments of equal length using a time-localised window function, e.g. Gaussian window or Hamming window, and then perform the DFT separately on each windowed segment of the signal, which together forms the time-frequency spectrum of the signal. In practice, the fast Fourier transform (FFT) is often used to perform the DFT of the STFT. The STFT requires that the signal to be analysed while stationary during a short time interval, i.e. the window time interval. To estimate the frequency content of the signal based on the STFT, the magnitude squared of the STFT, also called the spectrogram or sonogram, is used. The spectrogram is a graphical representation of a signal that estimates the energy distribution of the signal over the time-frequency domain. The spectrogram can be mathematically presented using the following equation:

$$SPEC_{x(n)}(n, w) = | STFT_{x(n)}(n, w)^2| \tag{5.3}$$

As an example, Figure 5.1 shows a spectrogram, i.e. the time-frequency spectrum, of roller bearing vibrations under normal and faulty conditions from Figure 3.9 with a Hamming window function of length 1333, 50% overlap between contiguous segments, and a maximum of 2048 points to compute the FFT. More accurate representations of signals with respect to time resolution can be achieved by reducing the window size, but that may lead to increased computation time. In addition, to increase the frequency resolution, a longer time interval is required, but this may invalidate the stationary assumption. In fact, the STFT uses windows with a fixed size that has to be chosen in advance. While monitoring the condition of rotating machines, the detection of faults from nonstationary signals that are masked with noises, and that sometimes are weak, may require more effective and flexible techniques. The STFT does not utilise window function with varying sizes for all the frequency components.

The application of the STFT and spectrograms in machine condition monitoring has been considered in various studies. For instance, Strangas et al. used four methods to identify developing electrical faults in electric drives; these include the STFT. In this study, the different fault conditions are classified utilising the linear-discriminant classifier and k-means classification (Strangas et al. 2008). Wang and Chen (2011) investigated the sensitivities of three time-frequency analysis techniques, including the STFT, for diagnosing the condition of rotating machines with a variable rotational speed. Belšak and Prezelj (2011) stated that the life cycle of a gear unit can be monitored more reliably with the use of appropriate spectrogram samples and a clear presentation of the pulsation of individual frequency components, which along with the average spectrum are used as measures for examining the condition of a gear unit. Banerjee and Das (2012) proposed and investigated a hybrid method for fault signal classification based on sensor data fusion by using a support vector machine (SVM) and STFT techniques. In this method, the STFT is used to separate the signal according to its frequency level and amplitude. In (Czarnecki 2016), an instantaneous frequency rate spectrogram (IFRS)

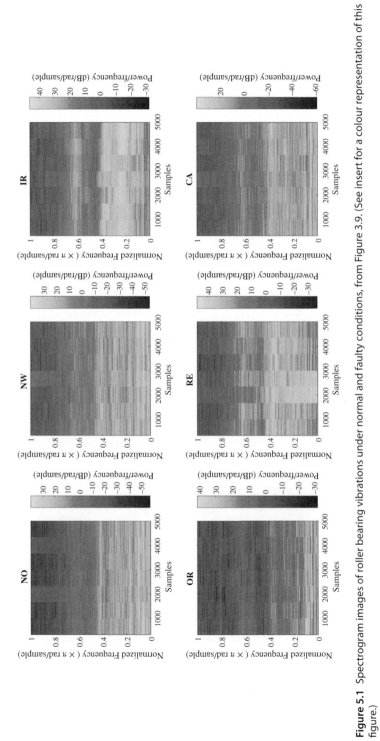

Figure 5.1 Spectrogram images of roller bearing vibrations under normal and faulty conditions, from Figure 3.9. (See insert for a colour representation of this figure.)

technique is presented for a joint time-frequency distribution. In this technique, the distribution is directly obtained by performing the STFT locally.

Moreover, Liu et al. proposed a rolling bearing fault diagnosis method that is based on the STFT and uses a stacked sparse autoencoder (SAE) to automatically extract the fault features. In this method, first spectrograms of the input signals are obtained using the STFT, then the SAE is employed to extract the fault features, and finally softmax regression is adopted to classify the bearing fault condition (Liu et al. 2016a). Although their method achieved interesting results for fault classification, Liu et al. stated that the STFT and spectrogram functions consumed a vast amount of memory due to their extensive matrix operations. In the same vein, in (He and He 2017), a deep learning–based technique for bearing fault diagnosis is proposed. In this technique, the raw data is transformed from the time domain to the frequency domain using the STFT to obtain a spectrum matrix. Then, based on the generated spectrum matrix, an optimised deep learning structure is built to diagnosis the bearing faults.

5.3 Wavelet Analysis

Wavelet analysis is another time-frequency domain analysis approach that decomposes the signal based on a family of 'wavelets'. The way wavelet analysis localises the signal's information in the time-frequency domain through variable spectrograms makes it a good alternative analysis method to the STFT method for the analysis of nonstationary signals. Unlike the window used with the STFT, the wavelet families have fixed shapes – e.g. Haar, Daubechies, symlets, Morlets, coiflets, etc. – but the wavelet function is scalable, which means the wavelet transformation is adaptable to a wide range of frequency- and time-based resolutions. The mother wavelet $\psi(t)$ can be expressed mathematically by the following equation,

$$\psi_{s,\tau}(t) = \frac{1}{\sqrt{s}} \psi \left(\frac{t - \tau}{s} \right) \tag{5.4}$$

where s represents the scaling parameter, τ is the transformation parameter, and t is the time. The original mother wavelet has $s = 1$ and $\tau = 0$. The three main transforms in wavelets analysis are the continuous wavelet transform (CWT), discrete wavelet transform (DWT), and wavelet packet transform (WPT) (Burrus et al. 1998). Readers who are interested in more details of the similarities and the differences between the STFT and WT are referred to (Daubechies 1990).

5.3.1 Wavelet Transform (WT)

The wavelet transform is an extension of the Fourier transform, which maps the original signal from the time domain to the time-frequency domain. The wavelet basis functions are obtained from the mother wavelet $\psi(t)$ through dilation and translation procedures. The obtained wavelet functions often have the same shape as the mother wavelet but different dimensions and places. The three main transforms in wavelets analysis are the CWT, DWT, and WPT. The subsequent subsections discuss those wavelet transformation methods in more detail.

5.3.1.1 Continuous Wavelet Transform (CWT)

The CWT of the time domain vibration signal (x), $W_{x(t)}(s, \tau)$ can be expressed using the following equation,

$$W_{x(t)}(s, \tau) = \frac{1}{\sqrt{s}} \int x(t)\psi^* \left(\frac{t - \tau}{s}\right) dt \tag{5.5}$$

where ψ^* represents the complex conjugate of $\psi(t)$ that is scaled and shifted using the s and τ parameters, respectively. The translation parameter τ is linked to the location of the wavelet window, as the window moved through the signal and scale parameter s is linked to the zooming action of the wavelets. The integral compares the local shape of the signal and the shape of the wavelet. Thus, the obtained coefficients generated using Eq. (5.5) describe the correlation between the waveform and the wavelet used at the several translations and scales. In other words, one may say that the obtained coefficients deliver the amplitudes of a series of wavelets over a choice of scales and translations. Following are the five easy steps utilised to compute the CWT of a signal (Saxena et al. 2016):

1. Select a wavelet, and match it to the segment at the start of the signal.
2. Calculate the CWT coefficients that measure how similar are the wavelet and the analysed segment of the signal are.
3. Shift the wavelet to the right by the translation parameter τ, and repeat steps 1 and 2. Calculate the CWT coefficients for all translations.
4. Scale the wavelet, and repeat steps 1–3.
5. Repeat step 4 for all scales.

The wavelet coefficients are often displayed in a scalogram, which is a function of both the scale and the wavelength. The most common wavelets used with the CWT are Morse wavelets, analytical Morlets, and bump wavelets. As an example, Figure 5.2 shows a scalogram of roller bearing vibrations under brand-new and inner race (IR) fault conditions from Figure 3.1. As can be seen in this figure, through the multiresolution ability of the wavelet coefficients, the differences in the vibration waveform of the IR fault and the normal condition, i.e. the brand-new condition, can be located.

Another significant aspect of the CWT is its ability to reconstruct the manipulated signal in the time frequency back into an approximation of the original signal $\hat{x}(t)$ using the inverse CWT. The inverse CWT reconstructs the original signal from its wavelet coefficients, which traditionally are presented in a two-dimensional integration, i.e. double-integral form, over the scale parameter s and the translation parameter τ. This can be expressed mathematically by the following equation,

$$\hat{x}(t) = \frac{1}{C_\psi} \int_{-\infty}^{\infty} d\tau \int_{-\infty}^{\infty} \frac{1}{s^2} W_{x(t)}(s, \tau)\psi_{s,\tau}(t) ds \tag{5.6}$$

where C_ψ is a constant based on the type of the selected wavelet.

The two main applications of the CWT are time-frequency analysis and filtering of time-localised frequency components. In an early example of research into machine diagnosis using the CWT, Lopez et al. presented an overview of the results obtained using wavelet/neural network system–based fault detection and identification of rotating machinery. The method is applied to a range of platforms including helicopter transmissions, turbo pumps, and gas turbines (Lopez et al. 1996). Furthermore, many authors have described the use of the CWT for detecting local faults in gears, rotors,

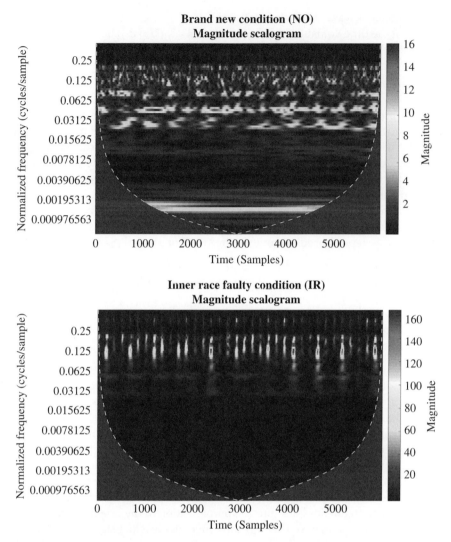

Figure 5.2 Magnitude scalogram images of roller bearing vibrations under normal and faulty conditions, from Figure 3.1. (See insert for a colour representation of this figure.)

and roller bearings. For example, Zheng et al. showed that the time average wavelet spectrum (TAWS) based on the CWT using a Morlet wavelet function can select features for effective gear fault identification. Based on this concept, they demonstrated a spectrum comparison method and an energy method based on TAWS that can detect the gear fault accurately (Zheng et al. 2002). Luo et al. (2003) proposed a method using online vibration analysis with fast CWT for condition monitoring of bearings and demonstrated its suitability for real-time applications. In this method, infinite impulse response (IIR) filters are used to compute the CWT. Zhang et al. (2006) presented a technique for machine health diagnosis using analytic wavelets that has the ability to extract defect features and construct the envelope of the signal in a single step. Hong and Liang proposed a new version of the Lempel-Ziv complexity as a bearing fault-severity measure based on the CWT results. In this method, the Lempel-Ziv complexity values are calculated from the wavelet coefficients at the best scale level. The authors demonstrated the efficiency of this method in measuring the severity of both inner- and outer-race faults (Hong and Liang 2009).

Furthermore, Wang et al. presented an approach for gear fault diagnosis based on complex Morlet CWT. In this method, a gear motion residual signal that represents the departure of the time synchronous average (TSA) signal from the average tooth-meshing vibration is analysed as source data due to its lower sensitivity to the changing load conditions (Wang et al. 2010a). Also, Rafiee et al. (2010) showed that complex Morlet wavelets have acceptable performance for both bearing and gear fault diagnosis. Konar and Chattopadhyay demonstrated that the CWT along with SVM approach can be used as a better alternative to the DWT/artificial neural network (ANN) based fault classification algorithm. Also, they showed that the choice of mother wavelet plays a crucial role in the fault-detection algorithm (Konar and Chattopadhyay 2011). Moreover, Kankar et al. introduced a method for rolling bearing fault diagnosis using the CWT. In this method, two wavelet selection criteria – the maximum energy to Shannon entropy ratio and the maximum relative wavelet energy – are utilised and compared to choose a suitable feature extraction. The bearing faults are classified using statistical features extracted from the wavelet coefficients as input to a machine learning classifier (Kankar et al. 2011). Saxena et al. (2016) demonstrated that CWT analysis is an effective tool for analysing bearing fault data and that CWT features can be used for visual inspection of bearing faults.

5.3.1.2 Discrete Wavelet Transform (DWT)

Discrete methods are often required for the computerised implementation and analysis process of wavelet transforms. The DWT of the time domain vibration signal (x), $W_{x(t)}(s, \tau)$, can be expressed using the following equation:

$$W_{x(t)}(s, \tau) = \frac{1}{\sqrt{2^j}} \int x(t)\psi^* \left(\frac{t - k2^j}{2^j} \right) dt \tag{5.7}$$

Here, the DWT represents the discrete form of the CWT, where $\psi_{s, \tau}(t)$ is discretised using dyadic scales, i.e. $s = 2^j$ and $\tau = k2^j$, where j and k are integers. In practice, the DWT is often implemented with a low-pass scaling filter $h(k)$ and a high-pass wavelet filter $g(k) = (-1)^k h(1 - k)$. These filters are created from scaling functions $\phi(t)$ and $\psi(t)$, which can be represented by the following equations:

$$\phi(t) = \sqrt{2} \sum_k h(k)\phi(2t - k) \tag{5.8}$$

$$\psi(t) = \sqrt{2} \sum_k g(k) \psi(2t - k) \tag{5.9}$$

The signal is decomposed into a set of low- and high-frequency signals using the wavelet filters such that

$$a_{j,k} = \sum_m h(2k - m)a_{j-1,m} \tag{5.10}$$

$$b_{j,k} = \sum_m g(2k - m)a_{j-1,m} \tag{5.11}$$

where the terms h and g are high-pass and low-pass filters derived from the wavelet function $\psi(t)$ and the scaling function $\phi(t)$; $a_{j,k}$ and $b_{j,k}$ are the coefficients that represent the low-frequency components and the high-frequency components of the signal, respectively. The analysis of a signal using the DWT technique produces few large-magnitude coefficients, but the noise in the signal results in smaller-magnitude coefficients, which makes it ideal for denoising and compressing signals. Briefly, the process of the DWT can be presented as follows.

To perform the DWT, also called multiresolution analysis, of a given discrete signal x, (i) the signal is filtered using special low-pass filter (L) and high-pass filter (H), e.g. Daubechies, coiflets, and symlets, producing two vectors of low and high sub-bands at the first level. The first vector is the approximation coefficient (A1), and the second is a detailed coefficient (D1) (see Figure 5.3). For the next level of the decomposition, (ii) the approximation coefficient, i.e. the low-pass sub-band, is further filtered using L and H filters, which produce a further approximation coefficient (A2) and detailed coefficient (D2) at the second level. The length of the coefficients in each level is half the number of the coefficients in the previous level (Mallat 1989; Mertins and Mertins 1999). Figure 5.3 illustrates the DWT with five decomposition levels. Figure 5.4 shows an original vibration signal from a cage fault from Figure 3.9 and its corresponding details of level 1 to level 5 decomposition using the sym6 wavelet, which is a member of the symlets wavelet family.

As far as the noise in vibration signals is concerned, one of the main applications of the DWT in machine fault diagnosis is denoising the vibration signal in the time domain as well as the frequency domain. One way to do this is by scaling the detailed coefficients using a threshold. Donoho and Johnstone reviewed and compared various proposals for the selection of thresholds that can be used for thresholding the wavelet coefficients. These include soft and hard thresholding (Donoho and Johnstone 1994). In both thresholding techniques, i.e. hard thresholding and soft thresholding, the coefficients with magnitude values (m) less than the thresholding value (τ) are set to zero. The only different between these two types of thresholding is based on how they deal

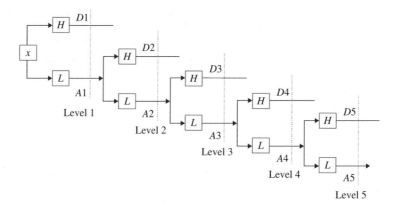

Figure 5.3 Illustration of the DWT with five decomposition levels.

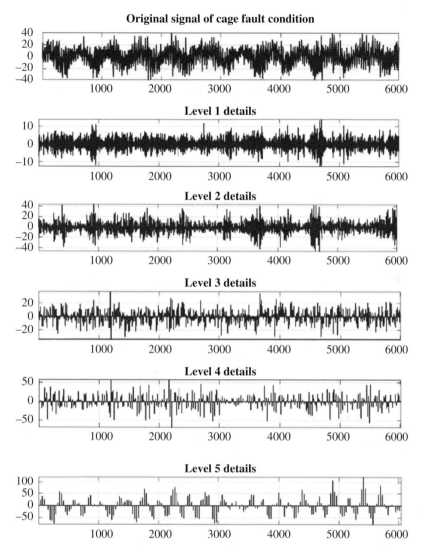

Figure 5.4 Vibration signal of a cage fault from Figure 3.9 and its corresponding details from level 1 to level 5.

with the coefficients that have magnitude values that are larger than τ; in hard thresholding, coefficients with magnitude values larger than τ are left unchanged, while in soft thresholding the coefficients are reduced by $(\tau - m)$. Figure 5.5 shows the result of denoising a vibration signal from a cage fault condition from Figure 3.9 using sym6 wavelet with hard thresholding (see Figure 5.5b) and soft thresholding (Figure 5.5c), with their corresponding signal-to-noise ratio (SNR).

Various types of wavelets have been widely used for analysing many rotating machine vibration signals. For example, Lin and Qu proposed a denoising technique based on wavelet analysis to extract a feature from mechanical vibration signals. In this method, the time-frequency resolution is adapted to different signals using the Morlet wavelet.

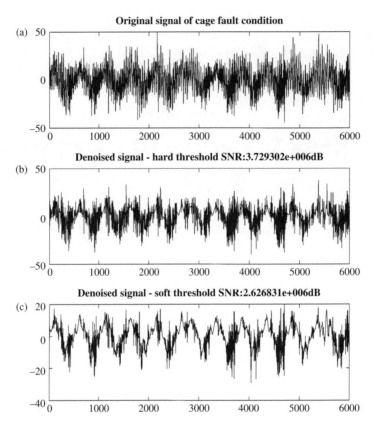

Figure 5.5 Example of wavelet denoising: (a) original vibration signal of a cage fault condition from Figure 3.9; (b) denoised signal using a hard threshold; (c) denoised signal using a soft threshold.

This is achieved by adjusting the value of the parameter β that controls the shape of the basic wavelet. By applying this method in rolling bearing diagnosis and gearbox diagnosis, they showed that this denoising method based on the Morlet wavelet offers more advantages than the soft-thresholding denoising method in feature extraction from machine vibration signals (Lin and Qu 2000). Prabhakar et al. (2002) showed that the DWT can be used as an effective tool for detecting single and multiple faults. Smith et al. proposed an approach to vibration analysis using wavelets for aircraft condition monitoring. In this approach, the characteristic features of vibration signals are extracted from noise utilising the Haar, Daubechies, and Morlet wavelets, and then the detection is performed using the signal's scalogram information. From their investigations, they showed that their suggested wavelet-based vibration-detection algorithms have been successfully applied to detect vibration signals. In addition, they have been able to recover key features of the vibration signal from noise using Haar, Daubechies, and Morlet wavelets (Smith et al. 2007). Chen et al. proposed a feature-extraction technique based on the overcomplete rational dilation discrete wavelet transform (ORDWT) for fault features confronted in gearbox fault diagnosis. In this method, the ORDWT is used to decompose the input vibration signal, while other auxiliary signal-processing techniques are used to process the reconstructed wavelet sub-bands of the vibration signals. The processing results showed that this method is able to detect various types of fault features (Chen et al. 2012). Li et al. introduced a method for diagnosing the condition of rotating machines using a wavelet transform with the ReverseBior 2.8 wavelet function and ant colony optimisation (ACO). The authors showed the effectiveness of this method by applying it to the task of diagnosing a fault in a bearing used in a centrifugal fan system (Li et al. 2012).

Moreover, Kumar and Singh proposed a technique based on signal decomposition using a symlet wavelet for measuring the outer race (OR) defect width of a roller bearing over a range of $0.5776 - 1.9614$ mm. Based on their experimental results and subsequent analysis, they demonstrated that decomposition of a vibration signal by a symlet with five levels is suitable for measuring OR defects of rolling bearings. In addition, the defect width is also verified using image analysis (Kumar and Singh 2013). Li et al. presented a multiscale slope feature-extraction method based on wavelet analysis, which include three main steps: (i) the DWT is used to obtain a series of detailed signals of vibration signals, (ii) the variances of the multiscale detailed signals are calculated, and finally (iii) the wavelet-based multiscale slope features are estimated based on the slope logarithmic variances. The authors showed the effectiveness of the suggested method in classifying different conditions of both bearings and gearboxes with high classification accuracy (Li et al. 2013). Recently, in (Ahmed et al. 2018), in their attempts to compress a large number of acquired vibration signals, the authors applied a thresholded Haar wavelet function to obtain sparse representations from roller bearing vibration signals. These sparse representations of the vibration signals are then used in a compressive sampling (CS) framework to obtain highly compressed measurements that possess the quality of the original signals.

5.3.2   Wavelet Packet Transform (WPT)

The WPT is an improvement of the DWT in which every signal detail obtained by the DWT is further decomposed into an approximation signal and a detail signal (Shen et al. 2013). Accordingly, the time domain vibration signal $x(t)$ can be decomposed using the following equations:

$$d_{j+1,2n} = \sum_m h(m - 2k)d_{j,n} \tag{5.12}$$

$$d_{j+1,2n+1} = \sum_m g(m - 2k)d_{j,n} \tag{5.13}$$

Here, m is the number of coefficients, and $d_{j,n}$, $d_{j+1,2n}$, and $d_{j+1,2n+1}$ are the wavelet coefficients at sub-bands n, $2n$, and $2n + 1$, respectively. The key advantages of the WPT over other types of wavelet transforms are the accurate and detailed representations of the decomposed signals. Furthermore, wavelet packet basis functions are localised in time, offering better signal approximation and decomposition (Saleh and Rahman 2005).

To perform the WPT of a given discrete signal x, (i) the signal is filtered using a special low-pass filter (L) and high-pass filter (H), producing two vectors of low and high sub-bands at the first level. The first vector is the approximation coefficient (A_L), and the second is a detailed coefficient (D_H) (see Figure 5.4). For the next level of the decomposition, (ii) both A_L and D_H are further filtered using L and H filters, which produces further approximation coefficients (AA_L and AD_H) and detailed coefficients

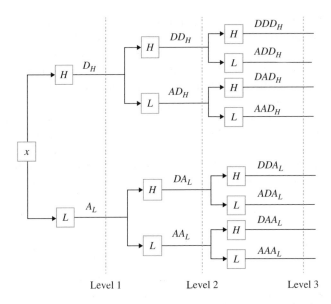

Figure 5.6 Illustration of the WPT with three decomposition levels.

(DA_L and DD_H) at the second level, and so on. Figure 5.6 illustrates the WPT with three decomposition levels.

Moving on now to consider the application of the WPT in machine fault diagnosis, in an early example of research into machine diagnosis using the WPT, Liu et al. (1997) developed a method based on wavelet packet theory along with a fault-dependent basis introduced to facilitate feature extraction. Moreover, Yen and Lin investigated the possibility of applying the WPT to the classification of vibration signals based on the idea that by using the WPT, a rich collection of time-frequency characteristics in a signal can be obtained. In this work, the authors detailed a feature-selection process that exploits signal class differences in the wavelet packet node energy, resulting in a reduced-dimensional feature space compared to the dimensions of the original time-series vibration signal. According to their experimental results, this method showed high classification accuracy for vibration monitoring (Yen and Lin 2000). Hu et al. (2007) proposed a method for fault diagnosis of rotating machinery based on an improved wavelet packet transform (IWPT) and showed its superiority in extracting the fault features of rolling element bearings. Sadeghian et al. (2009) proposed an algorithm for online detection of rotor bar breakage in induction motors using the WPT and neural networks. Xian and Zeng (2009) proposed using the WPT for extraction of features from vibration signals of rotating machinery in the time-frequency domain and hybrid SVM to classify the patterns inherent in the features extracted through the WPT of different fault conditions. In (Al-Badour et al. 2011), the authors used the CWT and WPT to analyse the monitored vibration signals of rotating machinery. Boskoski and Juricic proposed a method for detecting various mechanical faults in rotating machines under variable operating conditions. The fault detection is performed by calculating the Renyi entropy and Jensen-Renyi divergence of the envelope probability density function (PDF). Here, the envelope PDF is computed using wavelet transform coefficients (Boskoski and Juricic 2012).

Readers who are interested in more details of the application of wavelets in machine fault diagnosis may refer to several systematic reviews of wavelets for fault diagnosis of rotating machines (e.g. Peng and Chu 2004; Yan et al. 2014; Chen et al. 2016).

5.4 Empirical Mode Decomposition (EMD)

The EMD method was developed based on the assumption that any signal comprises different simple intrinsic modes of oscillations. Each intrinsic mode, linear or nonlinear, represents a simple oscillation that can be represented by an intrinsic mode function (IMF). The EMD is a nonlinear and adaptive data analysis technique that decomposes the time domain signal $x(t)$ into different scales or IMFs (Huang et al. 1998; Wu and Huang 2009; Huang 2014). These IMFs must fulfil the following equation conditions: (i) in the whole dataset, the number of extrema and the number of zero-crossings must either be equal or differ at most by one; (ii) at any point, the mean value of the envelope defined by the local maxima and the envelope defined by the local minima is zero. The process of EMD to decompose a signal $x(t)$ can be described as follows:

1. Identify all the local extrema, and then connect all the local maxima by a cubic spline as the upper envelope.
2. Repeat the procedure for the local minima to produce the lower envelope. The upper and lower envelopes should cover all the data between them.
3. The mean of the upper and lower envelope values is designated as m_1, and the difference between the signal $x(t)$ and m_1 is the first component, h_1, i.e.

$$h_1 = x(t) - m_1 \tag{5.14}$$

Ideally, if h_1 is an IMF, then h_1 is the first component of $x(t)$.

4. If h_1 is not an IMF, h_1 is treated as the original signal, and repeat the above steps; then

$$h_{11} = h_1 - m_{11} \tag{5.15}$$

5. After repeated shifting up to k times, h_{1k} becomes an IMF, such that:

$$h_{1k} = h_{1(k-1)} - m_{1k} \tag{5.16}$$

Then it is designated as

$$C_1 = h_{1k} \tag{5.17}$$

To stop the shifting process, a stoppage criterion must be selected. In (Huang et al. 1998), the stoppage criterion is determined by using a Cauchy type of convergence test. In this test, the normalised squared difference between two successive shifting operations is defined using the following equation:

$$SD_k = \frac{\displaystyle\sum_{t=0}^{T} |h_{k-1}(t) - h_k(t)|^2}{\displaystyle\sum_{t=0}^{T} h_{k-1}^2} \tag{5.18}$$

If SD_k is smaller than a predetermined value, the shifting process will be stopped. After the first IMF C_1 is found, which should contain the finest scale or the shortest period component of the signal, it can be separated from the original signal $x(t)$ such that

$$r_1 = x(t) - C_1 \tag{5.19}$$

The residue r_1 is treated as the new signal to which the same shifting process described will be repeated. This will result in obtaining r_1 such that

$$r_2 = r_1 - C_2 \tag{5.20}$$

This process can be repeated up to $r_j's$ where $r = 1, 2, 3, \ldots$; this can be expressed mathematically using the following equation:

$$r_j = r_{j-1} - C_j \tag{5.21}$$

Or, more efficiently,

$$x(t) = \sum_{j=1}^{n} c_j + r_n \tag{5.22}$$

where c_j is the jth IMF and r_n is the residue of data $x(t)$ after the extraction of the n IMFs. Figure 5.7 shows an example of EMD of a vibration signal with IMF 1 to IMF 8.

Based on EMD, Wu and Huang presented the ensemble empirical mode decomposition (EEMD) technique to overcome the problem of a single IMF consisting of waves with widely disparate scales, or a signal of a similar scale residing in different IMF, which is also called mode mixing. The implication of mode mixing is that EMD results can be sensitive to noise included in the signal (Wu and Huang 2009). The EEMD process can be described in the following steps: (i) add a white noise series to the signal of interest, (ii) decompose the signal with the added white noise in the IMFs, (iii) repeat with a different realisation of the white noise series each time, and finally (iv) obtain the (ensemble) means of the corresponding IMFs of the decomposition.

Many researchers have utilised EMD to analyse vibration signals for the purpose of machine fault diagnosis. For example, Junsheng et al. proposed an energy operator demodulation approach based on EMD to extract the instantaneous frequencies and amplitudes of the multicomponent amplitude-modulated and frequency-modulated (AM-FM) signals. They applied this method to machinery fault diagnosis, and the analysis results showed its efficiency to extract the characteristics of machinery fault vibrations (Junsheng et al. 2007). Wu and Qu introduced an improved slope-based method (ISBM) to restrain the end effect in EMD and showed that nonstationary and nonlinear time series can be decomposed efficiently and accurately into a set of IMFs and a residual trend. For the purpose of machine fault diagnosis, the IMFs derived from this method are used to extract features of faults and remove the interference from the background noise and some relevant components. Industrial case studies on large rotating machinery showed the effectiveness of this method to analyse nonstationary, nonlinear time series (Wu and Qu 2008). Gao et al. (2008) showed that IMFs sometimes fail to reveal signal characteristics due to the effect of noise. Hence, they proposed adaptive filtering features of EMD and a combined mode function (CMF). The results of a practical fault signal from the high-pressure cylinder of a power generator in a thermal electric plant showed that EMD and CMF can extract fault characteristics

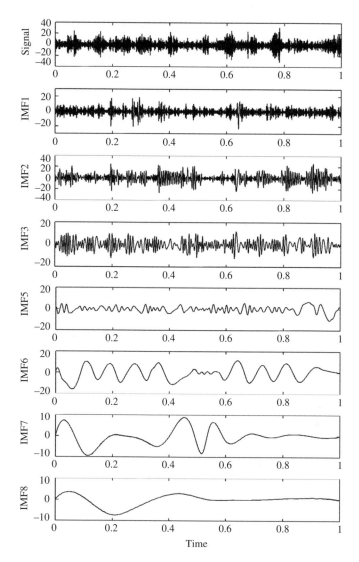

Figure 5.7 An example of EMD of a vibration signal with IMF 1 to IMF 8.

and identify fault patterns effectively (Gao et al. 2008). Lei et al. proposed a method based on EEMD to diagnose rotating machinery faults. Two vibration signals from a rub-impact fault in a power generator and an early rub-impact fault in a heavy oil catalytic cracking machine set were analysed utilising the proposed method to diagnose the faults. The results showed the effectiveness of this method in extracting the fault characteristics and identifying the faults (Lei et al. 2009). Bin et al. proposed a method combining the WPT, EMD, and backpropagation (BP) neural network for early fault diagnosis of rotating machinery. In this method, the acquired vibration signal is first decomposed using the WPT method, then the IMFs is obtained using the EMD method, and then the energy of the IMFs is proposed as an eigenvector to effectively express the failure feature; the results showed its effectiveness to eliminate the influence of false information and improve the accuracy of fault diagnosis (Bin et al. 2012).

Furthermore, Zhang and Zhou presented a procedure based on EEMD and optimised SVM for multi-fault diagnosis of rolling element bearings. Two types of features are extracted: EEMD energy entropy, which is used to specify whether the bearing has faults; and a singular value vector that is combined with optimised SVM to specify the fault type (Zhang and Zhou 2013). Dybała and Zimroz presented a fault-diagnosis method for rolling bearings based on EMD and CMFs. In this method, (i) the raw vibration signal is decomposed into a number of IMFs, (ii) CMFs are applied to combine neighbouring IMFs, and finally (iii) the signal is divided into three parts: noise-only IMFs, signal-only IMFs, and trend-only IMFs based on the Pearson correlation coefficient of each IMF and the empirically determined local mean of the original signal. To extract bearing fault features from the resultant signals, spectral analysis is used. The results of applying this method to raw vibration signals, which are generated by complex mechanical systems used in industry, showed that the proposed method can identify bearing faults at early stages of their development (Dybała and Zimroz 2014). Ali et al. suggested a feature-extraction method based on EMD energy entropy. They presented a mathematical analysis to select the most significant IMFs. With the selected features, an ANN is used to classify bearing defects. Experimental results indicate that the proposed method can reliably categorise bearing faults (Ali et al. 2015). In addition to these studies, a good and systematic review of EMD for fault diagnosis of rotating machines can be found in (Lei et al. 2013).

5.5 Hilbert-Huang Transform (HHT)

The Hilbert transform (HT) is a way to calculate the instantaneous frequency of a signal $x(t)$. This instantaneous frequency has the ability to divulge intra-wave frequency modulations, which can be used as a meaningful description of a signal function $x(t)$ (Boashash 1992 and Huang 2014). The Hilbert transform (HT) of a signal $x(t)$ can be described by its complex conjugate $y(t)$ such that

$$y(t) = \frac{P}{\pi} \int_{-\infty}^{+\infty} \frac{x(\tau)}{t - \tau} d\tau \qquad (5.23)$$

Here, P is the principal value of the singular integral. With the HT, the analytic signal $z(t)$, which is a complex signal whose imaginary part is the Hilbert transform of its real part, of $x(t)$ can be expressed mathematically in the following equation,

$$z(t) = x(t) + iy(t) = a(t)e^{i\varphi(t)} \qquad (5.24)$$

where $a(t)$ is the instantaneous amplitude of $x(t)$ and can be computed using Eq. (5.25),

$$a(t) = [x^2(t) + y^2(t)]^{\frac{1}{2}} \qquad (5.25)$$

and $\varphi(t)$ is the instantaneous phase of $x(t)$ that can be expressed mathematically using the following equation:

$$\varphi(t) = arctan\left(\frac{y(t)}{x(t)}\right) \qquad (5.26)$$

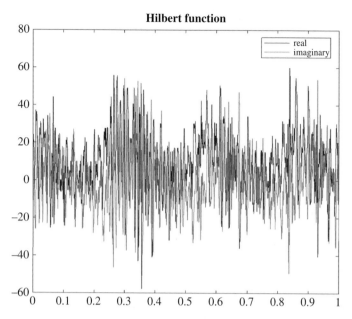

Figure 5.8 The real and imaginary signals of an outer race fault vibration signal produced using the Hilbert transform. (See insert for a colour representation of this figure.)

The instantaneous frequency, which is the time derivative of the instantaneous phase $\varphi(t)$, can be represented using the following equation:

$$w(t) = \frac{d\varphi(t)}{dt} \tag{5.27}$$

A simple way to achieve the Hilbert transform of a time domain signal $x(t)$ is by transforming into the frequency domain using the Fourier transform, shifting the phase angle of all components by $\pm\frac{\pi}{2}$, i.e. shifting by $\pm90°$, and then transforming back to the time domain. As an example, Figure 5.8 shows the real and imaginary signals of an OR fault vibration signal produced using the '*hilbert*' function in the Signal Processing Toolbox in MATLAB (https://uk.mathworks.com/help/matlab/index.html?s_tid=CRUX_lftnav). This function uses FFT to compute the Hilbert transform and returns the analytic signal $z(t)$ from $x(t)$. Another way to compute the analytic signal $z(t)$ is by decomposing a signal into IMFs using EMD, as described in Section 5.4, and then applying HT to all IMF components and computing the instantaneous frequency to all IMF; this technique is called HHT.

The vibration produced by rotating machines often contains nonlinear and nonstationary signals. As long as the Hilbert transform is appropriate for nonstationary and nonlinear signals, its application to mechanical vibration analysis has been studied in a large volume of published studies describing various techniques that can be used to analyse mechanical vibrations. For instance, Peng et al. proposed an improved HHT for vibration signal analysis and demonstrated that the improved HHT is a precise method for nonlinear and nonstationary signal analysis. In this method, the WPT is applied as a preprocessing step to decompose the signal into a set of narrowband signals before applying EMD. They showed that with the help of the WPT, each IMF produced from the EMD can correctly become monocomponent. Then, a screening process is performed to remove unrelated IMFs from the result (Peng et al. 2005b). Yu et al. introduced a method for roller bearing fault diagnosis using EMD and the Hilbert spectrum. In this method, (i) the vibration signal is decomposed using the WPT, (ii) the wavelet coefficients are analysed using HT, (iii) EMD is applied to the envelope signal to obtain IMFs, and finally (iv) some special IMFs are selected to obtain a local Hilbert marginal spectrum from which the faults can be diagnosed (Yu et al. 2005). Rai and Mohanty proposed a technique to estimate characteristic defect frequencies (CDFs) using HHT. In this technique, (i) the CDFs are determined using time domain EMD and HT, which generate monotonic IMFs; (ii) the generated IMFs are converted to the frequency domain using an FFT algorithm (Rai and Mohanty 2007).

Furthermore, Feldman (2011) presented a large number of examples devoted to illustrating key concepts of actual mechanical signals and demonstrated how the Hilbert transform offers advantages in machine fault diagnosis. Ming et al. found that the characteristic frequencies are quite clear in the envelope spectrum for further fault diagnosis. To compute this envelope spectrum, (i) the original signal is filtered using a cyclic Wiener filter, and (ii) the envelope spectrum is calculated using HT (Ming et al. 2011). Sawalhi and Randall investigated the defect vibration signal of a spalled bearing using vibration signatures of seeded faults at different speeds. This is achieved by considering the acceleration time signal responses resulting from a rolling element's entry into and exit from a typical spall. A method of joint treatment for the two events is proposed to extract the size of the spall. (i) The signal is pre-whitened using the autoregressive (AR) model. (ii) The pre-whitened signal is filtered using complex Morlet wavelets. (iii) The filtered signal with similar frequency content is selected, and its squared envelope signal is computed using HT and improved using EMD. (iv) The real cepstrum is utilised to determine the average separation of the two pulses (Sawalhi and Randall 2011). Soualhi et al. presented a technique that combines the HHT, the SVM, and the support vector regression (SVR). Experimental results using this method showed its suitability to improve the detection, diagnosis, and prognostics of bearing degradation (Soualhi et al. 2015).

5.6 Wigner-Ville Distribution

The Wigner distribution (WD) was first introduced in (Wigner 1932) and can be derived by generalising the relationship between the power spectrum and the autocorrelation function for a nonstationary, time-variant process. The WD for a signal $x(t)$ can be expressed mathematically by the following equation,

$$W_x(t,f) = \int_{-\infty}^{+\infty} x\left(t + \frac{\tau}{2}\right) x^*\left(t + \frac{\tau}{2}\right) e^{-2\pi ft} d\tau \qquad (5.28)$$

where $x^*(t)$ is the complex conjugate of $x(t)$.

The Wigner-Ville distribution (WVD) is defined as the WD for the analytic signal $z(t)$ such that,

$$W_z(t,f) = F_{\tau \to f}\left\{ z\left(t + \frac{\tau}{2}\right) z^*\left(t - \frac{\tau}{2}\right) \right\} \qquad (5.29)$$

Here, $z(t)$ can be represented mathematically by the following equation,

$$z(t) = x(t) + j\hat{x}(t) \tag{5.30}$$

where $\hat{x}(t)$ is the Hilbert transform of $x(t)$.

Various studies have assessed the efficacy of the WVD in machine fault diagnosis. For instance, Shin and Jeon investigated and applied the pseudo-Wigner-Ville distribution (PWVD) to analyse nonstationary signals during machinery condition monitoring, and they showed that the PWVD is ideal for portraying nonstationary time signals as well as stationary signals (Shin and Jeon 1993). Staszewski et al. presented a study of the WVD in gearbox condition monitoring and showed that the WVD is capable of detecting local tooth faults in spur gears. In this study, the WVD-based chosen features were good enough to separate the gearbox vibration data into the appropriate classes using neural network analysis (Staszewski et al. 1997). Stander et al. presented a vibration waveform normalisation approach that enables the use of the PWVD to indicate deteriorating fault conditions under fluctuating load conditions. Statistical parameters and various other features were extracted from the distribution (Stander et al. 2002). Li and Zhang proposed a method for fault diagnosis of bearings using the WVD based on EMD. In this method, (i) the original time series data is decomposed using EMD to obtain IMFs; (ii) the WVD is calculated for the selected IMF. The experimental results showed that the proposed method can be used effectively to diagnosis bearing faults (Li and Zhang 2006). Blodt et al. proposed a method that deals with the detection of mechanical load faults in induction motors during speed transients. In this method, the fault indicators are extracted from the WD for online condition monitoring. The experimental results demonstrated the practical feasibility of using time-frequency methods for online condition monitoring (Blodt et al. 2008).

Furthermore, Tang et al. proposed a wind turbine fault diagnosis method based on the Morlet wavelet transform and WVD. In this method, the raw signal is denoised using a Morlet wavelet transform process in which the appropriate scale parameter for the CWT is optimised using the cross-validation method (CVM). In addition, an auto terms window (ATW) function is used to suppress the cross terms in the WVD. The results indicate that the wind turbine gearbox fault feature is much clearer using the proposed method (Tang et al. 2010). Ibrahim and Albarbar compared the EMD and smoothed pseudo Wigner-Ville distribution (SPWVD) methods based on vibration signatures and energy calculation procedures for monitoring gearbox systems. This comparison showed that the calculation of energy using the EMD technique offers a more effective way to detect early faults than computations using the SPWVD method, and the computation of energy using the EMD technique is faster than the calculation done using the SPWVD method (Ibrahim and Albarbar 2011). Yang et al. proposed a data-driven fault-diagnosis method for wind turbine generator bearings based on the idea of sparse representation and shift-invariant dictionary learning to extract different impulsive components from vibration signals. In this method, the impulse signals at different locations with the same features can be represented by only one atom through a shift operation. Then, the shift-invariant dictionary is generated by taking all the possible shifts of a few short atoms. Based on the learned shift-invariant dictionary, the coefficients obtained can be sparse, with the extracted impulse signal being closer to the real signal. Finally, the time-frequency representation of the impulse component is achieved using both the WVD of every atom and the corresponding sparse coefficients (Yang et al. 2017).

5.7 Local Mean Decomposition (LMD)

The LMD is an adaptive analysis technique that decomposes a complicated signal into a set of product functions (PFs), each of which is the product of an envelope signal and a purely frequency modulated (FM) signal. Moreover, the complete time-frequency distribution of the original signal is derived. The LMD procedure for a time domain signal $x(t)$ can be described as follows:

1. Find all the local extremas $(n_1, n_2, n_3, \ldots)$
2. Compute the local mean value and the local envelope estimate of the maximum and minimum points of each half-wave oscillation of the signal. Thus the ith mean value (m_i) is given by Eq. (5.31):

$$m_i = \frac{(n_i + n_{i+1})}{2} \tag{5.31}$$

The ith envelope estimate (a_i) is given by Eq. (5.32):

$$a_i = \frac{|n_i - n_{i+1}|}{2} \tag{5.32}$$

3. Obtain the smoothed varying continuous local mean function $m_{11}(t)$ and the smoothed varying continuous envelope function $a_{11}(t)$ using a moving average (MA).
4. Subtract $m_{11}(t)$ from $x(t)$ to obtain the residual signal $h_{11}(t)$:

$$h_{11}(t) = x(t) - m_{11}(t) \tag{5.33}$$

5. Divide $h_{11}(t)$ by $a_{11}(t)$ to obtain $s_{11}(t)$:

$$s_{11}(t) = h_{11}(t)/a_{11}(t) \tag{5.34}$$

6. Compute envelope $a_{12}(t)$ of $s_{11}(t)$. If $a_{12}(t) \neq 1$, the process needs to be repeated for $s_{11}(t)$.
7. Compute a smoothed local mean $m_{12}(t)$ for $s_{11}(t)$, subtract it from $s_{11}(t)$ to obtain $h_{12}(t)$, and divide $h_{12}(t)$ by $a_{12}(t)$ to obtain $s_{12}(t)$. Repeat this process n times until a purely frequency modulated signal $s_{1n}(t)$ is obtained.
8. To obtain the envelope signal $a_1(t)$ of the first PF_1, multiply all the smoothed local envelopes during iteration:

$$a_1(t) = a_{11}(t)a_{12}(t)a_{13}(t) \ldots a_{1n}(t) \tag{5.35}$$

9. Compute the first PF_1 using $a_1(t)$ and the final frequency modulated $s_{1n}(t)$:

$$PF_1 = a_1(t)s_{1n}(t) \tag{5.36}$$

10. The smoothed version of the signal can be computed by subtracting the first PF_1 from the original signal:

$$u_1(t) = x(t) - PF_1 \tag{5.37}$$

This procedure is repeated m times; finally, $x(t)$ can be denoted as

$$x(t) = \sum_{i=1}^{m} PF_i(t) + u_m(t) \tag{5.38}$$

Here, m is the number of PFs.

Moving on now, various studies have assessed the efficacy of LMD in machine fault diagnosis. For instance, Wang et al. conducted a comparative study on LMD and EMD and their applications to diagnosing rotating machinery health. Their results reveal that LMD seems to be more suitable and have better performance than EMD for incipient fault detection (Wang et al. 2010b). In the same vein, Cheng et al. compared the performance of LMD and EMD in gear and roller bearing fault diagnosis. The analysis results demonstrated that the diagnosis approach based on LMD could identify gear and roller bearing conditions accurately and effectively (Cheng et al. 2012).

Liu and Han proposed a fault feature extraction method based on LMD and multiscale entropy (MSE). In this method, the multiscale entropies of each PF decomposed by LMD are used as the feature vectors. The analysis of practical bearing vibration signals showed the effectiveness of the proposed method (Liu and Han 2014).

Liu et al. proposed a hybrid fault-diagnosis method based on second-generation wavelet-denoising (SGWD) and LMD. In this method, (i) SGWD using neighbouring coefficients is employed as a noise remover in rotating machinery. (ii) LMD is used to decompose the denoised signals into several PFs, and (iii) the PF corresponding to the faulty feature signal is selected according to the correlation coefficients criterion. Finally, (iv) the frequency spectrum is analysed using FFT. Experimental results from the analysis of vibration signals collected from a gearbox and a real locomotive rolling bearing showed the effectiveness of this method (Liu et al. 2016b).

Li et al. introduced a fault-diagnosis method for rolling bearings based on LMD, improved multiscale fuzzy entropy (IMFE), Laplacian score (LS), and SVM. In their method, (i) LMD is used as a preprocessing step for the vibration signal, and (ii) the optimum PF is selected according to kurtosis features. (iii) The IMFE is used as a feature extractor to compute the multiscale fuzzy entropy of the optimum PF. (iv) LS is used to select the best scale factor according to their importance and distinguishability. Finally, (v) an improved SVM is adopted to deal with the classification problem. The experimental rolling bearing fault diagnosis showed the superiority of this method in identifying different categories and severities of faults in rolling bearings (Li et al. 2016).

Moreover, Feng et al. presented a systematic review of adaptive mode decomposition in two main types of algorithms and their application in signal analysis for machinery fault diagnosis. The first type are mono-component decomposition algorithms including EMD, LMD, intrinsic time-scale decomposition, local characteristic scale decomposition, Hilbert vibration decomposition, empirical wavelet transform, variation mode decomposition, nonlinear mode decomposition, and adaptive local iterative filtering. The second type are instantaneous frequency-estimation approaches including HHT, direct quadrature, and the normalised HT based on empirical AM-FM decomposition (Feng et al. 2017).

Yu and Lv proposed a method based on LMD and multilayer hybrid denoising (MHD) for feature extraction of early faults occurring in bearings. Based on the decomposed PFs by LMD, (i) an amplitude indication decision is used to select the effective PFs.

(ii) Wwavelet thresholding denoising (WTD) is used as a prefilter of a singular value decomposition (SVD) implementation that enables the preserved singular values to contain the most important information of the PFs. Finally, (iii) SVD is used to extract the most important principal features of the Hankel matrix of the PFs. The results of experimental analysis on vibration signal collected from rolling bearings showed that local mean decomposition-multilayer hybrid denoising (LMD-MHD) is effective for extracting weak features and performs well for bearing fault diagnosis (Yu and Lv 2017).

Li et al. introduced a method for early fault diagnosis in rotating machinery. In this method, (i) an improved LMD method called differential rational spline-based LMD is used to avoid the mode-mixing problem. (ii) With the obtained PFs, the Kullback-Leibler (K-L) divergence is applied to select the main PF components that contain the most fault information. Results of testing this method on experimental vibration signals showed its effectiveness in accurately decomposing the signals and in detecting early faults in gears and rolling bearings (Li et al. 2017).

5.8 Kurtosis and Kurtograms

As described in Chapter 3, the kurtosis is the fourth normalised central statistical moment, which measures whether the peak is higher or lower than the peak of the distribution. Based on the idea of the kurtosis in the time domain, spectral kurtosis (SK), as a frequency domain kurtosis (FDK) technique, was first introduced by Dwyer (1983). The FDK is defined as the ratio of the expected value of the fourth-order central moment of the STFT to the expected value of the square of the second-order central moment of the STFT (Dwyer 1984). This can be expressed mathematically using the following equation,

$$x_{FDK}(f) = \frac{E\{[x(q,f_p)]^4\}}{E\{[(x(q,f))^2]^2\}} \tag{5.39}$$

where

$$x(q,f) = \sqrt{\frac{h}{m}} \sum_{i=0}^{m-1} w_i x(i,q) exp(-jf_p i) \tag{5.40}$$

where h is the interval between successive observations of the process, $w_i = 1, f = 2\pi/m$, $q = 1, 2, \ldots n, i = p = 0, 1, 2, \ldots m$, and $j = \sqrt{-1}$. Also, $x(i, q)$ is the input signal and can be represented by the following equation:

$$x(i,q) = x[i + (q-1)m)n] \tag{5.41}$$

On the other hand, the simple idea of the SK technique is to utilise the kurtosis at each frequency as a measure to discover the presence of non-Gaussian components and to indicate the frequency bands in which these non-Gaussian components happen. SK is a representation of the kurtosis of each frequency component of a signal (de la Rosa et al. 2013). To compute the SK, the signal is first decomposed into the time-frequency domain, where the kurtosis values are defined for each frequency group. The SK of a signal $x(t)$, $x_{SK}(f)$ is defined as the fourth-order normalised cumulant. This can be computed using the following equation (Leite et al. 2015),

$$x_{SK}(f) = \frac{\langle |H(t,f)|^4 \rangle}{\langle |H(t,f)|^2 \rangle^2} - 2 \tag{5.42}$$

where $H(t,f)$ is the complex envelope function of $x(t)$ at frequency f obtained using the STFT algorithm:

$$H(t,f) = \int_{-\infty}^{+\infty} x(\tau) w(\tau - t) e^{-j2\pi f \tau} d\tau \tag{5.43}$$

The kurtogram algorithm (KUR) was first introduced by Antoni and Randall, and comes from the SK (Antoni and Randall 2006). The KUR computes the SK for several window sizes using a bandpass filter. In this bandpass filter, the central frequency f_c and the bandwidth can be determined with which to jointly maximise the kurtogram. Here, for all possible central frequencies and bandwidths, all possible window sizes should be considered. This will lead to increased computational cost in real applications. To overcome this problem, Antoni proposed a fast kurtogram based on the multirate filter-bank (MFB) structure and quasi-analytic filters. The fast KUR uses a series of digital filters rather than the STFT. Results from fast kurtogram showed the efficiency of this method: the results of the fast kurtogram can be computed more quickly than the kurtogram, and yet the results of the fast kurtogram are very similar to those results of kurtogram filters (Antoni 2007; Randall 2011). A more detailed explanation of the SK and KUR and their application for fault detection, diagnosis, and prognostics in rotating machines can be found in (Wang et al. 2016).

A considerable amount of literature has been published on the application of the SK and KUR in machine fault diagnosis. For instance, in 2006, Antoni and Randall published a paper in which they demonstrated how the SK can be elegantly and very efficiently applied to the vibratory monitoring of rotating machines. The basic idea is to use the high sensitivity of the SK for detecting and featuring incipient faults that produce impulse-like signals. Their methodology suggested two steps for how the SK can be used: (i) Use SK as a detection tool that precisely observes in which frequency band(s) the fault shows the best contrast form background noise; this step finds useful applications in surveillance protocols. (ii) Use the SK as a basis for designing ad hoc detection filters that can extract the mechanical signature of the fault; this step is of evident utility for diagnostic purposes (Antoni and Randall 2006).

In 2007, Sawalhi et al. introduced an algorithm for enhancing the surveillance capability of SK by using the minimum entropy deconvolution (MED) technique. The results showed that the use of the MED technique dramatically sharpens the pulses originating from the impacts of the balls with the spall and increases the kurtosis values to a level that reflects the severity of the fault. In addition, the algorithm was tested on signals collected from a gearbox for a bearing with a spalled OR. The results showed that the use of the MED along with SK analysis greatly improves the results of envelope analysis for making a complete diagnosis of the fault and trending its progression (Sawalhi et al. 2007).

Immovilli introduced a method for detection of generalised-roughness bearing faults using the SK energy of vibrations or current signals. In this method, the kurtogram of the SK was used to identify the bandwidth where the effect of the fault was stronger. Then, the energy of the signal in this bandwidth was used as a diagnostic feature. The experiments showed the effectiveness and reliability of this method for vibration monitoring (Immovilli et al. 2009).

Zhang and Randall proposed an approach for rolling element bearings using the fast KUR and a genetic algorithm to select an optimum bandpass filter. The experimental investigation showed the effectiveness of this method compared with traditional envelope analysis and with the fast KUR alone (Zhang and Randall 2009). Lei et al. proposed a method for fault diagnosis of rolling element bearings using an improved KUR method adopting the WPT as the filter of the KUR to overcome the shortcoming of the original KUR. The results of using this method on rolling bearing vibration signals verified the effectiveness of this method (Lei et al. 2011).

Furthermore, Wang et al. proposed an enhanced KUR for fault diagnosis of rolling element bearings where kurtosis values are calculated based on the power spectrum of the envelope of the signals extracted from wavelet packet nodes at different depths. The power spectrum of the envelope of the signals defines the sparse representation of the signals, and the kurtosis measures the protrusion of the sparse representation. The results showed the effectiveness of this method in the detection of various faults (Wang et al. 2013).

Liu et al. proposed gearbox fault feature extraction using an adaptive SK filtering technique to extract the signal transient based on a Morlet wavelet. In this method, the Morlet wavelet is used as a filter bank whose centre frequency is defined by wavelet correlation filtering. Different bandwidth filters in the filter bank are used to select the optimal filter, which maximises the SK for extracting the signal transients. The results of applying this method in the extraction of the signal transients that show the gear fault showed the effectiveness of this method in practical application (Liu et al. 2014a).

Tian et al. introduced an approach for detecting bearing faults and monitoring the degradation of bearings in electric motors. In this method, the fault features are extracted using SK and cross-correlation. Then, the features are combined to form a health index using principal component analysis (PCA) and a semi-supervised k-nearest neighbour (k-NN) distance measure. The results presented the effectiveness of this method in detecting incipient faults and diagnosing locations of faults as well as providing a health index that tracks the degradation of faults without missing intermittent faults (Tian et al. 2016).

Recently, Wang et al. presented a time-frequency analysis method for rotating machinery fault diagnosis based on ensemble local mean decomposition (ELMD) and fast KUR. In this method, (i) the ELMD process starts by decomposing the raw signal into PFs, and (ii) the PFs with the most fault information are selected according to kurtosis index; finally, (iii) the selected PF signals are further filtered by an optimal bandpass filter based on the fast KUR. The fault identification is inferred by the appearance of fault characteristic frequencies in the squared envelope spectrum of the filtered signal. The efficiency of this method in fault diagnosis for rotating machinery is demonstrated on gearbox and rolling bearing case analysis (Wang et al. 2018).

To conclude, a considerable amount of literature has been published on the use of frequency domain and time-frequency domain techniques for rotating machine fault diagnosis. Table 5.1 presents a summary of the frequency and time-frequency domain vibration analysis techniques that have been used in different studies of machine fault

Table 5.1 Summary of the frequency and time-frequency domain analysis techniques that have been used in different studies of rotating machine condition monitoring.

Studies	FFT	CWT	DWT	WPT	STFT	HHT	EMD	LMD	WVD	SK	KUR
Lin and Qu 2000; Peng et al. 2005a				✓		✓	✓				
Peter et al. 2001; Paliwal et al. 2014	✓	✓									
Sun and Tang 2002; Luo et al. 2003; Hong and Liang 2007; Li et al. 2007; Zhu et al. 2009; Su et al. 2010; Kankar et al. 2011; Li et al. 2011		✓									
Nikolaou and Antoniadis 2002; Ocak et al. 2007; Wang et al. 2015				✓							
Prabhakar et al. 2002; Lou and Loparo 2004; Purushotham et al. 2005; Tyagi 2008; Djebala et al. 2008; Xian 2010; Kumar and Singh 2013; Ahmed et al. 2016			✓								
Yu et al. 2005						✓	✓				
Junsheng et al. 2006; Yu and Junsheng 2006; Zhao et al. 2013; Dybała and Zimroz 2014							✓				
Li and Zhang 2006; Li et al. 2006							✓		✓		
Antoni and Randall 2006										✓	✓
Sawalhi et al. 2007										✓	
Rai and Mohanty. 2007; Pang et al. 2018	✓					✓					
Li et al. 2008			✓	✓							
Zhang and Randall 2009											✓

Table 5.1 (Continued)

Studies	FFT	CWT	DWT	WPT	STFT	HHT	EMD	LMD	WVD	SK	KUR
Immovilli et al. 2009; Wang and Liang 2011										✓	
Wensheng et al. 2010										✓	
Lei et al. 2011; Wang et al. 2013				✓			✓				✓
Zhou et al. 2011									✓		
Linsuo et al. 2011					✓				✓	✓	
Cheng et al. 2012; Liu and Han 2014; Tian et al. 2015; Li et al. 2016								✓			
Cozorici et al. 2012	✓	✓									
Jiang et al. 2013	✓			✓			✓				
Singhal and Khandekar 2013; Lin et al. 2016						✓					
Liu et al. 2014b					✓						
Liu et al. 2016a,b					✓						
Jacop et al. 2017	✓	✓									

diagnosis. According to these studies, these techniques can be used individually or in a mixture of several techniques to extract features from raw vibration data. There have also been many comprehensive review papers (Feng et al. 2013; Henao et al. 2014; Lee et al. 2014; Riera-Guasp et al. 2015; de Azevedo et al. 2016).

5.9 Summary

This chapter has introduced signal processing in the time-frequency domain and provided an explanation of several techniques that can be used to examine time-frequency characteristics of the time-indexed series signal, which can be obtained more effectively than the Fourier transform and its corresponding frequency spectrum features. Various techniques have been proposed to extract features from the raw vibration signals in the time-frequency domain. These include: (i) the STFT, which computes the DFT by decomposing a signal into shorter segments of equal length using a time-localised window function. (ii) Wavelet analysis decomposes the signal based on a family of 'wavelets'. Unlike the window used with the STFT, the wavelet function is scalable, which makes it adaptable to a wide range of frequencies and time-based resolution;

Table 5.2 Summary of publicly accessible software for some of the techniques introduced in this chapter.

Algorithm name	Platform	Package	Function
Spectrogram using STFT	MATLAB	Signal processing toolbox	spectrogram
Discrete-time analytic signal using the Hilbert transform			hilbert
Display wavelet family names	MATLAB	Wavelet Toolbox	Waveletfamilies ('f')
Display wavelet families with their corresponding properties			Waveletfamilies ('a')
Denoising of a one-dimensional signal using wavelets			wden
The continuous 1-D wavelet transform			cwt
The inverse continuous 1-D wavelet transform			icwt
Continuous wavelet transform	Python	Gregory R. Lee et al. (2018)	pywt.cwt
Discrete wavelet transform			pywt.dwt
Fast kurtogram	MATLAB	Antoni J. (2016)	Fast_Kurtogram
Visualise spectral kurtosis	MATLAB		kurtogram
Wigner-Ville distribution and smoothed pseudo Wigner-Ville distribution			wvd
Empirical mode decomposition			Emd
Spectral entropy of signal			pentropy

the three main transforms in wavelets analysis are the CWT, DWT, and WPT. (iii) The EMD decomposes the signal into different scales of IMFs. (iv) The HHT can be achieved by decomposing a signal into IMFs using EMD, applying HT to all IMF components, and computing the instantaneous frequency to all IMF. The HT can be achieved by transforming into the frequency domain using Fourier transform, shifting the phase angle of all components by $\pm\frac{\pi}{2}$, i.e. shifting by $\pm 90°$, and then transforming back to the time domain. (v) The WVD can be derived by generalising the relationship between the power spectrum and the autocorrelation function for nonstationary, time-variant process. And (vi) the SK can be achieved by decomposing the signal into the time-frequency domain where the kurtosis values are defined for each frequency group, while the KUR computes the SK for several window size using a bandpass filter. Most of the introduced techniques and their publicly accessible software are summarised in Table 5.2.

References

Ahmed, H.O.A., Wong, M.D., and Nandi, A.K. (2016). Effects of deep neural network parameters on classification of bearing faults. In: *IECON 2016-42nd Annual Conference of the IEEE Industrial Electronics Society*, 6329–6334. IEEE.

Ahmed, H.O.A., Wong, M.L.D., and Nandi, A.K. (2018). Intelligent condition monitoring method for bearing faults from highly compressed measurements using sparse over-complete features. *Mechanical Systems and Signal Processing* 99: 459–477.

Al-Badour, F., Sunar, M., and Cheded, L. (2011). Vibration analysis of rotating machinery using time–frequency analysis and wavelet techniques. *Mechanical Systems and Signal Processing* 25 (6): 2083–2101.

Ali, J.B., Fnaiech, N., Saidi, L. et al. (2015). Application of empirical mode decomposition and artificial neural network for automatic bearing fault diagnosis based on vibration signals. *Applied Acoustics* 89: 16–27.

Antoni, J. (2007). Fast computation of the kurtogram for the detection of transient faults. *Mechanical Systems and Signal Processing* 21 (1): 108–124.

Antoni, J. (2016). Fast kurtogram. Mathworks File Exchange Center. https://uk.mathworks .com/matlabcentral/fileexchange/48912-fast-kurtogram.

Antoni, J. and Randall, R.B. (2006). The spectral kurtosis: application to the vibratory surveillance and diagnostics of rotating machines. *Mechanical Systems and Signal Processing* 20 (2): 308–331.

Banerjee, T.P. and Das, S. (2012). Multi-sensor data fusion using support vector machine for motor fault detection. *Information Sciences* 217: 96–107.

Belšak, A. and Prezelj, J. (2011). Analysis of vibrations and noise to determine the condition of gear units. In: *Advances in Vibration Analysis Research*, 315–328. InTech.

Bin, G.F., Gao, J.J., Li, X.J., and Dhillon, B.S. (2012). Early fault diagnosis of rotating machinery based on wavelet packets—empirical mode decomposition feature extraction and neural network. *Mechanical Systems and Signal Processing* 27: 696–711.

Blodt, M., Bonacci, D., Regnier, J. et al. (2008). On-line monitoring of mechanical faults in variable-speed induction motor drives using the Wigner distribution. *IEEE Transactions on Industrial Electronics* 55 (2): 522–533.

Boashash, B. (1992). Estimating and interpreting the instantaneous frequency of a signal. II. Algorithms and applications. *Proceedings of the IEEE* 80 (4): 540–568.

Boskoski, P. and Juricic, D. (2012). Fault detection of mechanical drives under variable operating conditions based on wavelet packet Renyi entropy signatures. *Mechanical Systems and Signal Processing* 31: 369–381.

Brandt, A. (2011). *Noise and Vibration Analysis: Signal Analysis and Experimental Procedures*. Wiley.

Burrus, C.S., Gopinath, R.A., Guo, H. et al. (1998). *Introduction to Wavelets and Wavelet Transforms a Primer*, vol. 1. NJ: Prentice Hall.

Chen, B., Zhang, Z., Sun, C. et al. (2012). Fault feature extraction of gearbox by using overcomplete rational dilation discrete wavelet transform on signals measured from vibration sensors. *Mechanical Systems and Signal Processing* 33: 275–298.

Chen, J., Li, Z., Pan, J. et al. (2016). Wavelet transform based on inner product in fault diagnosis of rotating machinery: a review. *Mechanical Systems and Signal Processing* 70: 1–35.

Cheng, J., Yang, Y., and Yang, Y. (2012). A rotating machinery fault diagnosis method based on local mean decomposition. *Digital Signal Processing* 22 (2): 356–366.

Cohen, L. (1995). *Time-Frequency Analysis*, vol. 778. Prentice Hall.

Cozorici, I., Vadan, I., and Balan, H. (2012). Condition based monitoring and diagnosis of rotating electrical machines bearings using FFT and wavelet analysis. *Acta Electrotehnica* 53 (4): 350–354.

Czarnecki, K. (2016). The instantaneous frequency rate spectrogram. *Mechanical Systems and Signal Processing* 66: 361–373.

Daubechies, I. (1990). The wavelet transform, time-frequency localization and signal analysis. *IEEE Transactions on Information Theory* 36 (5): 961–1005.

de Azevedo, H.D.M., Araújo, A.M., and Bouchonneau, N. (2016). A review of wind turbine bearing condition monitoring: state of the art and challenges. *Renewable and Sustainable Energy Reviews* 56: 368–379.

de la Rosa, J.J.G., Sierra-Fernández, J.M., Agüera-Pérez, A. et al. (2013). An application of the spectral kurtosis to characterize power quality events. *International Journal of Electrical Power & Energy Systems* 49: 386–398.

Djebala, A., Ouelaa, N., and Hamzaoui, N. (2008). Detection of rolling bearing defects using discrete wavelet analysis. *Meccanica* 43 (3): 339–348.

Donoho, D.L. and Johnstone, I.M. (1994). Threshold selection for wavelet shrinkage of noisy data. In: *Engineering in Medicine and Biology Society, 1994. Engineering Advances: New Opportunities for Biomedical Engineers. Proceedings of the 16th Annual International Conference of the IEEE*, vol. 1, A24–A25. IEEE.

Dwyer, R. (1983). Detection of non-Gaussian signals by frequency domain kurtosis estimation. In: *Acoustics, Speech, and Signal Processing, IEEE International Conference on ICASSP'83*, vol. 8, 607–610. IEEE.

Dwyer, R. (1984). Use of the kurtosis statistic in the frequency domain as an aid in detecting random signals. *IEEE Journal of Oceanic Engineering* 9 (2): 85–92.

Dybała, J. and Zimroz, R. (2014). Rolling bearing diagnosing method based on empirical mode decomposition of machine vibration signal. *Applied Acoustics* 77: 195–203.

Feldman, M. (2011). Hilbert transform in vibration analysis. *Mechanical Systems and Signal Processing* 25 (3): 735–802.

Feng, Z., Liang, M., and Chu, F. (2013). Recent advances in time-frequency analysis methods for machinery fault diagnosis: a review with application examples. *Mechanical Systems and Signal Processing* 38 (1): 165–205.

Feng, Z., Zhang, D., and Zuo, M.J. (2017). Adaptive mode decomposition methods and their applications in signal analysis for machinery fault diagnosis: a review with examples. *IEEE Access* 5: 24301–24331.

Gabor, D. (1946). Theory of communication. Part 1: the analysis of information. *Journal of the Institution of Electrical Engineers-Part III: Radio and Communication Engineering* 93 (26): 429–441.

Gao, Q., Duan, C., Fan, H., and Meng, Q. (2008). Rotating machine fault diagnosis using empirical mode decomposition. *Mechanical Systems and Signal Processing* 22 (5): 1072–1081.

He, M. and He, D. (2017). Deep learning based approach for bearing fault diagnosis. *IEEE Transactions on Industry Applications* 53 (3): 3057–3065.

Henao, H., Capolino, G.A., Fernandez-Cabanas, M. et al. (2014). Trends in fault diagnosis for electrical machines: a review of diagnostic techniques. *IEEE Industrial Electronics Magazine* 8 (2): 31–42.

Hong, H. and Liang, M. (2007). Separation of fault features from a single-channel mechanical signal mixture using wavelet decomposition. *Mechanical Systems and Signal Processing* 21 (5): 2025–2040.

Hong, H. and Liang, M. (2009). Fault severity assessment for rolling element bearings using the Lempel–Ziv complexity and continuous wavelet transform. *Journal of Sound and Vibration* 320 (1–2): 452–468.

Hu, Q., He, Z., Zhang, Z., and Zi, Y. (2007). Fault diagnosis of rotating machinery based on improved wavelet package transform and SVMs ensemble. *Mechanical Systems and Signal Processing* 21 (2): 688–705.

Huang, N.E. (2014). *Hilbert-Huang Transform and Its Applications*, vol. 16. World Scientific.

Huang, N.E., Shen, Z., Long, S.R. et al. (1998). The empirical mode decomposition and the Hilbert spectrum for nonlinear and non-stationary time series analysis. *Proceedings of the Royal Society of London A: Mathematical, Physical and Engineering Sciences* 454 (1971): 903–995.

Ibrahim, G.R. and Albarbar, A. (2011). Comparison between Wigner–Ville distribution-and empirical mode decomposition vibration-based techniques for helical gearbox monitoring. *Proceedings of the Institution of Mechanical Engineers, Part C: Journal of Mechanical Engineering Science* 225 (8): 1833–1846.

Immovilli, F., Cocconcelli, M., Bellini, A., and Rubini, R. (2009). Detection of generalized-roughness bearing fault by spectral-kurtosis energy of vibration or current signals. *IEEE Transactions on Industrial Electronics* 56 (11): 4710–4717.

Jacop, A., Khang, H.V., Robbersmyr, K.G., and Cardoso, A.J.M. (2017). Bearing fault detection for drivetrains using adaptive filters based wavelet transform. In: *2017 20th International Conference on Electrical Machines and Systems (ICEMS)*, 1–6. IEEE.

Jiang, H., Li, C., and Li, H. (2013). An improved EEMD with multiwavelet packet for rotating machinery multi-fault diagnosis. *Mechanical Systems and Signal Processing* 36 (2): 225–239.

Junsheng, C., Dejie, Y., and Yu, Y. (2006). A fault diagnosis approach for roller bearings based on EMD method and AR model. *Mechanical Systems and Signal Processing* 20 (2): 350–362.

Junsheng, C., Dejie, Y., and Yu, Y. (2007). The application of energy operator demodulation approach based on EMD in machinery fault diagnosis. *Mechanical Systems and Signal Processing* 21 (2): 668–677.

Kankar, P.K., Sharma, S.C., and Harsha, S.P. (2011). Fault diagnosis of ball bearings using continuous wavelet transform. *Applied Soft Computing* 11 (2): 2300–2312.

Konar, P. and Chattopadhyay, P. (2011). Bearing fault detection of induction motor using wavelet and support vector machines (SVMs). *Applied Soft Computing* 11 (6): 4203–4211.

Kumar, R. and Singh, M. (2013). Outer race defect width measurement in taper roller bearing using discrete wavelet transform of vibration signal. *Measurement* 46 (1): 537–545.

Lee, G.R., Gommers, R., Wohlfahrt, K. et al., … 0-tree (2018). *PyWavelets/Pywt: PyWavelets v1.0.1* (Version v1.0.1). Zenodo. http://doi.org/10.5281/zenodo.1434616 (accessed 08 September 2018).

Lee, J., Wu, F., Zhao, W. et al. (2014). Prognostics and health management design for rotary machinery systems—reviews, methodology and applications. *Mechanical Systems and Signal Processing* 42 (1–2): 314–334.

Lei, Y., He, Z., and Zi, Y. (2009). Application of the EEMD method to rotor fault diagnosis of rotating machinery. *Mechanical Systems and Signal Processing* 23 (4): 1327–1338.

Lei, Y., Lin, J., He, Z., and Zi, Y. (2011). Application of an improved kurtogram method for fault diagnosis of rolling element bearings. *Mechanical Systems and Signal Processing* 25 (5): 1738–1749.

Lei, Y., Lin, J., He, Z., and Zuo, M.J. (2013). A review on empirical mode decomposition in fault diagnosis of rotating machinery. *Mechanical Systems and Signal Processing* 35 (1–2): 108–126.

Leite, V.C., da Silva, J.G.B., Veloso, G.F.C. et al. (2015). Detection of localized bearing faults in induction machines by spectral kurtosis and envelope analysis of stator current. *IEEE Transactions on Industrial Electronics* 62 (3): 1855–1865.

Li, F., Meng, G., Ye, L., and Chen, P. (2008). Wavelet transform-based higher-order statistics for fault diagnosis in rolling element bearings. *Journal of Vibration and Control* 14 (11): 1691–1709.

Li, H., Fu, L., and Zheng, H. (2011). Bearing fault diagnosis based on amplitude and phase map of Hermitian wavelet transform. *Journal of Mechanical Science and Technology* 25 (11): 2731–2740.

Li, H. and Zhang, Y. (2006). Bearing faults diagnosis based on EMD and Wigner-Ville distribution. In: *The Sixth World Congress on Intelligent Control and Automation, 2006, WCICA 2006*, vol. 2, 5447–5451. IEEE.

Li, H., Zheng, H., and Tang, L. (2006). Wigner-Ville distribution based on EMD for faults diagnosis of bearing. In: *International Conference on Fuzzy Systems and Knowledge Discovery*, 803–812. Berlin, Heidelberg: Springer.

Li, K., Chen, P., and Wang, H. (2012). Intelligent diagnosis method for rotating machinery using wavelet transform and ant colony optimization. *IEEE Sensors Journal* 12 (7): 2474–2484.

Li, L., Qu, L., and Liao, X. (2007). Haar wavelet for machine fault diagnosis. *Mechanical Systems and Signal Processing* 21 (4): 1773–1786.

Li, P., Kong, F., He, Q., and Liu, Y. (2013). Multiscale slope feature extraction for rotating machinery fault diagnosis using wavelet analysis. *Measurement* 46 (1): 497–505.

Li, Y., Liang, X., Yang, Y. et al. (2017). Early fault diagnosis of rotating machinery by combining differential rational spline-based LMD and K–L divergence. *IEEE Transactions on Instrumentation and Measurement* 66 (11): 3077–3090.

Li, Y., Xu, M., Wang, R., and Huang, W. (2016). A fault diagnosis scheme for rolling bearing based on local mean decomposition and improved multiscale fuzzy entropy. *Journal of Sound and Vibration* 360: 277–299.

Lin, H.C., Ye, Y.C., Huang, B.J., and Su, J.L. (2016). Bearing vibration detection and analysis using an enhanced fast Fourier transform algorithm. *Advances in Mechanical Engineering* 8 (10), p. 1687814016675080.

Lin, J. and Qu, L. (2000). Feature extraction based on Morlet wavelet and its application for mechanical fault diagnosis. *Journal of Sound and Vibration* 234 (1): 135–148.

Linsuo, S., Yazhou, Z., and Wenpeng, M. (2011). Application of Wigner-Ville-distribution-based spectral kurtosis algorithm to fault diagnosis of rolling bearing. *Journal of Vibration, Measurement and Diagnosis* 1: 010.

Liu, B., Ling, S.F., and Meng, Q. (1997). Machinery diagnosis based on wavelet packets. *Journal of Vibration and Control* 3 (1): 5–17.

Liu, H. and Han, M. (2014). A fault diagnosis method based on local mean decomposition and multi-scale entropy for roller bearings. *Mechanism and Machine Theory* 75: 67–78.

Liu, H., Huang, W., Wang, S., and Zhu, Z. (2014a). Adaptive spectral kurtosis filtering based on Morlet wavelet and its application for signal transients detection. *Signal Processing* 96: 118–124.

Liu, H., Li, L., and Ma, J. (2016a). Rolling bearing fault diagnosis based on STFT-deep learning and sound signals. *Shock and Vibration* 2016, 12 pages.

Liu, H., Wang, X., and Lu, C. (2014b). Rolling bearing fault diagnosis under variable conditions using Hilbert-Huang transform and singular value decomposition. *Mathematical Problems in Engineering* 2014, pp 1-10 .

Liu, Z., He, Z., Guo, W., and Tang, Z. (2016b). A hybrid fault diagnosis method based on second generation wavelet de-noising and local mean decomposition for rotating machinery. *ISA Transactions* 61: 211–220.

Lopez, J.E., Yeldham, I.A., and Oliver, K. (1996). *Overview of Wavelet/Neural Network Fault Diagnostic Methods Applied to Rotating Machinery*. Burlington, MA: AlphaTech Inc.

Lou, X. and Loparo, K.A. (2004). Bearing fault diagnosis based on wavelet transform and fuzzy inference. *Mechanical Systems and Signal Processing* 18 (5): 1077–1095.

Luo, G.Y., Osypiw, D., and Irle, M. (2003). On-line vibration analysis with fast continuous wavelet algorithm for condition monitoring of bearing. *Modal Analysis* 9 (8): 931–947.

Mallat, S.G. (1989). A theory for multiresolution signal decomposition: the wavelet representation. *IEEE Transactions on Pattern Analysis and Machine Intelligence* 11 (7): 674–693.

Mertins, A. and Mertins, D.A. (1999). *Signal Analysis: Wavelets, Filter Banks, Time-Frequency Transforms and Applications*. Wiley.

Ming, Y., Chen, J., and Dong, G. (2011). Weak fault feature extraction of rolling bearing based on cyclic Wiener filter and envelope spectrum. *Mechanical Systems and Signal Processing* 25 (5): 1773–1785.

Nikolaou, N.G. and Antoniadis, I.A. (2002). Rolling element bearing fault diagnosis using wavelet packets. *NDT & E International* 35 (3): 197–205.

Ocak, H., Loparo, K.A., and Discenzo, F.M. (2007). Online tracking of bearing wear using wavelet packet decomposition and probabilistic modeling: a method for bearing prognostics. *Journal of Sound and Vibration* 302 (4–5): 951–961.

Paliwal, D., Choudhur, A., and Govandhan, T. (2014). Identification of faults through wavelet transform vis-à-vis fast Fourier transform of noisy vibration signals emanated from defective rolling element bearings. *Frontiers of Mechanical Engineering* 9 (2): 130–141.

Pang, B., Tang, G., Tian, T., and Zhou, C. (2018). Rolling bearing fault diagnosis based on an improved HTT transform. *Sensors* 18 (4): 1203.

Peng, Z.K. and Chu, F.L. (2004). Application of the wavelet transform in machine condition monitoring and fault diagnostics: a review with bibliography. *Mechanical Systems and Signal Processing* 18 (2): 199–221.

Peng, Z.K., Peter, W.T., and Chu, F.L. (2005a). A comparison study of improved Hilbert–Huang transform and wavelet transform: application to fault diagnosis for rolling bearing. *Mechanical Systems and Signal Processing* 19 (5): 974–988.

Peng, Z.K., Peter, W.T., and Chu, F.L. (2005b). An improved Hilbert–Huang transform and its application in vibration signal analysis. *Journal of Sound and Vibration* 286 (1–2): 187–205.

Peter, W.T., Peng, Y.A., and Yam, R. (2001). Wavelet analysis and envelope detection for rolling element bearing fault diagnosis—their effectiveness and flexibilities. *Journal of Vibration and Acoustics* 123 (3): 303–310.

Prabhakar, S., Mohanty, A.R., and Sekhar, A.S. (2002). Application of discrete wavelet transform for detection of ball bearing race faults. *Tribology International* 35 (12): 793–800.

Purushotham, V., Narayanan, S., and Prasad, S.A. (2005). Multi-fault diagnosis of rolling bearing elements using wavelet analysis and hidden Markov model based fault recognition. *NDT & E International* 38 (8): 654–664.

Rafiee, J., Rafiee, M.A., and Tse, P.W. (2010). Application of mother wavelet functions for automatic gear and bearing fault diagnosis. *Expert Systems with Applications* 37 (6): 4568–4579.

Rai, V.K. and Mohanty, A.R. (2007). Bearing fault diagnosis using FFT of intrinsic mode functions in Hilbert–Huang transform. *Mechanical Systems and Signal Processing* 21 (6): 2607–2615.

Randall, R.B. (2011). *Vibration-based condition monitoring: industrial, aerospace and automotive applications*. John Wiley & Sons.

Riera-Guasp, M., Antonino-Daviu, J.A., and Capolino, G.A. (2015). Advances in electrical machine, power electronic, and drive condition monitoring and fault detection: state of the art. *IEEE Transactions on Industrial Electronics* 62 (3): 1746–1759.

Sadeghian, A., Ye, Z., and Wu, B. (2009). Online detection of broken rotor bars in induction motors by wavelet packet decomposition and artificial neural networks. *IEEE Transactions on Instrumentation and Measurement* 58 (7): 2253–2263.

Saleh, S.A. and Rahman, M.A. (2005). Modeling and protection of a three-phase power transformer using wavelet packet transform. *IEEE Transactions on Power Delivery* 20 (2): 1273–1282.

Sawalhi, N. and Randall, R.B. (2011). Vibration response of spalled rolling element bearings: observations, simulations and signal processing techniques to track the spall size. *Mechanical Systems and Signal Processing* 25 (3): 846–870.

Sawalhi, N., Randall, R.B., and Endo, H. (2007). The enhancement of fault detection and diagnosis in rolling element bearings using minimum entropy deconvolution combined with spectral kurtosis. *Mechanical Systems and Signal Processing* 21 (6): 2616–2633.

Saxena, M., Bannet, O.O., Gupta, M., and Rajoria, R.P. (2016). Bearing fault monitoring using CWT based vibration signature. *Procedia Engineering* 144: 234–241.

Shen, C., Wang, D., Kong, F., and Peter, W.T. (2013). Fault diagnosis of rotating machinery based on the statistical parameters of wavelet packet paving and a generic support vector regressive classifier. *Measurement* 46 (4): 1551–1564.

Shin, Y.S. and Jeon, J.J. (1993). Pseudo Wigner–Ville time-frequency distribution and its application to machinery condition monitoring. *Shock and Vibration* 1 (1): 65–76.

Singhal, A. and Khandekar, M.A. (2013). Bearing fault detection in induction motor using fast Fourier transform. In: *IEEE International Conference on Advanced Research in Engineering & Technology*, 190–194.

Smith, C., Akujuobi, C.M., Hamory, P., and Kloesel, K. (2007). An approach to vibration analysis using wavelets in an application of aircraft health monitoring. *Mechanical Systems and Signal Processing* 21 (3): 1255–1272.

Soualhi, A., Medjaher, K., and Zerhouni, N. (2015). Bearing health monitoring based on Hilbert–Huang transform, support vector machine, and regression. *IEEE Transactions on Instrumentation and Measurement* 64 (1): 52–62.

Stander, C.J., Heyns, P.S., and Schoombie, W. (2002). Using vibration monitoring for local fault detection on gears operating under fluctuating load conditions. *Mechanical Systems and Signal Processing* 16 (6): 1005–1024.

Staszewski, W.J., Worden, K., and Tomlinson, G.R. (1997). Time–frequency analysis in gearbox fault detection using the Wigner–Ville distribution and pattern recognition. *Mechanical Systems and Signal Processing* 11 (5): 673–692.

Strangas, E.G., Aviyente, S., and Zaidi, S.S.H. (2008). Time–frequency analysis for efficient fault diagnosis and failure prognosis for interior permanent-magnet AC motors. *IEEE Transactions on Industrial Electronics* 55 (12): 4191–4199.

Su, W., Wang, F., Zhu, H. et al. (2010). Rolling element bearing faults diagnosis based on optimal Morlet wavelet filter and autocorrelation enhancement. *Mechanical Systems and Signal Processing* 24 (5): 1458–1472.

Sun, Q. and Tang, Y. (2002). Singularity analysis using continuous wavelet transform for bearing fault diagnosis. *Mechanical Systems and Signal Processing* 16 (6): 1025–1041.

Tang, B., Liu, W., and Song, T. (2010). Wind turbine fault diagnosis based on Morlet wavelet transformation and Wigner-Ville distribution. *Renewable Energy* 35 (12): 2862–2866.

Tian, J., Morillo, C., Azarian, M.H., and Pecht, M. (2016). Motor bearing fault detection using spectral kurtosis-based feature extraction coupled with K-nearest neighbor distance analysis. *IEEE Transactions on Industrial Electronics* 63 (3): 1793–1803.

Tian, Y., Ma, J., Lu, C., and Wang, Z. (2015). Rolling bearing fault diagnosis under variable conditions using LMD-SVD and extreme learning machine. *Mechanism and Machine Theory* 90: 175–186.

Tyagi, C.S. (2008). A comparative study of SVM classifiers and artificial neural networks application for rolling element bearing fault diagnosis using wavelet transform preprocessing. *Neuron* 1: 309–317.

Wang, D., Peter, W.T., and Tsui, K.L. (2013). An enhanced Kurtogram method for fault diagnosis of rolling element bearings. *Mechanical Systems and Signal Processing* 35 (1–2): 176–199.

Wang, H. and Chen, P. (2011). Fuzzy diagnosis method for rotating machinery in variable rotating speed. *IEEE Sensors Journal* 11 (1): 23–34.

Wang, L., Liu, Z., Miao, Q., and Zhang, X. (2018). Time-frequency analysis based on ensemble local mean decomposition and fast kurtogram for rotating machinery fault diagnosis. *Mechanical Systems and Signal Processing* 103: 60–75.

Wang, X., Makis, V., and Yang, M. (2010a). A wavelet approach to fault diagnosis of a gearbox under varying load conditions. *Journal of Sound and Vibration* 329 (9): 1570–1585.

Wang, Y., He, Z., and Zi, Y. (2010b). A comparative study on the local mean decomposition and empirical mode decomposition and their applications to rotating machinery health diagnosis. *Journal of Vibration and Acoustics* 132 (2): 021010.

Wang, Y. and Liang, M. (2011). An adaptive SK technique and its application for fault detection of rolling element bearings. *Mechanical Systems and Signal Processing* 25 (5): 1750–1764.

Wang, Y., Xiang, J., Markert, R., and Liang, M. (2016). Spectral kurtosis for fault detection, diagnosis and prognostics of rotating machines: a review with applications. *Mechanical Systems and Signal Processing* 66: 679–698.

Wang, Y., Xu, G., Liang, L., and Jiang, K. (2015). Detection of weak transient signals based on wavelet packet transform and manifold learning for rolling element bearing fault diagnosis. *Mechanical Systems and Signal Processing* 54: 259–276.

Wensheng, S., Fengtao, W., Zhixin, Z. et al. (2010). Application of EMD denoising and spectral kurtosis in early fault diagnosis of rolling element bearings. *Journal of Vibration and Shock* 29 (3): 18–21.

Wigner, E.P. (1932). On the quantum correction for thermodynamic equilibrium. *Physical Review* 40: 749–759.

Wu, F. and Qu, L. (2008). An improved method for restraining the end effect in empirical mode decomposition and its applications to the fault diagnosis of large rotating machinery. *Journal of Sound and Vibration* 314 (3–5): 586–602.

Wu, Z. and Huang, N.E. (2009). Ensemble empirical mode decomposition: a noise-assisted data analysis method. *Advances in Adaptive Data Analysis* 1 (1): 1–41.

Xian, G.M. (2010). Mechanical failure classification for spherical roller bearing of hydraulic injection molding machine using DWT–SVM. *Expert Systems with Applications* 37 (10): 6742–6747.

Xian, G.M. and Zeng, B.Q. (2009). An intelligent fault diagnosis method based on wavelet packer analysis and hybrid support vector machines. *Expert Systems with Applications* 36 (10): 12131–12136.

Yan, R., Gao, R.X., and Chen, X. (2014). Wavelets for fault diagnosis of rotary machines: a review with applications. *Signal Processing* 96: 1–15.

Yang, B., Liu, R., and Chen, X. (2017). Fault diagnosis for a wind turbine generator bearing via sparse representation and shift-invariant K-SVD. *IEEE Transactions on Industrial Informatics* 13 (3): 1321–1331.

Yen, G.G. and Lin, K.C. (2000). Wavelet packet feature extraction for vibration monitoring. *IEEE Transactions on Industrial Electronics* 47 (3): 650–667.

Yu, D., Cheng, J., and Yang, Y. (2005). Application of EMD method and Hilbert spectrum to the fault diagnosis of roller bearings. *Mechanical Systems and Signal Processing* 19 (2): 259–270.

Yu, J. and Lv, J. (2017). Weak fault feature extraction of rolling bearings using local mean decomposition-based multilayer hybrid denoising. *IEEE Transactions on Instrumentation and Measurement* 66 (12): 3148–3159.

Yu, Y. and Junsheng, C. (2006). A roller bearing fault diagnosis method based on EMD energy entropy and ANN. *Journal of Sound and Vibration* 294 (1–2): 269–277.

Zhang, L., Gao, R.X., and Lee, K.B. (2006). Spindle health diagnosis based on analytic wavelet enveloping. *IEEE Transactions on Instrumentation and Measurement* 55 (5): 1850–1858.

Zhang, X. and Zhou, J. (2013). Multi-fault diagnosis for rolling element bearings based on ensemble empirical mode decomposition and optimized support vector machines. *Mechanical Systems and Signal Processing* 41 (1–2): 127–140.

Zhang, Y. and Randall, R.B. (2009). Rolling element bearing fault diagnosis based on the combination of genetic algorithms and fast kurtogram. *Mechanical Systems and Signal Processing* 23 (5): 1509–1517.

Zhao, S., Liang, L., Xu, G. et al. (2013). Quantitative diagnosis of a spall-like fault of a rolling element bearing by empirical mode decomposition and the approximate entropy method. *Mechanical Systems and Signal Processing* 40 (1): 154–177.

Zheng, H., Li, Z., and Chen, X. (2002). Gear fault diagnosis based on continuous wavelet transform. *Mechanical Systems and Signal Processing* 16 (2–3): 447–457.

Zhou, Y., Chen, J., Dong, G.M. et al. (2011). Wigner–Ville distribution based on cyclic spectral density and the application in rolling element bearings diagnosis. *Proceedings of the Institution of Mechanical Engineers, Part C: Journal of Mechanical Engineering Science* 225 (12): 2831–2847.

Zhu, Z.K., He, Z., Wang, A., and Wang, S. (2009). Synchronous enhancement of periodic transients on polar diagram for machine fault diagnosis. *International Journal of Wavelets, Multiresolution and Information Processing* 7 (4): 427–442.

Part III

Rotating Machine Condition Monitoring Using Machine Learning

6

Vibration-Based Condition Monitoring Using Machine Learning

6.1 Introduction

As described in Chapter 1, machine condition monitoring (MCM) is an important part of condition-based maintenance (CBM), which is regarded as an efficient maintenance approach that can help in avoiding unnecessary maintenance tasks during the time-based maintenance (TBM) approach as well as the high cost of corrective maintenance, particularly for large-scale applications of rotating machines. Moreover, several studies have demonstrated the economic advantages of CBM in several applications of rotating machines (e.g. McMillan and Ault 2007; Verma et al. 2013; Van Dam and Bond 2015; Kim et al. 2016). In CBM, the decision to undertake maintenance is made based on the machine's current health condition, which can be recognised using a condition monitoring (CM) system. When a defect occurs in the machine, an accurate CM technique allows early detection of faults and correct identification of the type of defects. Accordingly, the more accurate and sensitive the CM system, the more precise the maintenance decision can be, and the more time is available to plan and do maintenance before a machine breakdown occurs.

The main goal of MCM is to avoid catastrophic machine failure that may cause secondary damage, downtime, potential safety incidents, lost production, and higher costs associated with repairs. CM techniques in rotating machinery involve the practice of monitoring measurable data, e.g. vibration, acoustic emission, electric current, etc., to classify changes in machine condition. Of these techniques, vibration-based CM has been extensively studied and has become a well-accepted technique for planned maintenance management. Typically, different fault conditions produce different forms of vibration spectra. As a result, vibration analysis in principle allows us to examine the inner parts of the operating machine and analyse its health condition based on the produced vibration signals without physically opening it. Additionally, various characteristic features can be detected from vibration signals, which makes it one of the best choices for MCM.

This chapter describes vibration-based MCM using machine learning algorithms. The first part of the chapter gives an overview of the vibration-based MCM process, describes the fault-detection and -diagnosis problem framework, and describes the types of learning that can be applied to vibration data. The second part defines the main problems of learning from vibration data for the purpose of fault diagnosis and describes techniques to prepare vibration data for analysis to overcome the aforementioned problems.

Condition Monitoring with Vibration Signals: Compressive Sampling and Learning Algorithms for Rotating Machines,
First Edition. Hosameldin Ahmed and Asoke K. Nandi.
© 2020 John Wiley & Sons Ltd. Published 2020 by John Wiley & Sons Ltd.

6.2 Overview of the Vibration-Based MCM Process

6.2.1 Fault-Detection and -Diagnosis Problem Framework

In vibration-based CM, by analysing the physical features of the acquired vibration signals, one is able to correctly categorise the acquired vibration signal as the corresponding condition, which is generally a multiclass classification problem. As shown in Figure 6.1, a simple vibration-based CM system consists of three key steps:

(1) *Data acquisition.* A sensor (e.g. velocity sensor or accelerometer) is mounted on the component of interest to collect input vibration measurements, i.e. raw data that can be transmitted, stored, and processed.
(2) *Vibration data analysis.* This step includes preprocessing, filtering, feature extraction, and selection of the vibration data acquired in step 1.
(3) *Machine health diagnosis.* This involves detection and identification of a fault by using a classifier to discriminate the data signals into different classes utilising the extracted features, while prognosis aims to predict the residual life of a machine before a breakdown takes place (Nandi and Jack 2004).

In step 1 of the vibration-based fault-detection and -diagnosis framework, real vibration data are collected using sensors such as accelerometers. With regard to the type of vibration data used in the field of machine health CM, Lei et al. (2018) stated that most of the research publications in the field of machine fault and prognosis used data acquired from accelerated degradation test beds instead of real industrial equipment. This is due to the fact it is difficult to collect high-quality run-to-failure data of machinery for academic research, for the following reasons:

- Machinery normally undergoes a long-term degradation procedure from health to failure that may take several months. Hence, it is expensive to collect all the run-to-failure data from a rotating machine with a long-term degradation process.
- Since an unforeseen machine failure may lead to unexpected machine downtime, accidents, and injuries, machines are not allowed to run to failure.
- Monitoring data from machines, such as wind turbine gearboxes, automotive gearboxes, and aircraft engines, are usually mixed with lots of interferences from the outside environment. Moreover, various monitoring data are taken during the out-of-service period, e.g. the breakdown or the restart time, which presents different behaviours compared with monitoring data acquired under the in-service period. These often decrease the quality of the data.
- Limited run-to-failure data sources from a few military and commercial institutions are available only to a few academics privileged to cooperate with these institutions.

Figure 6.1 The overall framework of vibration-based machine condition monitoring. (See insert for a colour representation of this figure.)

To develop and validate a machine fault diagnosis method, one may also utilise simulated data, which is often generated using a model that simulates the rotating machine under different operating and fault conditions. For instance, Li et al. (2000) presented a technique for motor rolling bearing fault diagnosis using a neural network and time-frequency domain bearing vibration analysis. In this study, computer-simulated vibration data are first utilised to study and design the neural network motor bearing fault-diagnosis algorithm. Then, real bearing vibration data acquired in real-time are applied to perform initial testing and validation of the proposed algorithm. Moreover, Ocak and Loparo (2004) presented two separate algorithms for estimating the running speed and key bearing frequencies, which are required for failure detection and diagnosis, of an induction motor using vibration data. In this study, a linear vibration model is developed for generating simulated vibration data. Then, the simulated data is used to validate the performance of the two presented algorithms. Furthermore, Fan and Zuo (2006) proposed a method for the detection of gear faults based on the Hilbert and wavelet packet transforms. In this study, both simulated data and real data collected from a gearbox are used to demonstrate the effectiveness of the proposed method.

In both types of vibration data, i.e. real measured or simulated vibration data, we often have many signals recorded over a time period from one machine or several machines. These vibration measurements often represent both healthy and faulty conditions. To design a vibration-based fault-diagnosis algorithm, these vibration measurements collected under variable conditions are frequently gathered in datasets. For example, the main bearing vibration dataset used in the case studies presented in this book is acquired from experiments on a test rig that simulates an environment running roller bearings. In these experiments, several interchangeable faulty roller bearings are inserted in the test rig to represent the types of faults that can normally happen in roller bearings. The test rig (Figure 6.2) used to acquire the bearing vibration dataset consists of a 12 V DC electric motor driving the shaft through a flexible coupling. The shaft is supported by two plummer bearing blocks, where a series of damaged bearings are inserted. Two accelerometers are used to measure the resultant vibrations in the horizontal and vertical planes. The output from the accelerometers is fed back via a charge amplifier to a Loughborough Sound Images DSP32 analog-to-digital converter (ADC) card utilising a low-pass filter by means of a cut-off of 18 kHz. The sampling rate is 48 kHz, giving slight oversampling. Six health conditions of roller bearings have been recorded: two normal conditions, i.e. brand-new (NO) and worn but undamaged (NW); and four faulty conditions, i.e. inner race (IR), outer race (OR), rolling element (RE), and cage (CA). Table 6.1 presents an explanation of the corresponding characteristics of these bearing health conditions.

The data were recorded using 16 different speeds in the range 25–75 rev s^{-1}. At each speed, 10 time series were recorded for each condition, i.e. 160 examples per condition. This resulted in a total of 960 examples with 6000 data points to work with. Figure 6.3 illustrates some typical time-series plots for the six different aforementioned conditions (Ahmed and Nandi 2018).

As described in Chapter 3, in reality, the acquired vibration signal typically comprises a large collection of responses from several sources in a rotating machine and some background noise. Thus, the direct use of the collected vibration signals for diagnosing rotating machine faults is challenging. In place of processing the raw vibration signals,

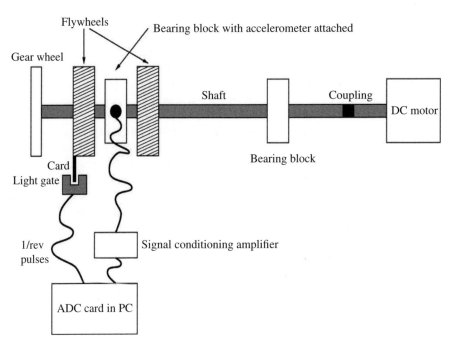

Figure 6.2 The test rig used to collect vibration data from bearings used in the case studies presented in this book.

Table 6.1 The characteristics of bearing health conditions in the acquired bearing dataset.

Condition	Characteristic
NO	The bearing is brand new and in perfect condition.
NW	The bearing has been in service for some period of time but is in good condition.
IR	Inner race fault. This fault is created by cutting a small groove in the raceway of the inner race.
OR	Outer race fault. This fault is created by cutting a small groove in the raceway of the outer race.
RE	Roller element fault. This fault is created by using an electrical etcher to mark the surface of the balls, simulating corrosion.
CA	Cage fault. This fault is created by removing the plastic cage from one of the bearings, cutting away a section of the cage so that two of the balls are not held at a regular space and free to move.

one may compute certain attributes of the raw vibration signal that can represent the signal in essence. In the machine learning community, these attributes are also known as *characteristics*, *signatures*, or *features*. Hence, in the second step of the vibration-based fault-detection and -diagnosis framework, several types of methods need to be adopted in a cascade of steps starting from raw vibration datasets and ending at final mature sets

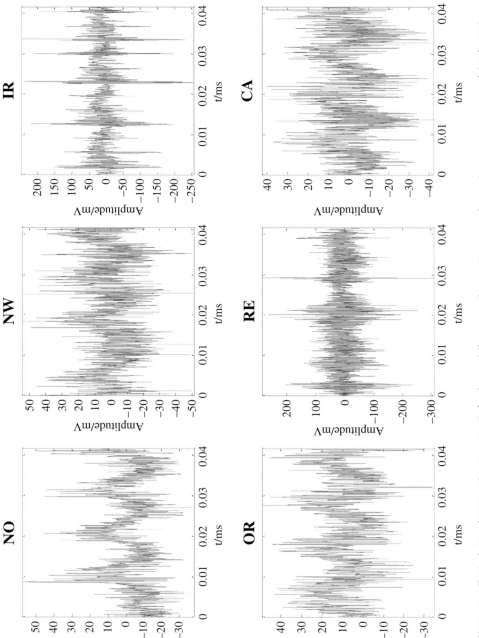

Figure 6.3 Typical time domain vibration signals for the six different conditions. (See insert for a colour representation of this figure.)

of results. These include vibration analysis techniques that have the ability to obtain useful information about machine conditions from the raw vibration datasets, which can be successfully used for fault diagnosis.

The following section of this chapter moves on to describe several different types of learning that can be applied to vibration data.

6.3 Learning from Vibration Data

In Chapter 3, we have mentioned that visual inspection is not dependable in the rotating machine CM field for many reasons that can be summarised as follows:

(1) Not all waveform signals from rotating machines provide clear visual differences.
(2) In practice, we deal with a large collection of vibration signals that often contain background noise.
(3) Sometimes we deal with low-amplitude signals measured in a noisy background.
(4) The demand for early detection of faults makes manual inspection of all the collected vibration signals impractical.

Hence, a learning algorithm is required to automatically detect meaningful patterns in the vibration data. The literature on machine fault diagnosis has highlighted various machine learning algorithms (Liu et al. 2018; Martin-Diaz et al. 2018; Zhao et al. 2019). Each of these machine learning algorithms has three main components, as follows (Domingos 2012):

1. *Representation.* A classifier must be represented in some formal language that a computer can handle. On the other hand, selecting a representation for a learning algorithm is equivalent to selecting the set of classifiers that may be able to learn. This set is known as the *hypothesis space* of the learning algorithm. Examples of representations include instances, hyperplanes, decision trees, sets of rules, neural networks, and graphical models.
2. *Evaluation.* An evaluation function, also called an *objective function*, is required to differentiate good classifiers from bad ones. Examples include accuracy, error rate, precision, recall, likelihood, posterior probability, margin, information gain, and Kullback-Leibler (K-L) divergence.
3. *Optimisation.* A method to search among the classifiers for the highest-scoring one is required. The selection of the optimisation technique is the key to the effectiveness of the learning algorithm. Examples of optimisation methods include combinatorial optimisation, e.g. greedy search; and continuous optimisation that includes constraint methods (e.g. linear programming and quadratic programming) and unconstrained methods (e.g. gradient descent, conjugate gradient, and Quasi-Newton methods).

Various classification techniques can be used to classify different vibration type based on the features provided. If the vibration signals' features are carefully devised and the parameters of the classifiers are carefully tuned, it is possible to achieve high classification accuracies. Part IV of this book introduces some widely used state-of-the art classifiers, decision trees, random forests, multinomial logistic regression, support vector machines, and artificial neural networks (ANNs) that have already been used for classification of vibration signal status.

In machine learning, there are several different types of learning that can be applied to vibration data. The following subsections briefly describe these types of learning.

6.3.1 Types of Learning

6.3.1.1 Batch vs. Online Learning

Batch learning, also called *offline learning*, is performed when we have access to all the vibration data. In this type of learning, all the data are used together for learning and optimising the model parameters. In *online learning*, we do not have access to all the data. Instead, the observations arrive one at a time, and the model is trained using one observation at a time (Suthaharan 2016).

6.3.1.2 Instance-Based vs. Model-Based Learning

The main task of machine learning algorithms in machine fault diagnosis is to make a prediction about the machine's health. This means, given a number of training observations, the algorithm must be capable to generalise to new observations. There are two main learning techniques to achieve this generalisation: (i) instance-based learning and (ii) model-based learning. In the instance-based learning technique, the algorithm requires a measure of similarity between new observations and known observations, e.g. k-nearest neighbour (*k*-NN) algorithms (Aha et al. 1991; Géron 2017; Liu et al. 2018). In model-based learning, to achieve a generalisation from a set of given observations, we need to build a model of these observations that we can employ to make predictions: e.g. fuzzy logic, ANNs, and case-based reasoning (CBR) (Murphey et al. 2006).

6.3.1.3 Supervised Learning vs. Unsupervised Learning

Supervised learning and unsupervised learning are defined based on the class definition. In supervised learning, the classes are known and class boundaries are well defined in the given training data sources. In other words, supervised learning is based on training a data example from a data source with its corresponding classification label already assigned. In unsupervised learning, also known as *clustering*, the classes or class boundaries are not available, i.e. not known. In unsupervised learning, the class boundaries are statistical and not defined exactly (Suthaharan 2016).

6.3.1.4 Semi-Supervised Learning

Traditionally, supervised classifiers need only labelled data to train. However, in many real applications, we often have datasets that include a large number of unlabelled examples and a small number of labelled examples; this is because labelled data are often difficult, expensive, or time-consuming to get, as they require the efforts of experienced human annotators. On the other hand, unlabelled data may be relatively easy to acquire; nevertheless, there have been few methods to utilise them. Semi-supervised learning addresses this limitation by means of utilising a large amount of unlabelled data together with the labelled data, to build better classifiers (Zhu 2006).

6.3.1.5 Reinforcement Learning

The key objective in solving a machine learning problem is to produce intelligent programs or intelligent agents via the process of learning and adapting to a changed environment. Reinforcement learning is a branch of artificial intelligence (AI) in which learners or software agents learn from direct interaction with the environment. The agent observes the information from the environment in the form of state spaces and action spaces. The agent can also learn even if a complete model or complete information about the environment is not available. An agent gets feedback about its actions as a reward. The received reward is assigned to the new state. Here, the reinforcement learning agent has to maximise the rewards received during interactions with the environment to find the optimal policy to solve a specific task (Sutton and Barto 2011; Kulkarni 2012).

Reinforcement learning is different from supervised learning, which learns from examples provided by a knowledgeable external supervisor. This is an important type of learning; however, by itself it is not adequate for learning from interactions. Reinforcement learning combines the fields of dynamic programming and supervised learning to produce a machine learning algorithm that is very close to techniques used by human learning.

6.3.1.6 Transfer Learning

As mentioned earlier, semi-supervised learning algorithms address the problem that many real applications often have datasets that include a large number of unlabelled examples and a small number of labelled examples, by using a large amount of unlabelled data together with a small amount of labelled data. However, most semi-supervised learning algorithms assume that the distributions of the labelled and unlabelled data are the same. Transfer learning, in contrast, allows the domains, tasks, and distributions used in training and testing to be different. In this learning technique, labelled data from a different domain is used to enhance the learning process. As described by Pang and Yang (2010), based on different conditions between the source and the target domain tasks, transfer learning can be categorised into three main subsettings, as follows:

(1) *Inductive transfer learning.* The source task is different from the target task, regardless of whether the source and the target are in the same domain. In this type of transfer learning, in the target domain, we need a dataset with its corresponding class labels to induce an objective predictive model $f_T(.)$ for use in the target domain. In the source domain, data can be labelled or unlabelled. If a lot of the labelled data in the source domain are available, the inductive transfer learning setting is similar to the multitask learning setting. But when the available data in the source are not labelled, the inductive transfer learning setting is similar to the self-taught learning setting, in which the label spaces between the source and target domains may be different.

(2) *Transductive transfer learning.* The source and target tasks are the same, while the source and target domains are different. In this type of transfer learning, no labelled data are available in the target domain, while a lot of labelled data are available in the source domain. There are two settings for transductive transfer learning: (i) when the feature spaces between the source and target domains are different, i.e. $X_S \neq X_T$; and (ii) when $X_S = X_T$, but the marginal probability distributions of the input data of the source and the target are different, i.e. $P(X_S) \neq P(X_T)$.

(3) *Unsupervised transfer learning.* The target task is different from but related to the source task. This type of transfer learning focuses on solving unsupervised learning tasks in the target domain. In this case, the available data in both the source and the target domains are not labelled.

Several studies have used transfer learning in the field of machine fault diagnosis (e.g. Shen et al. 2015; Wang et al. 2016; Zhang et al. 2017; Wen et al. 2017; Han et al. 2018).

6.3.2 Main Challenges of Learning from Vibration Data

6.3.2.1 The Curse of Dimensionality

Current data analysis has to deal with a tremendous amount of data. Due to recent advances in sensor technologies, data are more and more easily collected and stored. The huge increase in the amount of data is found not only in the number of observations, i.e. signal examples acquired over time, but also in the number of attributes in each measured observation (Verleysen and François 2005). The measured data are often gathered into vectors whose dimension correspond to the number of measured attributes. As the dimensions increase, each example can be represented as a vector in a high-dimensional space. Data analysis tools based on learning principles infer knowledge, or information, from available learning examples. Successful development of learning algorithms requires enough data for learning, and the number of learning examples should grow exponentially with the dimensions. This exponential growth is the first consequence of what is known as the *curse of dimensionality*. Therefore, it is the primary issue in designing classifiers when the number of attributes (N) is much larger than the number of examples (L), i.e. ($N >> L$). This problem can be addressed by reducing the dimensionality of the data.

As described in Section 6.2.1, the first step of a vibration-based CM system is acquiring vibration data. In this step, a sensor (e.g. velocity sensor or accelerometer) is mounted to the component of interest to collect input vibration measurements, i.e. raw data that can be transmitted, stored, and processed. Sampling theorems, including the Shannon-Nyquist theorem, are the core of the current sensing systems. However, a Nyquist sampling rate that is at least twice the highest frequency contained in the signal is high for some modern developing applications, e.g. industrial rotating machines (Eldar 2015). One aspect of much of the literature on using the Nyquist sampling rate is that it may result in measuring a large amount of data that need to be transmitted, stored, and processed. Moreover, in applications that include wideband, it is often very costly to collect samples at the necessary rate. It is clear that acquiring a large amount of data requires large storage and time for signal processing, and this also may limit the number of machines that can be monitored remotely across wireless sensor networks (WSNs) due to bandwidth and power constraints.

For this reason, it is currently becoming essential to developing new CM methods that not only have the ability to achieve accurate detection and identification of machine health conditions but also have the capability to address two main challenges:

(1) Cost of learning from a large amount of vibration data, including transmission costs, computation costs, and power needed for computation.
(2) Demand for early fault detection.

6.3.2.2 Irrelevant Features

As mentioned earlier, a collected vibration signal typically contains a large collection of responses from several sources in the rotating machine along with background noise, which can be caused by sources of data and imperfection in the technologies used to collect the data. Hence, to make sensible deductions automatically using a machine learning–based classifier is not an easy task, owing to the curse of dimensionality. Instead of processing raw signals, the common approach is to compute certain attributes of the raw signals that can represent the original signals. These features can be categorised into three types: (i) strongly relevant, (ii) weakly relevant, and (iii) irrelevant features (John et al. 1994). In classification problems, irrelevant features can never contribute to classification accuracy. On the other hand, relevant features, whether strongly or weakly relevant, can contribute to classification accuracy, i.e. removal of relevant feature can result in a change in the classification results.

6.3.2.3 Environment and Operating Conditions of a Rotating Machine

In reality, many structures are subject to various environmental and operating conditions, e.g. temperature and humidity that affect their dynamical behaviour. Changing environmental and operating conditions can affect measured signals, and these ambient variations of the system can mask subtle changes in the system's vibration signal caused by damage (Sohn 2006; Deraemaeker et al. 2008). Various environmental and operating conditions have an effect on the accuracy of measured vibration signals. These environmental conditions include wind, temperature, base strains, humidity, and magnetic fields; operating conditions include ambient loading conditions, operational speed, and mass loading. This problem can be addressed by normalising vibration datasets.

6.3.3 Preparing Vibration Data for Analysis

6.3.3.1 Normalisation

As just described, various environmental and operating conditions affect the accuracy of measured vibration signals. This problem can be addressed by normalising the acquired vibration dataset. Commonly used normalisation processes employed with measured vibration data include the following (Farrar et al. 2001; Kantardzic 2011):

1. The mean value of a measured time series is often subtracted from that signal to remove DC offsets from the signal.
2. Division by the standard deviation of the signal is performed to normalise the varying amplitudes in the signal.
3. Min-max normalisation performs a linear transformation on the original data. Suppose that min_a and max_a are the minimum and maximum values of the measured signal. Min-max normalisation maps a value x_i to $\acute{x}_i$ in the range [$new - min_a$, new-max_a]:

$$\acute{x}_i = \left(\frac{(x_i - min_a)}{(max_a - min_a)} \right) * (new - min_a - \text{new-}max_a) + new - min_a \qquad (6.1)$$

4. In Z-score normalisation, the values of a measured signal are normalised based on the mean and standard deviation of the measured signal x,

$$\acute{x}_i = \frac{(x_i - \bar{x})}{\sigma_x} \qquad (6.2)$$

where $\bar{x}$ and σ_x are the mean and the standard deviation of the measured signal.

5. Normalisation by decimal scaling moves the decimal point of values of the measured signal:

$$\hat{x}_i = \frac{x_i}{10^k} \tag{6.3}$$

6.3.3.2 Dimensionality Reduction

As mentioned earlier, the curse of dimensionality can be addressed by reducing the data's dimensionality. This can be achieved by computing certain features of the raw vibration signals. However, extracting useful features from a large and noisy vibration dataset is not an easy task. A reasonable approach to tackle the challenges of dealing with these types of data is to learn a subspace from the high-dimensional data, i.e. map high-dimensional data into a low-dimensional representation. Various techniques have been proposed to learn subspace features from high-dimensional vibration signals, including feature-extraction and feature-selection techniques. The following subsections describe briefly these techniques.

Feature Extraction Feature extraction is a dimensionality reduction procedure that transforms high-dimensional input into a reduced set of features. The feature-transformation techniques aim to reduce the high-dimensional input data to a low-dimensional set that are linear or nonlinear combinations of the original signals. Various techniques have been proposed to learn subspace features from raw vibration signals. These include:

- Linear subspace learning (LSL), e.g. principal component analysis (PCA), independent component analysis (ICA), linear discriminant analysis (LDA), canonical correlation analysis (CCA), and partial least squares (PLS)
- Nonlinear subspace learning (NLSL), e.g. kernel principal component analysis (KPCA), kernel linear discriminant analysis (KLDA), kernel independent component analysis (KICA), isometric feature mapping (ISOMAP), diffusion maps (DM), Laplacian eigenmaps (LE), local linear embedding (LLE), Hessian-based local linear embedding (HLLE), local tangent space alignment analysis (LTSA), maximum variance unfolding (MVU), and stochastic proximity embedding (SPE).

These techniques will be described in detail in Chapters 7 and 8, respectively.

Feature Selection Feature selection, also called *subset selection*, techniques aim to select a subset of features that can sufficiently represent the characteristic of the original features. In view of that, this will reduce the computational cost and may remove irrelevant and redundant features and consequently improve learning performance. In vibration-based machinery fault diagnosis, the main task is to categorise the acquired vibration signal correctly as a corresponding machine condition, which is generally a multiclass classification problem. Hence, the aim of feature selection in MCM is to select a subset of features that are able to discriminate between instances that belong to different classes. In other words, the main task of feature selection is to select a subset of features that are the most relevant to the classification problem. Based on their relationship with learning algorithms, feature-selection techniques can be categorised into three main groups:

- Filter models, e.g. Fisher score (FS), Laplacian score (LS), relief algorithms, Pearson correlation coefficient (PCC), information gain (IG), gain ratio (GR), mutual information (MI), Chi-squared (Chi-2), etc.
- Wrapper models, which can be categorised into sequential selection algorithms and heuristic search algorithms.
- Embedded models, e.g. LASSO, Elastic Net, the Classification and Regression Tree (CART), C4.5, and SVM–Recursive Feature Elimination (SVM-RFE).

These techniques will be described in detail in Chapter 9.

6.4 Summary

The main aim of MCM is to avoid catastrophic machine failure that may cause secondary damage, downtime, safety incidents, production loss, and higher costs associated with repairs. CM techniques for rotating machinery involve the practice of monitoring measurable data, e.g. vibrations, acoustic emissions, electric currents, etc., to classify changes in machine health. Of these monitoring techniques, vibration-based CM has become a well-accepted technique for planned maintenance management. This chapter has described vibration-based MCM using machine learning algorithms. An overview of the vibration-based MCM process and fault-detection and -diagnosis problem framework have been described. Moreover, several different types of learning that can be applied to vibration data have been described. These include batch learning, online learning, instance-based learning, model-based learning, supervised learning, unsupervised learning, semi-supervised learning, reinforcement, and transfer learning. Furthermore, the main problems of learning from vibration data for the purpose of fault diagnosis and the techniques to prepare vibration data for analysis to overcome these problems have been described. These techniques include normalisation techniques and dimensionality-reduction techniques, which include feature-extraction and feature-selection techniques.

References

Aha, D.W., Kibler, D., and Albert, M.K. (1991). Instance-based learning algorithms. *Machine Learning* 6 (1): 37–66.

Ahmed, H. and Nandi, A.K. (2018). Compressive sampling and feature ranking framework for bearing fault classification with vibration signals. *IEEE Access* 6: 44731–44746.

Deraemaeker, A., Reynders, E., De Roeck, G., and Kullaa, J. (2008). Vibration-based structural health monitoring using output-only measurements under changing environment. *Mechanical Systems and Signal Processing* 22 (1): 34–56.

Domingos, P. (2012). A few useful things to know about machine learning. *Communications of the ACM* 55 (10): 78–87.

Eldar, Y.C. (2015). *Sampling Theory: Beyond Bandlimited Systems*. Cambridge University Press.

Fan, X. and Zuo, M.J. (2006). Gearbox fault detection using Hilbert and wavelet packet transform. *Mechanical Systems and Signal Processing* 20 (4): 966–982.

Farrar, C.R., Sohn, H., and Worden, K. (2001). *Data Normalization: A Key for Structural Health Monitoring* (No. LA-UR-01-4212). NM, US: Los Alamos National Lab.

Géron, A. (2017). *Hands-on Machine Learning with Scikit-Learn and TensorFlow: Concepts, Tools, and Techniques to Build Intelligent Systems.* O'Reilly Media, Inc.

Han, T., Liu, C., Yang, W. et al. (2018). Deep transfer network with joint distribution adaptation: a new intelligent fault diagnosis framework for industry application. arXiv:1804.07265.

John, G.H., Kohavi, R., and Pfleger, K. (1994). Irrelevant features and the subset selection problem. In: *Machine Learning Proceedings 1994*, 121–129. San Francisco, CA: Morgan Kaufmann.

Kantardzic, M. (2011). *Data Mining: Concepts, Models, Methods, and Algorithms.* Wiley.

Kim, J., Ahn, Y., and Yeo, H. (2016). A comparative study of time-based maintenance and condition-based maintenance for optimal choice of maintenance policy. *Structure and Infrastructure Engineering* 12 (12): 1525–1536.

Kulkarni, P. (2012). *Reinforcement and Systemic Machine Learning for Decision Making*, vol. 1. Wiley.

Lei, Y., Li, N., Guo, L. et al. (2018). Machinery health prognostics: a systematic review from data acquisition to RUL prediction. *Mechanical Systems and Signal Processing* 104: 799–834.

Li, B., Chow, M.Y., Tipsuwan, Y., and Hung, J.C. (2000). Neural-network-based motor rolling bearing fault diagnosis. *IEEE Transactions on Industrial Electronics* 47 (5): 1060–1069.

Liu, R., Yang, B., Zio, E., and Chen, X. (2018). Artificial intelligence for fault diagnosis of rotating machinery: a review. *Mechanical Systems and Signal Processing* 108: 33–47.

Martin-Diaz, I., Morinigo-Sotelo, D., Duque-Perez, O., and Romero-Troncoso, R.J. (2018). An experimental comparative evaluation of machine learning techniques for motor fault diagnosis under various operating conditions. *IEEE Transactions on Industry Applications* 54 (3): 2215–2224.

McMillan, D. and Ault, G.W. (2007). Quantification of condition monitoring benefit for offshore wind turbines. *Wind Engineering* 31 (4): 267–285.

Murphey, Y.L., Masrur, M.A., Chen, Z., and Zhang, B. (2006). Model-based fault diagnosis in electric drives using machine learning. *IEEE/ASME Transactions on Mechatronics* 11 (3): 290–303.

Nandi, A.K. and Jack, L.B. (2004). Advanced digital vibration signal processing for condition monitoring. *International Journal of COMADEM* 7 (1): 3–12.

Ocak, H. and Loparo, K.A. (2004). Estimation of the running speed and bearing defect frequencies of an induction motor from vibration data. *Mechanical Systems and Signal Processing* 18 (3): 515–533.

Pan, S.J. and Yang, Q. (2010). A survey on transfer learning. *IEEE Transactions on Knowledge and Data Engineering* 22 (10): 1345–1359.

Shen, F., Chen, C., Yan, R., and Gao, R.X. (2015). Bearing fault diagnosis based on SVD feature extraction and transfer learning classification. In: *Prognostics and System Health Management Conference (PHM), 2015*, 1–6. IEEE.

Sohn, H. (2006, 1851). Effects of environmental and operational variability on structural health monitoring. *Philosophical Transactions of the Royal Society A: Mathematical, Physical and Engineering Sciences* 365: 539–560.

Suthaharan, S. (2016). Machine learning models and algorithms for big data classification. *Integrated Series in Information Systems* 36: 1–12.

Sutton, R.S. and Barto, A.G. (2011). *Reinforcement Learning: An Introduction*. Cambridge, U.K: Cambridge Univ. Press.

Van Dam, J. and Bond, L.J. (2015). Economics of online structural health monitoring of wind turbines: cost benefit analysis. *AIP Conference Proceedings* 1650 (1): 899–908.

Verleysen, M. and François, D. (2005). The curse of dimensionality in data mining and time series prediction. In: *International Work-Conference on Artificial Neural Networks*, 758–770. Berlin, Heidelberg: Springer.

Verma, N.K., Khatravath, S., and Salour, A. (2013). Cost benefit analysis for condition-based maintenance. In: *2013 IEEE Conference on Prognostics and Health Management (PHM)*, 1–6.

Wang, J., Xie, J., Zhang, L., and Duan, L. (2016). A factor analysis based transfer learning method for gearbox diagnosis under various operating conditions. In: *International Symposium on Flexible Automation (ISFA)*, 81–86. IEEE.

Wen, L., Gao, L., and Li, X. (2017). A new deep transfer learning based on sparse auto-encoder for fault diagnosis. *IEEE Transactions on Systems, Man, and Cybernetics: Systems* 99: 1–9.

Zhang, R., Tao, H., Wu, L., and Guan, Y. (2017). Transfer learning with neural networks for bearing fault diagnosis in changing working conditions. *IEEE Access* 5: 14347–14357.

Zhao, R., Yan, R., Chen, Z. et al. (2019). Deep learning and its applications to machine health monitoring. *Mechanical Systems and Signal Processing* 115: 213–237.

Zhu, X. (2006). *Semi-Supervised Learning Literature Survey. Computer Science*, vol. 2(3), 4. University of Wisconsin-Madison.

7

Linear Subspace Learning

7.1 Introduction

As described in Chapter 6, the core objective of vibration-based machinery condition monitoring is to correctly categorise the acquired vibration signal into the corresponding machine health condition, which is generally a multiclass classification problem. In practice, the collected vibration signal usually contains a large collection of responses from several sources in the rotating machine and some background noise. Thus the direct usage of the acquired vibration signal in rotating machine fault diagnosis is challenging both via manual inspection and via automatic monitoring. As described in Chapter 3, a common alternative to processing the raw vibration signal is to compute certain features of the raw signal that can describe the signal in essence. In the machine learning community, these features are also called *characteristics*, *signatures*, or *attributes*. However, extracting useful features from such a large amount of noisy vibration data is not an easy task. A reasonable approach to tackle the challenges involved in dealing with too many samples is to learn a subspace from the high-dimensional data, i.e. map a high-dimensional dataset into a low-dimensional representation.

Various techniques have been proposed to learn subspace features from a large number of vibration signals in rotating machine fault diagnosis. These include linear subspace learning (LSL), nonlinear subspace learning (NLSL), and feature-selection techniques. There are two main objectives of learning reduced features from the high-dimensional original data: (i) to reduce computational complexity; and (ii) to enhance classification accuracy. For a complete view of the field, this chapter introduces LSL techniques that can be used to learn features from a large amount of vibration signals. LSL has been extensively used in several areas of information processing, such as data mining, dimensionality reduction, and pattern recognition. The basic idea of LSL is to map a high-dimensional feature space to a lower-dimensional feature space through linear projection.

The problem of LSL can be formulated mathematically as follows. Given a dataset $X = [x_1, x_2, \ldots, x_L]$ for training, where L is the number of observations and each example x_l is a $n \times 1$ vector in a vector space $\mathbb{R}^n$. The objective of LSL techniques is to find a linear transformation, i.e. projection, $W \in \mathbb{R}^{n \times m}$ to map the n-dimensional sampled vibration signal to a lower-dimensional feature space, using the following equation:

$$\hat{x}_l = W^T x_l \tag{7.1}$$

Condition Monitoring with Vibration Signals: Compressive Sampling and Learning Algorithms for Rotating Machines, First Edition. Hosameldin Ahmed and Asoke K. Nandi.
© 2020 John Wiley & Sons Ltd. Published 2020 by John Wiley & Sons Ltd.

Here, $l = 1, 2 \ldots L$, $\hat{x}_l$ is the transformed feature vector with reduced dimensions (the extracted features) where $\hat{x}_l \in \mathbb{R}^{m \times 1}$ ($m < n$), and W is a transformation matrix. For the purpose of classification, these features are often used as input to a classifier, e.g. a support vector machine (SVM) and artificial neural network (ANN). Numerous LSL techniques have been used in machine fault diagnosis to extract a new low-dimensional feature space of the collected vibration data, which is usually a linear combination of the original high-dimensional feature space of the acquired vibration data. Of these techniques, principal component analysis (PCA) (Wold et al. 1987), independent component analysis (ICA), and linear discriminant analysis (LDA) are amongst the most common techniques that have been used in machine fault diagnosis. Canonical correlation analysis (CCA) and partial least squares (PLS) are recently considered in many applications of fault detection. The subsequent subsections discuss these types of techniques in more detail. The other two types of feature learning techniques, i.e. NLSL and feature-selection techniques, will be covered in detail in Chapter 8 and Chapter 9, respectively.

7.2 Principal Component Analysis (PCA)

PCA is an orthogonal linear feature projection algorithm that aims to find all the components (eigenvectors) in descending order of significance where the first few principal components (PCs) possess most of the variance of the original data. PCA can be used to form a low-dimensional feature vector from a high-dimensional data. To reduce the dimensionality of the data by means of PCA, one ignores the least significant of these components from PCA. There are two main techniques to accomplish PCA: (i) using eigenvector decomposition and (ii) using singular value decomposition (SVD) (Shlens 2014):

7.2.1 PCA Using Eigenvector Decomposition

The procedure for PCA using eigenvector decomposition involves the following steps:

1. Calculate the mean vector of the data.
2. Compute the covariance matrix of the data.
3. Obtain the eigenvalues and eigenvectors of the covariance matrix.

Suppose that the input dataset $X = [x_1 x_2 \ldots x_L]$ has L observations and n-dimensional space (Table 7.1). PCA transforms X to $\hat{X} = [\hat{x}_1, \hat{x}_2, \ldots, \hat{x}_L]$ in a new m-dimensional space.

The main target is to reduce the dimensionality from n to m ($m \ll n$) of vibration samples using Eq. (7.1). To compute the transformation matrix W, PCA uses eigenvalues and eigenvectors such that

$$\lambda V_1 = C_X V_2 \tag{7.2}$$

where λ is an eigenvalue, V_1 is an eigenvector, and C_X is the corresponding covariance matrix of data X that can be calculated using the following equation:

$$C_X = \frac{1}{L} \sum_{i=1}^{L} (x_i - \bar{x})(x_i - \bar{x})^T \tag{7.3}$$

Table 7.1 Input dataset $X_{L \times n}$.

	Measurement 1	Measurement 2	...	Measurement n
Observation 1			...	
Observation 2			...	
$\vdots$	$\vdots$	$\vdots$	$\vdots$	$\vdots$
Observation L			...	

Here, $\bar{x}$ can be computed using the following equation:

$$\bar{x} = \frac{1}{L} \sum_{i=1}^{L} x_i \tag{7.4}$$

PCA aims to find all the components (eigenvectors) in descending order of significance where the first few principal components possess most of the variance of the original data. Hence, it can be employed to form a low-dimensional feature vector. To reduce the dimensionality of the data X by means of PCA using eigenvectors, one ignores the least-significant components from the principal components, i.e. columns of V, such that

$$\hat{X} = W_1^T X \tag{7.5}$$

where $\hat{X} \in R^{L \times m}$ is the obtained data matrix with a reduced dimensionality, and W_1 is the projection matrix in which each column vector is composed of the corresponding eigenvectors of the m largest eigenvalues ($m \ll n$) of the covariance matrix C.

7.2.2 PCA Using SVD

A different technique to compute the transformation matrix W of PCA is SVD, which is a matrix decomposition technique that breaks a data matrix into three essential parts (Golub and Reinsch 1970; Moore 1981) such that,

$$X = U \sum V_2^T \tag{7.6}$$

where $X \in R^{L \times n}$ is an input matrix, $U \in R^{L \times L}$ is an orthogonal matrix, $\sum \in R^{L \times n}$ is a diagonal matrix, and $V_2 \in R^{n \times n}$ is an orthogonal matrix. The diagonal components of $\sum$ are the singular values of X. The L columns of U are the left-singular vectors, and the n columns of V are the right-singular vectors of X. The relationship between singular vectors/values and eigenvectors/eigenvalues can be described as follows (Lu et al. 2013):

- The columns of U, i.e. the left-singular vectors of X, are the eigenvectors of XX^T.
- The columns of V, i.e. the right-singular vectors of X, are the eigenvectors of $X^T X$.
- The diagonal components of $\sum$, which are the nonzero singular values of X, are the square roots of the nonzero eigenvalues of both XX^T and $X^T X$.

To reduce the dimensionality of the data X by means of PCA using SVD, one ignores the least-significant components from the principal components, i.e. columns of V_2, such that

$$\hat{X} = W_2^T X \tag{7.7}$$

where $\hat{X} \in R^{L \times m}$ is the obtained data matrix with a reduced dimensionality, and W_2 is the projection matrix that is composed of the first m selected columns ($m \ll n$) of V_2.

7.2.3 Application of PCA in Machine Fault Diagnosis

The application of PCA in vibration signal analysis as a dimensionality reduction technique has been widely employed in various studies of machine fault diagnosis. For instance, Baydar and Ball (2001) proposed a method for gear failure detection using a wavelet transform, improving its capability with PCA. In this method, PCA is used as a preprocessing step before the wavelet transform. The authors used the phase and the magnitude information of the wavelet to reveal fault symptoms in gearboxes. The experimental results using vibration signals acquired from a gearbox showed that the wavelet transform on its own is a very useful tool in gear diagnostics and can indicate progressing fault conditions in gearboxes. It was also shown that using PCA as a preprocessing of the vibration signal could improve the detection capability of the wavelet to make a more dependable decision.

Li et al. (2003) proposed a method that used PCA analysis to reduce the dimensionality of the feature space and obtain an efficient subspace for diagnosing gear faults. In this method, 10 time domain–based statistical parameters, including standard deviation, mean value, maximum peak value, root mean square, skewness, kurtosis, crest factor, clearance factor, impulse factor, and shape factor, were extracted in the time domain. PCA used SVD to determine the principal components of the original feature space. Typical raw signals under different conditions, including normal, a cracked tooth, and a broken tooth, acquired from a gearbox at a 12.5 KHz sampling rate were used to validate the proposed method. The experimental results showed that none of the 10 extracted features can be used to identify the cracked-tooth condition versus the normal condition. By applying PCA to obtain the principal components of the original features, the results showed that the proposed method is sensitive to different working conditions of gearbox, and it is able to identify industrial gearbox defects.

Malhi and Gao (2004) introduced a feature-selection approach based on PCA method to select the most representative features for the classification of faulty components and fault severity in three types of roller bearings. In this study, a set of 13 features extracted in time, frequency, and wavelet-domain these include peak value, root mean square, crest factor, rectified skew, and kurtosis from time-domain; bearing pass frequency of outer race (BPFO), bearing pass frequency of inner race (BPFI), ball spin frequency (BSF), and power in fault frequency range in frequency-domain. The PCA is used to reduce the input feature dimensions. To investigate the effectiveness of the selected features, i.e. the selected principal components, in machine fault classification, two scenarios of supervised and unsupervised classification procedure were used. The experimental results showed that three features selected by the PCA technique achieved better results with approximately 1% error compared to the use of the 13 original features as inputs to a classifier that resulted in 9% classification error. In general, the PCA-based selected features showed the capability to improve the accuracy of the classification scheme, for both supervised classification using feedforward neural network (FFNN) and radial basis function (RBF) networks (to be described in Chapter 12), and unsupervised competitive learning.

Serviere and Fabry (2005) proposed a method for diagnosing electromechanical systems using PCA and blind source separation (BSS) of modulated sources. In this method, modified PCA was utilised as a first step in the separation process to filter out the noise and whiten the observations. Experimental vibration data acquired from a complex electromechanical system was used to validate this method, and the results proved that the proposed method is available to denoise such sources as well as sinusoidal ones. Sun et al. (2007) presented a method for diagnosing rotating machine faults using the combination of the C4.5 decision tree algorithm (to be described in Chapter 10) and PCA. In this study, 18 features (including 7 features extracted in the time domain and 11 features extracted in the frequency domain) were extracted from the acquired signals. Then, the PCA model was used to reduce the dimensionality of these features. With these reduced features, C4.5 was used to deal with the fault-diagnosis problem. To validate the proposed method, six types of running conditions (normal, unbalance, rotor radial rub, oil whirl, shaft crack, and a combined condition of unbalance and radial rub) were used. The results demonstrated the effectiveness of the C4.5- and PCA-based diagnosis method for rotating machinery faults.

Moreover, He et al. (2009) presented a low-dimensional principal components (PCs) representation technique for representing and classifying machine health conditions. In this technique, the PCA method was used to automatically extract the PC representations from the time domain and frequency domain statistical features of the acquired signals. The experimental results of an internal-combustion engine sound analysis and automobile gearbox vibration signal analysis demonstrated the effectiveness of the proposed method. Furthermore, to view and detect incipient failures in large-size low-speed rolling bearings, Žvokelj et al. (2010) proposed a data-driven multivariate and multiscale monitoring method called ensemble empirical mode decomposition based multiscale PCA (EEMD-MSPCA). In this method, the PCA multivariate monitoring technique and ensemble empirical mode decomposition (EEMD) are combined. To validate the proposed method for the task of monitoring bearing conditions and filtering signals, simulated as well as real vibration and acoustic emission (AE) signals of roller bearings were used. The experimental results demonstrated that the use of AE and vibration signals and the proposed EEMD-MSPCA method is an effective combination for fault detection.

Trendafilova (2010) proposed a method for roller bearing fault detection and classification based on pattern recognition and PCA of measured vibration signals. In this method, the measured signals are preprocessed using a wavelet transform, and a modified PCA is used to reduce the number of wavelet coefficients and extract the relevant features. Four different bearings are used to study four conditions: no fault (NO), an inner race fault (IR), an outer race fault (OR), and a rolling element fault (RE) with very small notches. A simple pattern-recognition technique based on the 1-nearest neighbour method was applied, and it showed that between 94% and 96% of the faults were correctly detected and classified. De Moura et al. (2011) introduced a fault-diagnosis method based on the combination of PCA and ANN with two fluctuation analysis methods – detrended-fluctuation analysis (DFA) and rescaled-range analysis (RSA) – for roller bearings using vibration signals. In this method, first, the acquired vibration signals are preprocessed by DFA and RSA techniques and investigated using ANN and PCA in a total of four approaches. To validate this method, three different levels of bearing fault severities (0.15, 0.50, and 1.00 mm wide) along with a standard no-fault class were studied and compared. The vibration signals of roller bearings were collected under different frequency and load conditions. The experimental results showed the efficiency of the proposed combinations of techniques in discriminating vibration signals collected from bearings with four different fault severities. In particular, the classifier based on PCA presented performance slightly inferior to the one implemented by the neural network, but with a lower computational cost.

Furthermore, Dong and Luo (2013) proposed a technique for predicting the bearing degradation process based on PCA and optimised least square-support vector machine (LS-SVM). In this method, (i) the time domain, frequency domain, and time-frequency domain feature-extraction techniques are used to extract features from the collected vibration signals. Then, PCA is used to reduce the feature dimension and extract typical sensitive features. Finally, the LS-SVM is used to deal with the classification problem. Moreover, particle swarm optimisation (PSO) utilised to select the LS-SVM parameters. The experimental data sets used to validate this method are generated from bearing run-to-failure tests under constant load conditions. The experimental results proved the effectiveness of the proposed method. Gharavian et al. (2013) presented a method for a linear feature extraction for fault diagnosis of an automobile gearbox using vibration signals. In this method, first continuous wavelet transform (CWT) using a Morlet wavelet is applied over the acquired vibration signals. Then, to prevent the curse of dimensionality problem, Fisher discriminant analysis (FDA) is employed. With the reduced features, Gaussian mixture model (GMM) and K nearest neighbour (KNN) classifiers are individually examined. The fault-diagnosis results are compared to the results of PCA–based features using the same classifiers, i.e. GMM and KNN. The experimental results showed that FDA-based features achieved higher accuracy and a lower-cost monitoring system for a multispeed gearbox compared to PCA-based features.

Yang and Wu (2015) presented a method for diagnosing gear deterioration using EEMD and PCA techniques. In this method, features of the gear fault in the acquired vibration signals are extracted using EEMD and the marginal Hilbert spectrum analysis. Then, the extracted features are ordered using PCA, where the high-priority features in the PCs space represent the majority of the dynamical characteristics of the defective gears. ANN is used to classify the selected PCs. To validate the proposed method, six conditions of gear faults – normal, slight worn tooth, severe worn tooth, broken tooth, slight gear unbalance, and severe gear unbalance – are considered. The experimental results showed the effectiveness of the proposed method. Zuber and Bajrić (2016) studied the application of ANN and PCA on vibration signals for automated fault classification of roller element bearings. To validate this method, several types of roller element bearing faults at different levels of load were studied. The experimental results demonstrated the effectiveness of the proposed method. Song et al. (2019) introduced a method called dynamic principal component analysis (DPCA) for rotating machinery based on time-frequency domain analysis and PCA to enhance demodulation performance under low signal-to-noise ratio (SNR). In this method, a periodic modulation wave signal is extracted using time-frequency analysis and PCA. Simulation analysis and two real application cases were used to validate the proposed method. The experimental results of applying the proposed method to the vibration signals of the pump and acoustic signals of the propeller showed the effectiveness of the proposed method.

Moreover, Ahmed and Nandi (2019) proposed a three-stage hybrid method, compressive sampling with correlated principal and discriminant components (CS-CPDC), for roller bearing fault diagnosis based on compressed measurements. In this method, PCA and LDA features are combined via CCA in a three-step process to transform the characteristic space of the compressed measurements into a low-dimensional space of correlated and discriminant attributes (this method will be described in detail in Chapter 16). Wang et al. (2018) developed an algorithm for wind turbine fault detection and identification using PCA-based variable selection. In this method, a variable-selection algorithm based on PCA with multiple selection criteria is used to select a set of variables. Three performance measures – cumulative percentage variance, average correlation, and percentage entropy – are applied to evaluate different aspects of the algorithm with respect to the variable selection. To validate the proposed method, supervisory control and data acquisition system (SCADA) data with different types of faults was used to evaluate the T selection technique. With the selected variables, ANN was adopted to deal with the prediction problem. The experimental results showed the effectiveness of the proposed method in wind turbine fault diagnosis.

7.3   Independent Component Analysis (ICA)

ICA is a linear transformation algorithm that aims to transform an observed multidimensional random vector into components that are statistically as independent from each other as possible. Also, it can be defined as an algorithm that aims to find individual signals from mixtures of signals (Comon 1994; Hyvärinen et al. 2001). ICA is very closely linked to the BSS technique described in Chapter 3. The basic assumption in the standard ICA model is that the observed data X is a mix of sources or a vector of independent components (s_i) such that

$$X = AS \tag{7.8}$$

where $X = [x_1 x_2 ... x_L] \in R^{L \times n}$, $S = [s_1 s_2 ... s_L] \in R^{L \times k}$ is the independent components matrix ($k \leq n$), and $A \in R^{k \times n}$ is a mixing matrix that is invertible and square in the simplest case. This statistical model describes how the observed data are generated by a process of mixing the components of S. Here, the mixing matrix A is supposed to be unidentified, and the independent components of S cannot be directly obtained. So, the aim is to estimate both A and S utilising X. The basic assumption for ICA is that the components of S are statistically independent and that they have non-Gaussian distributions. However, Hyvärinen and Oja (2000) stated that in the basic model we do not assume these distributions are known.

The estimated matrix A, which is assumed to be square, can be used to recover the independent components from the data by using the inverse matrix W of the mixing matrix A such that

$$\hat{S} = A^{-1}X = WX \tag{7.9}$$

The estimation of W is based on cost functions, also called *objective functions* or *contrast functions*. Solutions W are found at the minima or maxima of these functions. To utilise the non-Gaussianity in ICA estimation, there are several measures that can be used to measure the non-Gaussianity of a random variable, e.g. Kurtosis and Negentropy (Hyvärinen and Oja 1997; Hyvärinen 1999a,b). Moreover, the minimisation of mutual information (MI) and maximisation of likelihood estimation are widely used in the estimation of the ICA model (Bell and Sejnowski 1995; Hyvärinen 1999a,b).

7.3.1 Minimisation of Mutual Information

The MI of the components of vector s can be expressed mathematically in Eq. (7.10),

$$MI(s) = \sum_i H(s_i) - H(s) \qquad (7.10)$$

where H represents Shannon's entropy, $H(s) = - \int p(s)log\,p(s)ds$, and $p(s)$ is the joint probability density of the components of vector s. The value of MI is normally non-negative and zero if and only if the variables are statistically independent. For an original random vector x with an invertible linear transformation $s = Wx$, the MI for s in terms of x can be defined in Eq. (7.11),

$$MI(s) = \sum_i H(s_i) - H(x) - log \mid det\,W \mid \qquad (7.11)$$

The use of MI can also be motivated using Kullback-Leibler divergence as a distance between two probability densities (Hyvärinen 1999b) such that,

$$\delta(pd_1, pd_2) = \int pd_1(s)log\,\frac{pd_1(s)}{pd_2(s)}ds \qquad (7.12)$$

where pd_1 and pd_2 are two probability densities. If s_i in Eq. (7.10) arc independent, one may measure the independence of the s_i as the Kullback-Leibler divergence between the real density $pd(s)$ and the estimated density $\widehat{pd}(s) = pd_1(s_1), pd_{21}(s_2), \dots, pd_n(s_n)$.

7.3.2 Maximisation of the Likelihood

The likelihood (Lh) can be formulated in the ICA model estimation, and this can be represented by the following equation,

$$Lh = \sum_{t=1}^{T} \sum_{i=1}^{n} f_i(W_i^T x(t)) + T\,ln \mid det\,W \mid \qquad (7.13)$$

where f_i are the density functions of the s_i, and the $x(t), t = 1, 2, \dots, T$ are the realisation of x (Hyvärinen 1999a,b).

Various ICA model algorithms have been proposed based on the type of objective functions described here. Of these methods, fast ICA algorithms, first developed by Hyvärinen and Oja (1997), is based on the maximisation of non-Gaussianity. In fact, fast ICA is one of the most commonly used ICA algorithms. Here, negentropy is employed to measure the non-Gaussianity such that,

$$N(x) = H(x_{Gauss}) - H(x) \qquad (7.14)$$

Here, x_{Gauss} is a Gaussian random variable, and x is a random variable. The objective function in fast ICA is based on marginal entropy maximisation such that,

$$J(x) = (E[G(x)] - E[G((x_{Gauss})])^2 \qquad (7.15)$$

where G is non-quadratic function and x_{Gauss} is a Gaussian variable with zero mean and unit variance. By considering the sphered data v in place of x and using the Kurtosis contrast function, the fast ICA algorithm can be represented mathematically using the following equation:

$$w \leftarrow E[z(w^T z)^3] - 3w \qquad (7.16)$$

Based on Eq. (7.16), after each iteration, the weight w is updated, and the final w provides the independent components using the linear combination $w^T z$. A more recently developed version has the ICA method evolve using the reconstruction cost and score matching to learn overcomplete features (Hyvärinen 2005; Le et al. 2011).

Due to the fact that ICA methods converge better with sphered data, it is often useful to sphere or whiten the data x before performing ICA. The sphering or whitening can be accomplished using standard PCA described earlier. Another preprocessing step usually used before performing ICA is to make the observed signals zero mean and decorrelate them.

7.3.3 Application of ICA in Machine Fault Diagnosis

The application of ICA in vibration signal analysis has been widely employed in various studies of machine fault diagnosis. For instance, Ypma et al. (1999) introduced a method for fault detection in rotating machines using ICA and support vector data description. This method is a fusion of multichannel measurement of machine vibration using ICA, followed by a description of the admissible domain, i.e. the part of the feature space that indicates normal machine operation, with support vector domain description (SVDD). A number of induced faults of a small submersible (loose foundation, imbalance, OR failure of the uppermost ball bearing) are used to validate the proposed method. The experimental results showed that the combination of measurement channels with ICA achieved improved results in fault detection. Tian et al. (2003) introduced a method for gearbox fault diagnosis using ICA in the frequency domain and wavelet filtering. In this method, fast Fourier transform (FFT) is first applied to the input vibration signals, and then ICA is used to find the unmixing matrix W that can be used to estimate the separated signals S. The experimental result using vibration signals from a gearbox simulator showed that ICA in the frequency domain is a promising technique for fault diagnosis of a rotating machine.

Moreover, Zuo et al. proposed the use of wavelet transform to preprocess the vibration data collected from a single sensor to obtain multiple data series at different scales of the wavelet transform. Then, these multiple data series are used as input to ICA or PCA for detection of a single independent source (Zuo et al. 2005). This method is applied to impulse detection using a simulated signal series and a real signal series collected from a gearbox with a gear tooth fault. The experimental results showed that ICA in combination with wavelet transform worked better than PCA in combination with wavelet transform for impulse identification. Li et al. (2006) presented a fault-recognition method, called independent component analysis-factorial hidden Markov model (ICA-FHMM), for speed-up and slow-down process of rotating machine. In this method, ICA is used to extract features from the multichannel vibration data, and the factorial hidden Markov model (FHMM) is a classifier to identify the faults of the speed-up and slow-down process in rotating machinery. Four running conditions (unbalance, rubbing, oil whirl, and pedestal looseness) performed on a Bently rotor kit are used to validate the proposed method. The experimental results showed the effectiveness of the proposed method in fault recognition. In addition, this method is compared to another method called independent component analysis-hidden Markov model (ICA-HMM) that combines ICA and the hidden Markov model (HMM), and the results showed that ICA-FHMM is faster than ICA-HMM.

Widodo et al. (2007) studied the application of ICA and SVMs to detect and diagnosis faults of the induction motor. ICA is employed to extract features from the original signals. PCA is also used to extract features from the original signals for comparison with ICA does. With the extracted features, SVM with a sequential minimal optimisation algorithm is used to deal with the classification problem. Vibration data from six induction motors are used to validate the proposed method. The experimental results showed that the combination of ICA and SVMs is an effective technique for fault diagnosis of induction motors. Jiao and Chang (2009) proposed a method called independent component analysis-multilayer perceptron (ICA-MLP) for fault diagnosis of rotor systems using ICA and a multilayer perceptron (MLP). In this method, ICA is used to extract features from multichannel vibration measurements acquired at different rotating speeds and/or loads. With these extracted features, the MLP classifier is used to deal with classification problem. The experimental results demonstrated the efficiency of the proposed method in fault diagnosis of the rotor system. Li et al. (2010) introduced a technique for gear multi-fault diagnosis of rotating machines using ICA and fuzzy K-nearest neighbour (FKNN). In this process, ICA is used to separate characteristic vibration signals and inference vibration signals from the parallel time series obtained from multichannel accelerometers. Then, the wavelet transform and autoregressive (AR) model are performed to extract features from the original feature vector. ICA is used again to reduce the dimensionality of the feature space. With these reduced features, FKNN is employed to identify the conditions of the gears of interest. The experimental results showed the effectiveness of the proposed method in gear fault diagnosis.

Furthermore, Wang et al. (2011) studied the application of constraint-independent component analysis (cICA) in machine fault diagnosis. In this study, cICA is used to extract the desired faulty signal based on some prior mechanical information as a reference. Also, the methods of creating a reference of cICA for machine fault diagnostics are discussed. Simulated and real vibration data of bearings from normal to final failure are used to validate the proposed method. The experimental results demonstrated the effectiveness of the proposed method in machine fault diagnosis. Guo et al. (2014) introduced a feature-extraction scheme for rolling element bearing fault diagnosis using the envelope extraction and ICA. In this scheme, envelope extraction is used to obtain impulsive components corresponding to the faults from the rolling element bearing and reduce the dimension of the vibration sources in the sensor-picked signals. Then, ICA is used to separate envelopes according to the independence of vibrations sources. In this way, the vibration features related to the rolling element bearing faults can be separated from the disturbances and clearly exposed by the envelope spectrum. Simulation and experimental results showed the effectiveness of the proposed method in fault detection in rolling element bearings in a rotating machine. Wang et al. proposed a method to identify compound faults from measured mixed signals, based on the EEMD method and ICA technique. In this method, (i) a vibration signal is decomposed into intrinsic mode functions (IMFs) using the EEMD method to obtain multichannel signals. (ii) According to a cross-correlation criterion, the corresponding IMF is selected as the input matrix of ICA. Finally, (iii) the compound faults can be separated effectively by executing the ICA method, which makes the fault features more easily extracted and more clearly identified. Experimental results showed the effectiveness of the proposed method in separating compound faults, which works not only for the OR defect but also for the roller defect and the unbalance fault of the experimental system (Wang et al. 2014).

Recently, Žvokelj et al. (2016) proposed a multivariate and multiscale statistical process monitoring method for detecting incipient failures in large slewing bearings. In this method, ICA and EEMD are combined to adaptively decompose signals into different time scales; thus it can handle multiscale system dynamics. This method is called EEMD-MSICA and has the ability to denoise multivariate signal. In addition, EEMD-MSICA in combination with envelope analysis can be used as a diagnostic tool. A synthetic signal, as well as real signals acquired through conducting an accelerated run-to-failure lifetime experiment from a slewing bearing test stand, are used to validate the proposed method. The experimental results showed that the EEMD-MSICA method can effectively extract information from high-dimensional datasets containing a large number of multiscale features. Thus, it is a reliable fault-detection, diagnosis, and signal-denoising tool.

7.4 Linear Discriminant Analysis (LDA)

Different from PCA, searching for the most important components of samples, LDA aims to find discriminant components that distinguish different class samples (Balakrishnama and Ganapathiraju 1998). In fact, LDA collects the samples from the same class and expands the margin of samples from different classes. The Fisher LDA analysis method (Welling 2005; Sugiyama 2006) considers maximising the Fisher criterion function $J(W)$, i.e. the ratio of the between the class scatter (S_B) to the within-class scatter (S_w) such that

$$J(W) = \frac{|W^T S_B W|}{|W^T S_w W|} \tag{7.17}$$

where

$$S_B = \frac{1}{L} \sum_{i=1}^{c} l_i (\mu^i - \mu)(\mu^i - \mu)^T \tag{7.18}$$

$$S_w = \frac{1}{L} \sum_{i=1}^{c} \sum_{j=1}^{l_i} (x_j^i - \mu^i)(x_j^i - \mu^i)^T \tag{7.19}$$

where L is the total number of observations and c is the number of classes. $x \, \varepsilon \, R^{L \times n}$ is the training dataset, x_1^i represents the dataset belong to the cth class, n_i is the number of measurements of the ith class, μ^i is the mean vector of class i, and μ is the mean vector of all training datasets such that,

$$\mu^i = \frac{1}{L_c} \sum_{i \in c} x_i \tag{7.20}$$

$$\mu = \frac{1}{L} \sum_{i} x_i \tag{7.21}$$

where L_c is the number of cases in class c.

The total scatter (S_T) can be represented using the following equation:

$$S_T = S_w + S_B \tag{7.22}$$

Then the objective function $J(W)$ in Eq. (7.17) can be rewritten as Eq. (7.23):

$$J(W) = \frac{|W^T S_T W|}{|W^T S_w W|} - 1 \tag{7.23}$$

Here, the objective function can be described as maximising the total scatter of the data while minimising the within scatter of the classes. LDA projects the space of the original data onto a $(c-1)$–dimension space by finding the optimal projection matrix W that maximises the $J(W)$ in Eq. (7.24) such that,

$$\widehat{W} = \arg \max_W J(W) \tag{7.24}$$

Here, $\widehat{W}$ is composed of the selected eigenvectors $(\widehat{w}_1, \dots, \widehat{w}_{m2})$ with the first $n2$ largest eigenvalues $(n2 = c - 1)$. Also, the problem of maximising the objective function J can be transformed into a constrained optimisation problem using the following equation:

$$min_w - \frac{1}{2} W^T S_B W|$$
$$\text{s.t} \quad W^T S_w W| = 1 \tag{7.25}$$

7.4.1 Application of LDA in Machine Fault Diagnosis

The application of LDA in vibration signal analysis has been widely employed in various studies of machine fault diagnosis. For instance, Lee et al. (2005) proposed a diagnosis algorithm for induction motor fault detection using LDA. In this method, (i) PCA is used to reduce the input dimension, and then (ii) LDA used to extract features for each fault. With these extracted features, a distance measure between the predefined fault vectors and the input vector is used for diagnosis. Various experiments under noisy conditions are used to validate the proposed method. The experimental results showed the effectiveness of the proposed method in induction motor fault diagnosis. Park et al. (2007) proposed a feature-extraction method and fusion algorithm, which constructed PCA and LDA for induction motor fault detection. In this method, after extracting features using PCA and LDA from the original signals, the reference data is used to produce matching values. In the diagnostic step, two matching values obtained by PCA and LDA are combined with a probability model, and a faulted signal is finally diagnosed. The simulation results of various noisy conditions demonstrated the effectiveness of the proposed method in induction motor fault detection.

Moreover, Jakovljevic et al. (2012) presented a process for detecting broken rotor bars in induction motor using vibration signal analysis. In this method, PCA is used for dimensionality reduction and ICA is employed as a data-classification technique. Real vibration signals with 16 data points are acquired using 2 accelerometers, 66 from a faulty motor and 44 from the normal (i.e. healthy) motor, and used to validate the proposed method. After the data acquisition and feature extraction, the 16-dimensional data points are reduced to 2-dimensional and 3-dimensional data points. The experimental results showed that the 3-dimensional reduction achieved 97% fault-classification accuracy. Harmouche et al. (2014) proposed a method for bearing ball faults using spectral features and LDA. In this method, FFT is used to obtain the spectral features of the feature space. Then, LDA is applied on the dataset that excludes the data coming from OR faults and IR fault conditions, i.e. only samples corresponding to normal and ball faults are used to build the LDA model. Experimental vibration data with the mentioned health conditions are used to validate the proposed method. The experimental results showed the effectiveness of the proposed method in the diagnosis of bearing faults. Real vibration signals acquired from roller bearings with six health conditions – brand new (NO), worn (NW), IR fault (IR), OR (OR) fault, rolling element (RE) fault, and cage (CA) fault – are used to validate the proposed methods. The experimental results showed that the features extracted using the CS and LDA achieved 100% classification accuracy using 20% of the data.

Furthermore, Ciabattoni et al. (2015) proposed an LDA-based technique for motor bearing fault detection. In this technique, an LDA-based algorithm, called Δ-LDA, is designed to overcome the problem of a between-class scatter matrix trace very close to zero by using the difference of covariance matrices. The proposed method is tested in an experimental fault-diagnosis case study. The experimental results showed the effectiveness of Δ-LDA in motor bearing fault classification. Recently, Ahmed et al. (2017) presented a strategy for bearing fault classification based on compressive sampling (CS). Under this strategy, compressed vibration signals are first obtained by resampling the acquired bearing vibration signals in the time domain with a random Gaussian matrix using different compressed sensing sampling rates. Then, three approaches have been chosen to process these compressed data for the purpose of bearing fault classification: these include using the data directly as the input of classifier, and extracting features from the data using linear feature-extraction methods, namely, unsupervised PCA and LDA.

7.5   Canonical Correlation Analysis (CCA)

The CCA was first introduced by Hotelling (1936). Different from PCA and LDA, which encompass only one dataset, CCA is a statistical method for finding linear combinations of two datasets that can maximise their correlation (Hardoon et al. 2004). For example, let $(y_1, y_2) \in R^{m1}, R^{m2}$ be two vectors with covariance C_{11} and C_{22}, respectively, and cross-covariance C_{12}. CCA finds linear combinations of y_1 and y_2 vectors $(\acute{w}_1 y_1, \acute{w}_2 y_2)$ that are maximally correlated such that the following objective function is maximised,

$$(W_1, W_2) = arg\max_{W_1, W_2} (\acute{w}_1 y_1, \acute{w}_2 y_2) \tag{7.26}$$

$$= arg\max_{W_1, W_2} \frac{\acute{w}_1 C_{12} w_2}{\sqrt{\acute{w}_1 C_{11} w_1 \acute{w}_2 C_{22} w_2}} \tag{7.27}$$

where C_{11}, C_{22}, and C_{12} can be computed using the following equations:

$$C_{11} = \frac{1}{m} y_1 \acute{y}_1 \tag{7.28}$$

$$C_{22} = \frac{1}{m} y_2 \acute{y}_2 \tag{7.29}$$

$$C_{12} = \begin{bmatrix} C_{11} & C_{12} \\ C_{21} & C_{22} \end{bmatrix} \tag{7.30}$$

Based on the theory of CCA, the solution of the objective function in (7.27) is not affected by scaling w_1 or w_2 individually or together: for example, by scaling the numerator and denominator of (7.27) by α the solution of this scaled objective function will be equivalent to the solution before scaling. This can be mathematically given by the following equation:

$$\frac{\alpha \acute{w}_1 C_{12} w_2}{\sqrt{\alpha^2 \acute{w}_1 C_{11} w_1 \acute{w}_2 C_{22} w_2}} = \frac{\acute{w}_1 C_{12} w_2}{\sqrt{\acute{w}_1 C_{11} w_1 \acute{w}_2 C_{22} w_2}} \tag{7.31}$$

Accordingly, the optimization problem in (7.27) can be rewritten as Eq. (7.32):

$$\arg \max_{W_1, W_2} \acute{w}_1 C_{12} w_2$$
$$\text{s.t} \quad \acute{w}_1 C_{11} w_1 = 1, \acute{w}_2 C_{22} w_2 = 1. \tag{7.32}$$

To find w_1 and w_2 that maximise the correlation between y_1 and y_2, we assume that with these equations w_1 can be derived from w_2 or vice versa. Hence, the corresponding Lagrangian can be mathematically given by Eq. (7.33):

$$L(\lambda, w_1, w_2) = \acute{w}_1 C_{12} w_2 - \frac{\lambda_1}{2}(\acute{w}_1 C_{11} w_1 - 1) - \frac{\lambda_2}{2}(\acute{w}_2 C_{22} w_2 - 1) \tag{7.33}$$

Then, by taking derivatives in respect to w_1 and w_2, the following equations can be obtained:

$$\frac{\partial f}{\partial w_1} = C_{12} w_2 - \lambda_1 C_{11} w_1 = 0 \tag{7.34}$$

$$\frac{\partial f}{\partial w_2} = C_{21} w_1 - \lambda_2 C_{22} w_2 = 0 \tag{7.35}$$

Then, with $\acute{w}_1$ multiplied by Eq. (7.34), $\acute{w}_2$ multiplied by Eq. (7.35), and subtracting the resultant equations, Eq. (7.36) can be obtained:

$$\acute{w}_1 C_{12} w_2 - \acute{w}_1 \lambda_1 C_{11} w_1 - \acute{w}_2 C_{21} w_1 + \acute{w}_2 \lambda_2 C_{22} w_2 = 0$$
$$\acute{w}_2 \lambda_2 C_{22} w_2 - \acute{w}_1 \lambda_1 C_{11} w_1 = 0 \tag{7.36}$$

The constraints of Eqs. (7.32) and (7.36) indicate that $\lambda_2 - \lambda_1 = 0$. Hence, let $\lambda = \lambda_1 = \lambda_2$ and assume that C_{22} is invertible. Then, from Eq. (7.35), w_2 can be represented using the following equation:

$$w_2 = \frac{C_{22}^{-1} C_{21} w_1}{\lambda} \tag{7.37}$$

By substituting in Eq. (7.34), we obtain,

$$\frac{C_{12} C_{22}^{-1} C_{21} w_1}{\lambda} - \lambda C_{11} w_1 = 0 \tag{7.38}$$

More efficiently,

$$C_{12} C_{22}^{-1} C_{21} w_1 = \lambda^2 C_{11} w_1 \tag{7.39}$$

Equation (7.39) is a generalised eigen-problem that can be solved to obtain w_1 and then obtain w_2 using Eq. (7.37).

Recently, CCA has been explored for fault detection in many applications. For example, Chen et al. (2016a,b) considered the application of CCA to deal with the detection of incipient multiplicative faults in industrial processes. Jiang et al. (2017) proposed data-driven distributed fault detection for large-scale processes based on a genetic algorithm (GA)-regularised CCA. Liu et al. (2018) presented a mixture of variational CCA for nonlinear and quality-relevant process monitoring. Moreover, Chen et al. (2018) considered the application of generalised CCA and randomised algorithms in fault detection of non-Gaussian processes. Furthermore, in machine fault detection and diagnosis, Ahmed and Nandi (2019) proposed a three-stage hybrid method, CS-CPDC, for diagnosing bearing faults based on compressed measurements. In this method, PCA and LDA features are combined via CCA in a three-step process to transform the characteristic space of the compressed measurements into a low-dimensional space of correlated and discriminant attributes (this method will be described in detail in Chapter 16).

7.6 Partial Least Squares (PLS)

PLS (which also goes by the name *projection to latent structures*) is a method of modelling relationships between sets of observed variables by means of latent variables. It was first proposed by Hermon Wold (1966a,b), who introduced nonlinear estimation by iterative least squares (NILES) procedures. Then, NILES with ordinary least squares (OLS) was employed as an iterative estimation technique to estimate principal components of PCA. Since then, variants of PLS have been proposed, e.g. nonlinear iterative partial least squares (NIPALS), partial least squares path modelling (PLS-PM), and partial least squares regression (PLS-R) (Wold 1966a,b, 1973; Geladi and Kowalski 1986; Wold et al. 2001). The simple idea of PLS is to construct new predictor variables, latent variables, as linear combinations of the original variables in a matrix X of features, i.e. descriptor variables, and a matrix Y of response variables (classes). Briefly, we present a mathematical description of the PLS procedure.

Let $X \in R^{L \times n}$ represent a data matrix with L examples and n-dimensional space of feature vectors. Similarly, let $Y \in R^{L \times c}$ be a matrix that represents the class labels. The PLS procedure for dimension reduction can be described as follows:

(1) The PLS method decomposes the zero-mean matrix $X \in R^{L \times n}$ and the zero mean $Y \in R^{L \times c}$ into

$$X = TP^T + E \tag{7.40}$$

$$Y = UQ^T + F \tag{7.41}$$

where $T \in R^{L \times p}$ and $U \in R^{L \times p}$ are matrices comprising p extracted latent vectors; the matrix $P \in R^{n \times p}$ and the vector $q \in R^{1 \times p}$ denote the loadings; and the matrices $E \in R^{L \times n}$ and $F \in R^{L \times c}$ are the residuals.

(2) The PLS method creates a latent subspace composed of a set of weight vectors $W = \{w_1, w_2, \ldots, w_p\}$ using the NIPALS method (Wold 1973). The sample covariance between latent vectors T and U can be mathematically represented using Eq. (7.42):

$$[cov(t_i, u_i)] = \max_{|W_i| = |s_i| = 1} [cov(Xw_i, Ys_i)]^2 \tag{7.42}$$

Here, t_i and u_i are the ith column of matrix T and U, respectively. To find the score vectors t_i and u_i, the NIPALS method begins with random initialisation of y-space vector u_i and repeats a sequence of the following steps until convergence (Rosipal and Krämer 2005):

a) $w_i = X^T u_i/(u_i^T u_i)$ d) $s_i = Y^T t_i/(t_i^T t_i)$

b) $||w_i|| \rightarrow 1$ e) $||s_i|| \rightarrow 1$

c) $t_i = Xw_i$ f) $u_i = Ys_i$

After these steps, matrices X and Y are deflated by subtracting their rank-one approximations based on the calculated t_i and u_i. Different types of deflation define several variants of PLS. For example, (Wold 1975).

PLS methods are being considered in a large diversity of applications, e.g. fault diagnosis in chemical processes (Chiang et al. 2000; Russell et al. 2012), appearance-based person recognition (Schwartz and Davis 2009), key performance indicator-related fault diagnosis (Yin et al. 2015), etc.

PLS has been used in various studies of machine fault diagnosis. For instance, to address the nonstationary and nonlinear characteristics of vibration signals generated by roller bearings, Cui et al. (2017) proposed a fault-diagnosis method based on SVD, empirical mode decomposition (EMD), and variable predictive model-based class discrimination (VPMCD). In this method, PLS is used to estimate the model parameters of VPMCD. The experimental results showed that the proposed method can effectively identify a faulty bearing. Moreover, Lv et al. (2016) proposed a method for wind turbine fault diagnosis based on multivariate statistics. In this method, using the original data, PLS is used to build the relationship between input and output variables and obtain the monitoring model. Computer simulation results demonstrated the effectiveness of reducing the dimensions of data and realising the fault diagnosis. Moreover, Fuente and colleagues proposed a method for fault detection and identification based on multivariate statistical techniques. In this method, the PLS is utilised for detecting faults and subsequently Fisher's discriminant analysis (FDA) is used to identify the faults (Fuente et al. 2009).

7.7 Summary

This chapter has presented commonly appropriate methods for LSL that can be used to reduce a large amount of collected vibration data to a few dimensions without significant loss of information. Various linear techniques have been proposed to learn subspaces from the high-dimensional data. Of these techniques, PCA, ICA, and LDA are amongst the most commonly used in machine fault diagnosis. CCA and PLS have been recently considered in many application of fault detection including machine fault detection. Most of the introduced techniques and their publicly accessible software are summarised in Table 7.2.

Table 7.2 Summary of Some of the introduced techniques and their publically accessible software.

Algorithm name	Platform	Package	Function
Eigenvalues and eigenvectors	MATLAB	Mathematics Toolbox, Linear Algebra	eig
Singular value decomposition			svd
Principal component analysis of raw data		Statistics and Machine Learning Toolbox – Dimensionality Reduction and Feature Extraction	pca
Feature extraction by using reconstruction ICA			rica
Canonical correlation			canoncorr

References

Ahmed, H. and Nandi, A. (2019). Three-stage hybrid fault diagnosis for rolling bearings with compressively-sampled data and subspace learning techniques. *IEEE Transactions on Industrial Electronics* 66 (7): 5516–5524.

Ahmed, H.O.A., Wong, M.D., and Nandi, A.K. (2017). Compressive sensing strategy for classification of bearing faults. In: *2017 IEEE International Conference on Acoustics, Speech and Signal Processing (ICASSP)*, 2182–2186. IEEE.

Balakrishnama, S. and Ganapathiraju, A. (1998). Linear discriminant analysis-a brief tutorial. *Institute for Signal and Information Processing* 18: 1–8.

Baydar, N. and Ball, A.D. (2001). Detection of gear failures using wavelet transform and improving its capability by principal component analysis. In: *The 14th International Congress on Condition Monitoring and Diagnostic Engineering Management* (eds. A.G. Starr and R.B.K.N. Rao), 411–418. Manchester: Elsevier Science Ltd.

Bell, A.J. and Sejnowski, T.J. (1995). An information-maximization approach to blind separation and blind deconvolution. *Neural Computation* 7 (6): 1129–1159.

Chen, Z., Ding, S.X., Peng, T. et al. (2018). Fault detection for non-Gaussian processes using generalized canonical correlation analysis and randomized algorithms. *IEEE Transactions on Industrial Electronics* 65 (2): 1559–1567.

Chen, Z., Ding, S.X., Zhang, K. et al. (2016a). Canonical correlation analysis-based fault detection methods with application to alumina evaporation process. *Control Engineering Practice* 46: 51–58.

Chen, Z., Zhang, K., Ding, S.X. et al. (2016b). Improved canonical correlation analysis-based fault detection methods for industrial processes. *Journal of Process Control* 41: 26–34.

Chiang, L.H., Russell, E.L., and Braatz, R.D. (2000). Fault diagnosis in chemical processes using Fisher discriminant analysis, discriminant partial least squares, and principal component analysis. *Chemometrics and Intelligent Laboratory Systems* 50 (2): 243–252.

Ciabattoni, L., Cimini, G., Ferracuti, F. et al. (2015). A novel LDA-based approach for motor bearing fault detection. In: *2015 IEEE 13th International Conference on Industrial Informatics (INDIN)*, 771–776. IEEE.

Comon, P. (1994). Independent component analysis, a new concept? *Signal Processing* 36 (3): 287–314.

Cui, H., Hong, M., Qiao, Y., and Yin, Y. (2017). Application of VPMCD method based on PLS for rolling bearing fault diagnosis. *Journal of Vibroengineering* 2308, 19 (1): 160–175.

De Moura, E.P., Souto, C.R., Silva, A.A., and Irmao, M.A.S. (2011). Evaluation of principal component analysis and neural network performance for bearing fault diagnosis from vibration signal processed by RS and DF analyses. *Mechanical Systems and Signal Processing* 25 (5): 1765–1772.

Dong, S. and Luo, T. (2013). Bearing degradation process prediction based on the PCA and optimized LS-SVM model. *Measurement* 46 (9): 3143–3152.

Fuente, M.J., Garcia-Alvarez, D., Sainz-Palmero, G.I., and Villegas, T. (2009). Fault detection and identification method based on multivariate statistical techniques. In: *ETFA 2009: IEEE Conference on Emerging Technologies & Factory Automation, 2009*, 1–6. IEEE.

Geladi, P. and Kowalski, B.R. (1986). Partial least-squares regression: a tutorial. *Analytica Chimica Acta* 185: 1–17.

Gharavian, M.H., Ganj, F.A., Ohadi, A.R., and Bafroui, H.H. (2013). Comparison of FDA-based and PCA-based features in fault diagnosis of automobile gearboxes. *Neurocomputing* 121: 150–159.

Golub, G.H. and Reinsch, C. (1970). Singular value decomposition and least squares solutions. *Numerische Mathematik* 14 (5): 403–420.

Guo, Y., Na, J., Li, B., and Fung, R.F. (2014). Envelope extraction based dimension reduction for independent component analysis in fault diagnosis of rolling element bearing. *Journal of Sound and Vibration* 333 (13): 2983–2994.

Hardoon, D.R., Szedmak, S., and Shawe-Taylor, J. (2004). Canonical correlation analysis: an overview with application to learning methods. *Neural Computation* 16 (12): 2639–2664.

Harmouche, J., Delpha, C., and Diallo, D. (2014). Linear discriminant analysis for the discrimination of faults in bearing balls by using spectral features. In: *2014 International Conference on Green Energy*, 182–187. IEEE.

He, Q., Yan, R., Kong, F., and Du, R. (2009). Machine condition monitoring using principal component representations. *Mechanical Systems and Signal Processing* 23 (2): 446–466.

Hotelling, H. (1936). Relations between two sets of variates. *Biometrika* 28 (3/4): 321–377.

Hyvärinen, A. (1999a). Fast and robust fixed-point algorithms for independent component analysis. *IEEE Transactions on Neural Networks* 10 (3): 626–634.

Hyvärinen, A., 1999b. Survey on independent component analysis.

Hyvärinen, A. (2005). Estimation of non-normalized statistical models by score matching. *Journal of Machine Learning Research* 6 (Apr): 695–709.

Hyvärinen, A., Hoyer, P.O., and Inki, M. (2001). Topographic independent component analysis. *Neural Computation* 13 (7): 1527–1558.

Hyvärinen, A. and Oja, E. (1997). A fast fixed-point algorithm for independent component analysis. *Neural Computation* 9 (7): 1483–1492.

Hyvärinen, A. and Oja, E. (2000). Independent component analysis: algorithms and applications. *Neural Networks* 13 (4–5): 411–430.

Jakovljevic, B.B., Kanovic, Z.S., and Jelicic, Z.D. (2012). Induction motor broken bar detection using vibration signal analysis, principal component analysis and linear discriminant analysis. In: *2012 IEEE International Conference on Control Applications (CCA)*, 1686–1690. IEEE.

Jiang, Q., Ding, S.X., Wang, Y., and Yan, X. (2017). Data-driven distributed local fault detection for large-scale processes based on the GA-regularized canonical correlation analysis. *IEEE Transactions on Industrial Electronics* 64 (10): 8148–8157.

Jiao, W. and Chang, Y. (2009). ICA-MLP classifier for fault diagnosis of rotor system. In: *IEEE International Conference on Automation and Logistics, 2009: ICAL'09*, 997–1001. IEEE.

Le, Q.V., Karpenko, A., Ngiam, J., and Ng, A.Y. (2011). ICA with reconstruction cost for efficient overcomplete feature learning. In: *Advances in Neural Information Processing Systems*, 1017–1025. https://papers.nips.cc/paper/4467-ica-with-reconstruction-cost-for-efficient-overcomplete-feature-learning.pdf.

Lee, D.J., Park, J.H., Kim, D.H., and Chun, M.G. (2005). Fault diagnosis of induction motor using linear discriminant analysis. In: *International Conference on Knowledge-Based and Intelligent Information and Engineering Systems*, 860–865. Berlin, Heidelberg: Springer.

Li, W., Shi, T., Liao, G., and Yang, S. (2003). Feature extraction and classification of gear faults using principal component analysis. *Journal of Quality in Maintenance Engineering* 9 (2): 132–143.

Li, Z., He, Y., Chu, F. et al. (2006). Fault recognition method for speed-up and speed-down process of rotating machinery based on independent component analysis and Factorial Hidden Markov Model. *Journal of Sound and Vibration* 291 (1–2): 60–71.

Li, Z.X., Yan, X.P., Yuan, C.Q., and Li, L. (2010). Gear multi-faults diagnosis of a rotating machinery based on independent component analysis and fuzzy k-nearest neighbor. In: *Advanced Materials Research*, vol. 108, 1033–1038. Trans Tech Publications.

Liu, Y., Liu, B., Zhao, X., and Xie, M. (2018). A mixture of variational canonical correlation analysis for nonlinear and quality-relevant process monitoring. *IEEE Transactions on Industrial Electronics* 65 (8): 6478–6486.

Lu, H., Plataniotis, K.N., and Venetsanopoulos, A. (2013). *Multilinear Subspace Learning: Dimensionality Reduction of Multidimensional Data*. Chapman and Hall/CRC.

Lv, F., Zhang, Z., Zhai, K., and Ju, X. (2016). Research on fault diagnosis method of wind turbine based on partial least square method. In: *2016 2nd IEEE International Conference on Computer and Communications (ICCC)*, 925–928. IEEE.

Malhi, A. and Gao, R.X. (2004). PCA-based feature selection scheme for machine defect classification. *IEEE Transactions on Instrumentation and Measurement* 53 (6): 1517–1525.

Moore, B. (1981). Principal component analysis in linear systems: controllability, observability, and model reduction. *IEEE Transactions on Automatic Control* 26 (1): 17–32.

Park, W.J., Lee, S.H., Joo, W.K., and Song, J.I. (2007). A mixed algorithm of PCA and LDA for fault diagnosis of induction motor. In: *International Conference on Intelligent Computing*, 934–942. Berlin, Heidelberg: Springer.

Rosipal, R. and Krämer, N. (2005). Overview and recent advances in partial least squares. In: *International Statistical and Optimization Perspectives Workshop "Subspace, Latent Structure and Feature Selection"*, 34–51. Berlin, Heidelberg: Springer.

Russell, E.L., Chiang, L.H., and Braatz, R.D. (2012). *Data-Driven Methods for Fault Detection and Diagnosis in Chemical Processes*. Springer Science & Business Media.

Schwartz, W.R. and Davis, L.S. (2009). Learning discriminative appearance-based models using partial least squares. In: *2009 XXII Brazilian Symposium on Computer Graphics and Image Processing (SIBGRAPI)*, 322–329. IEEE.

Serviere, C. and Fabry, P. (2005). Principal component analysis and blind source separation of modulated sources for electro-mechanical systems diagnostic. *Mechanical Systems and Signal Processing* 19 (6): 1293–1311.

Shlens, J. (2014). A tutorial on principal component analysis. arXiv preprint arXiv:1404.1100.

Song, Y., Liu, J., Chu, N. et al. (2019). A novel demodulation method for rotating machinery based on time-frequency analysis and principal component analysis. *Journal of Sound and Vibration* 442: 645–656.

Sugiyama, M. (2006). Local fisher discriminant analysis for supervised dimensionality reduction. In: *Proceedings of the 23rd international conference on Machine learning*, 905–912. ACM.

Sun, W., Chen, J., and Li, J. (2007). Decision tree and PCA-based fault diagnosis of rotating machinery. *Mechanical Systems and Signal Processing* 21 (3): 1300–1317.

Tian, X., Lin, J., Fyfe, K.R., and Zuo, M.J. (2003). Gearbox fault diagnosis using independent component analysis in the frequency domain and wavelet filtering. In: *2003 IEEE International Conference on Acoustics, Speech, and Signal Processing, 2003. Proceedings. (ICASSP'03)*, vol. 2, 245–248. IEEE.

Trendafilova, I. (2010). An automated procedure for detection and identification of ball bearing damage using multivariate statistics and pattern recognition. *Mechanical Systems and Signal Processing* 24 (6): 1858–1869.

Wang, H., Li, R., Tang, G. et al. (2014). A compound fault diagnosis for rolling bearings method based on blind source separation and ensemble empirical mode decomposition. *PLoS One* 9 (10): e109166.

Wang, Y., Ma, X., and Qian, P. (2018). Wind turbine fault detection and identification through PCA-based optimal variable selection. *IEEE Transactions on Sustainable Energy* 9 (4): 1627–1635.

Wang, Z., Chen, J., Dong, G., and Zhou, Y. (2011). Constrained independent component analysis and its application to machine fault diagnosis. *Mechanical Systems and Signal Processing* 25 (7): 2501–2512.

Welling, M. (2005). Fisher linear discriminant analysis. Department of Computer Science, University of Toronto. https://www.ics.uci.edu/~welling/teaching/273ASpring09/Fisher-LDA.pdf.

Widodo, A., Yang, B.S., and Han, T. (2007). Combination of independent component analysis and support vector machines for intelligent faults diagnosis of induction motors. *Expert Systems with Applications* 32 (2): 299–312.

Wold, H. (1966a). Estimation of principal components and related models by iterative least squares. In: *Multivariate Analysis*, 391–420. New York: Academic Press.

Wold, H. (1966b). Nonlinear estimation by iterative least squares procedures. In: *Festschrift for J. Neyman: Research Papers in Statistics* (ed. F.N. David), 411–444. Wiley.

Wold, H. (1973). Nonlinear iterative partial least squares (NIPALS) modelling: some current developments. In: *Multivariate Analysis–III*, 383–407. Academic Press.

Wold, H. (1975). Path models with latent variables: the NIPALS approach. In: *Quantitative sociology*, 307–357. Academic Press.

Wold, S., Esbensen, K., and Geladi, P. (1987). Principal component analysis. *Chemometrics and Intelligent Laboratory Systems* 2: 37–52.

Wold, S., Sjöström, M., and Eriksson, L. (2001). PLS-regression: a basic tool of chemometrics. *Chemometrics and Intelligent Laboratory Systems* 58 (2): 109–130.

Yang, C.Y. and Wu, T.Y. (2015). Diagnostics of gear deterioration using EEMD approach and PCA process. *Measurement* 61: 75–87.

Yin, S., Zhu, X., and Kaynak, O. (2015). Improved PLS focused on key-performance-indicator-related fault diagnosis. *IEEE Transactions on Industrial Electronics* 62 (3): 1651–1658.

Ypma, A., Tax, D.M., and Duin, R.P. (1999). Robust machine fault detection with independent component analysis and support vector data description. In: *Neural Networks for Signal Processing IX, 1999. Proceedings of the 1999 IEEE Signal Processing Society Workshop*, 67–76. IEEE.

Zuber, N. and Bajrić, R. (2016). Application of artificial neural networks and principal component analysis on vibration signals for automated fault classification of roller element bearings. *Eksploatacja i Niezawodność* 18 (2): 299–306.

Zuo, M.J., Lin, J., and Fan, X. (2005). Feature separation using ICA for a one-dimensional time series and its application in fault detection. *Journal of Sound and Vibration* 287 (3): 614–624.

Žvokelj, M., Zupan, S., and Prebil, I. (2010). Multivariate and multiscale monitoring of large-size low-speed bearings using ensemble empirical mode decomposition method combined with principal component analysis. *Mechanical Systems and Signal Processing* 24 (4): 1049–1067.

Žvokelj, M., Zupan, S., and Prebil, I. (2016). EEMD-based multiscale ICA method for slewing bearing fault detection and diagnosis. *Journal of Sound and Vibration* 370: 394–423.

8

Nonlinear Subspace Learning

8.1 Introduction

As mentioned previously in this book, in practice, a collected vibration signal usually contains a large collection of responses from several sources in the rotating machine and some background noise. This makes the direct usage of the acquired vibration signal in rotating machine fault diagnosis – either by manual inspection or automatic monitoring – challenging, owing to the curse of dimensionality. As an alternative to processing raw vibration signals, the common method is to compute certain features of the raw vibration signal that can describe the signal in essence. As described in Chapter 7, a reasonable approach to tackle the challenges involved in dealing with too many samples is to learn a subspace of the high-dimensional data: i.e. map a high-dimensional dataset into a low-dimensional representation. Many high-dimensional datasets have a nonlinear nature. Thus, many linear subspace learning methods were introduced in the previous chapter, which can be used to linearly learn a low-dimensional representation of a high-dimensional dataset. However, linear subspaces may be inefficient for some datasets that have a nonlinear structure.

Various nonlinear subspace learning techniques have been proposed to learn subspace features from a large amount of vibration signals in rotating machines fault diagnosis. These include kernel principal component analysis (KPCA), kernel linear discriminant analysis (KLDA), kernel independent component analysis (KICA), isometric feature mapping (ISOMAP), diffusion maps (DMs), Laplacian eigenmaps (LE), local linear embedding (LLE), Hessian-based local linear embedding (HLLE), local tangent space alignment analysis (LTSA), maximum variance unfolding (MVU), and stochastic proximity embedding (SPE). The subsequent subsections discuss these types of techniques and their application in machine fault diagnosis in more detail.

8.2 Kernel Principal Component Analysis (KPCA)

KPCA is the nonlinear extension of linear principal component analysis (PCA), described in the previous chapter. In fact, it is the reconstruction of linear PCA in a high-dimensional space, created using a kernel function (Schölkopf et al. 1998). The basic idea of KPCA is to first map the input space into a feature space via nonlinear mapping and then compute the principal components in that feature space. KPCA computes the principal eigenvectors of the kernel matrix, rather than those of the

Condition Monitoring with Vibration Signals: Compressive Sampling and Learning Algorithms for Rotating Machines, First Edition. Hosameldin Ahmed and Asoke K. Nandi.
© 2020 John Wiley & Sons Ltd. Published 2020 by John Wiley & Sons Ltd.

covariance matrix. The reformulation of PCA in kernel space is straightforward since a kernel matrix is similar to the product of the data points in the high-dimensional space that is constructed using the kernel function.

As described in the previous chapter, linear PCA relies on eigenvectors of the sample covariance matrix XX^T, which is a dot product between vectors. To map data into the higher-dimensional space $\emptyset : R^2 \rightarrow R^3$ using a kernel function, the dot product in the new feature space can be represented using the following equation:

$$K(x_i, x_j) = \emptyset(x_i)\emptyset(x_j) \tag{8.1}$$

The main goal of the kernel function is to find a certain way to map the data into higher-dimensional space. Typical examples of commonly used kernel functions are a polynomial kernel function and Gaussian kernel function that can be defined using the following equations, respectively,

$$K(x, y) = (X^T y + c)^d \tag{8.2}$$

$$K(x, y) = e^{\left(\frac{-|x-y|^2}{2\sigma^2}\right)} \tag{8.3}$$

As described in the previous chapter, to compute the transformation matrix W, linear PCA uses eigenvalues and eigenvectors such that

$$\lambda V_1 = C_X V_2 \tag{8.4}$$

where λ is an eigenvalue, V_1 is an eigenvector, and C_X is the corresponding covariance matrix of data X that can be calculated using the following equation:

$$C_X = \frac{1}{L} \sum_{i=1}^{L} (x_i - \overline{x})(x_i - \overline{x})^T \tag{8.5}$$

Here, L is the number of observations and $\overline{x}$ can be computed using the following equation:

$$\overline{x} = \frac{1}{L} \sum_{i=1}^{L} x_i \tag{8.6}$$

To extend linear PCA and enable it to produce nonlinear subspaces, we compute the principal eigenvectors of the kernel matrix in Eq. (8.1), rather than those of the covariance matrix in Eq. (8.5). This extension to linear PCA is called KPCA, and its procedure can be described as follows:

(1) Map the input data into a higher-dimensional feature space using kernel function K such that,

$$\emptyset : X \rightarrow H, X \in R^n \tag{8.7}$$

where $X \in R^n$ is the input data and $H \in R^N$ is the higher-dimensional feature space $(n \ll N)$.

(2) Compute the covariance of the data in the feature space H such that,

$$C_H = \frac{1}{L} \sum_{i=1}^{L} \emptyset(x_i)\emptyset(x_i)^T \tag{8.8}$$

(3) Compute the principal components by solving the eigenvalues and eigenvectors such that,

$$\lambda V_1 = C_H V_2 \tag{8.9}$$

$$\lambda V_1 = \frac{1}{L} \sum_{i=1}^{L} \emptyset(x_i)\emptyset(x_i)^T V_2 \tag{8.10}$$

Then, the eigenvector V_1 can be represented as follows:

$$V_1 = \frac{1}{\lambda L} \sum_{i=1}^{L} \emptyset(x_i)\emptyset(x_i)^T V_2 = \sum_{i=1}^{L} \frac{\emptyset(x_i)^T V_2}{\lambda L}\emptyset(x_i) \tag{8.11}$$

Let $\sum_{i=1}^{L} \frac{\emptyset(x_i)^T V_2}{\lambda L} = \alpha_i$. Then the eigenvector V_1 can be represented as follows:

$$V_1 = \sum_{i=1}^{L} \alpha_i \emptyset(x_i) \tag{8.12}$$

Substituting Eqs. (8.12) and (8.1) in Eq. (8.10) yields Eq. (8.13),

$$\lambda \alpha_i = K\alpha_j \tag{8.13}$$

To normalise the data in the feature space defined by the kernel function with a zero mean we need to centre the high-dimensional feature space by subtracting the mean of the data in the feature space defined by the kernel function K such that,

$$\widehat{\emptyset}(x_K) = \emptyset(x_i)^T - \frac{1}{L} \sum_{i=1}^{L} \emptyset(x_K) \tag{8.14}$$

Then, the normalised kernel matrix can be represented as follows:

$$\widehat{K}(x_i, x_j) = \widehat{\emptyset}(x_i)\widehat{\emptyset}(x_j) \tag{8.15}$$

More efficiently,

$$\widehat{K} = K - 21_L K + 1_L K 1_L \tag{8.16}$$

where

$$1_L = \frac{1}{L} \begin{bmatrix} 1 & \cdots & 1 \\ \vdots & \ddots & \vdots \\ 1 & \cdots & 1 \end{bmatrix} \tag{8.17}$$

Finally, the low-dimensional representation y_i is found by solving the eigen problem $\lambda \alpha_i = \widehat{K}\alpha_j$ and projecting the data to each new dimension j such that,

$$y_i = \sum_{i=1}^{L} \alpha_{ij} K(x_i, x), \text{ for } j = 1, 2, \ldots, m. \tag{8.18}$$

Other linear subspace learning methods that used kernel functions to learn nonlinear subspace include linear discriminant analysis (LDA), independent component analysis (ICA), and canonical component analysis (CCA) extended to KLDA, KICA, and kernel canonical component analysis (KCCA), respectively. Because of the limitations on space, we cannot detail these methods here. Readers who are interested may be referred to (Mika et al. 1999; Lai and Fyfe 2000; Bach and Jordan 2002).

8.2.1 Application of KPCA in Machine Fault Diagnosis

KPCA is one of the most widely used nonlinear dimensionality reduction techniques in machine fault diagnosis. There have been several studies in the literature reporting the application of KPCA in machine fault diagnosis. For instance, Qian et al. (2006) presented an approach for early detection of cracked gear teeth based on KPCA. In this study, the authors analysed a collection of gear vibration signals using PCA and KPCA methods and identified normal and abnormal conditions of gears. The results verified that both PCA and KPCA are able to recognise gear conditions. However, for monitoring the running condition of the gear, the KPCA method is more effective than the PCA method. Widodo and Yang (2007) studied the application of nonlinear feature extraction and support vector machine (SVM) for fault diagnosis in induction motors. In this study, the authors used a PCA and ICA procedure and adopted the kernel trick to nonlinearly map the data into a feature space. Then, SVMs are employed to classify induction motor faults. Three direction vibration signals and three-phase current signals are used to validate the proposed method. The result of fault classification using SVMs showed that nonlinear feature extraction can improve the performance of a classifier.

He et al. (2007) proposed a technique for gearbox condition monitoring based on KPCA. In this technique, first, statistical features of the vibration signals in the time domain and frequency domain are extracted to represent the characteristics of the gear's condition. The statistical features in the time domain include absolute mean, maximum peak value, root mean square, square root value, variance, kurtosis, crest factor, and shape factor. Also, wavelet packet analysis is used to eliminate noises and enhance machine signals. In the frequency domain, the fast Fourier transform (FFT) is used to transform the machine signal's in the frequency domain. The frequency spectra of the secondary signals are divided into eight equal bands; then, the spectral energies of these eight bands are taken as the frequency-domain statistical features. Then, two subspace structures are considered for gearbox condition monitoring. The first subspace, called KPCA1, is constructed from all training data with all health conditions. The second subspace, called KPCA2, is constructed using training data from each health state. Vibration signals from an automobile gearbox are used to validate the proposed methods. The results demonstrated the effectiveness of both subspace methods in gearbox condition monitoring.

Dong et al. (2017) introduced a method for diagnosing bearing faults based on KPCA and an optimised k-nearest neighbour model. In this method, first, the collected vibration signals are decomposed using local mean decomposition (LMD) and then the Shannon entropy of the decomposed product functions (PFs) is calculated. Then, KPCA is used to reduce the feature dimensions. Finally, the particle swarm optimisation (PSO)-based optimised Kohonen neural network (KNN) model is utilised to deal with the fault-diagnosis problem. Two case studies of roller bearings are used to examine the efficiency of the proposed method. The results demonstrated the effectiveness of the proposed method in bearing fault diagnosis.

8.3 Isometric Feature Mapping (ISOMAP)

ISOMAP is a nonlinear dimensionality-reduction technique based on the idea of viewing the problem as a combination of the main algorithmic features, computational efficiency, global optimality, and asymptotic convergence guarantee features of PCA and multidimensional scaling (MDS), with the flexibility to learn a wide class of nonlinear manifolds (Tenenbaum et al. 2000; Samko et al. 2006). The ISOMAP algorithm takes as input the distance $d_X(i, j)$ between all pairs i, j from N data points in the high-dimensional input space X. The algorithm outputs coordinate vectors Y_i in a lower d-dimensional Euclidean space Y that best represents the intrinsic geometry of the data. The procedure for the ISOMAP technique can be described in three main steps as follows,

1. Construct a neighbourhood graph. Define the overall data points of graph G by connecting points i and j if [as measured by $d_X(i, j)$] they are closer than ε ($\varepsilon - Isomap$), or if i is one of the K nearest neighbours of j (K-isomap). Set the edge lengths equal to $d_X(i, j)$.
2. Compute the shortest paths. Initialise $d_G(i, j) = d_X(i, j)$ if i and j are linked by an edge; $d_G(i, j) = \infty$ otherwise. Then, for each value of $k = 1, 2, \ldots, N$ in turn, recall all entries $d_G(i, j)$ by

$$\min\{d_G(i, j), d_G(i, k) + d_G(k, j)\} \qquad (8.19)$$

   The matrix of final values $D_G(i, j) = \{d_G(i, j)\}$ will contain the shortest distances between all pairs of points in G.
3. Construct d-dimensional embedding. This step applies the MDS algorithm to the matrix of graph distances $D_G(i, j) = \{d_G(i, j)\}$, constructing an embedding of the data in a d-dimensional Euclidean space that best preserves the manifold's estimated intrinsic geometry. The coordinate vectors Y_i for points in Y are selected to minimise the cost function

$$E = \|\tau(D_G) - \tau(D_Y)\|_{L2} \qquad (8.20)$$

   Here, D_Y represents the matrix of Euclidean distances $\{d_y(i, j) = \|y_i - y_j\|\}$; $\|.\|_{L2}$ is the matrix norm; and τ is and operator that converts distances to inner products that uniquely describe the geometry of the data in a form that supports efficient optimisation.

Samko et al. (2006) proposed an algorithm for the automatic selection of the optimal values of K and ε in the first step of ISOMAP. This algorithm consists of four steps as follows,

1. Select the interval of possible values of K, $K_{opt} \in [K_{min}, K_{max}]$. Here, K_{min} is the minimal value of K, with which the neigborhood graph (from the second step of ISOMAP) is connected, and K_{max} can be selected as the largest value of K in the following equation,

$$\frac{2 * P}{N} \leq K + 2 \qquad (8.21)$$

   where P is the number of edges and N is the number of nodes in the neighbourhood graph from the second step in the ISOMAP algorithm.
   With regard to ε, select ε_{max} as $\max(d_X(i, j))$.
2. Compute the cost function $E(K)$ in Eq. (8.19) for each $K \in [K_{min}, K_{max}]$.
3. Compute all minima of $E(K)$ and the corresponding K that compose the set I_K of initial candidates for the optimal value.

4. Run the ISOMAP algorithm for each $K \in I_K$, and determine K_{opt} using the following equation:

$$K_{opt} = \arg \min_K (1 - \rho^2_{D_x D_y})$$
(8.22)

Here, D_x and D_y are the matrices of the Euclidean distances between pairs of points in the input and the output spaces, respectively, and ρ represents the standard linear correlation coefficient taken over all entries of D_x and D_y.

8.3.1 Application of ISOMAP in Machine Fault Diagnosis

Various studies have assessed the efficacy of ISOMAP in machine fault diagnosis. For example, Xu and Chen (2009) proposed a method for electromechanical equipment fault prediction based on ISOMAP and artificial neural network (ANN) techniques. In this method, the ISOMAP algorithm is applied to extract and reduce the dimensionality of the operation data of electromechanical equipment. The reduced data is used as input of a neural network for fault prediction. Vibration signals at the front and rear bearings of a flue gas turbine are used to validate the proposed method. The results showed the suitability of the proposed method in fault prediction of electromechanical equipment. Moreover, Benkedjouh et al. (2013) proposed the introduced the utilisation of the ISOMAP algorithm to perform nonlinear feature reduction, combined with nonlinear support vector regression (SVR) to construct health indicators that can be used to estimate the health state of roller bearings and predict their remaining useful life (RUL). This method is divided into two steps: an offline step and an online step. The offline step is employed for learning the bearings' degradation models by utilising the ISOMAP and SVR methods. The online step uses the models learned in the offline step to examine the current health condition of new tested bearings and to predict their RUL. Real vibration signals from three degraded bearings are used to test the proposed method. Eight features are first extracted from the raw vibration signals of each degraded bearing using the wavelet packet decomposition technique. Then, ISOMAP is used to reduce these eight features to one feature, called the *health indicator*. The input parameter of ISOMAP is estimated using an optimisation approach. The results showed the effectiveness of the proposed method in modelling the evaluation of the degradations and predicting the RUL of the bearings.

Furthermore, Zhang, et al. (2013) introduced a fault-diagnosis technique for rotating machines based on ISOMAP. In this technique, ISOMAP is used to capture the nonlinear intrinsic manifold structure of the high-dimensional signal space and then map it into a lower-dimensional manifold space in which the fault classification is performed using a classifier. The number of neighbours k for ISOMAP is determined empirically by selecting the value k that provides the best classification result as an optimal value of k from a range of values of k. Two vibration datasets – rotor bed data, with unbalance, misalignment, and rub impact faults; and rolling bearing data, with a normal condition, inner race (IR) fault, and outer race (OR) fault – are used to verify the fault-classification performance of the proposed method. Three classifiers – minimum-distance classifier, KNN, and SVM with a radial basis function (RBF) kernel – are employed in the reduced dimension space to deal with the classification problem. The results showed the effectiveness of the proposed method in diagnosing rotating machine faults.

Yin et al. (2016) proposed a model for assessing machine health based on the wavelet packet transform (WPT), ISOMAP, and deep belief network (DBN). In this model, first, features are extracted from the original vibration signal using the time domain, frequency domain, and WPT. Then, ISOMAP is used to reduce the dimensionality of the extracted features. The obtained low-dimensional features are used as input for the DBN to evaluate the performance status of the machine. Vibration signals acquired from roller bearings are used to validate this method. The results demonstrated the effectiveness of the proposed method in roller bearing health assessment compared to other methods considered by the authors in this study.

8.4 Diffusion Maps (DMs) and Diffusion Distances

To find meaningful geometric descriptions of datasets, Coifman and Lafon (2006) proposed a framework based on diffusion processes. In this framework, eigenfunctions of Markov matrices are used to define a random walk on the data to obtain new descriptions of datasets through a family of mappings called *diffusion maps*, which embed the data points into a Euclidean space in which the usual distance describes the relationship between pairs of points in terms of connectivity. This defines a useful distance between points in the data called the *diffusion distance*. Different geometric representations of the dataset can be obtained by iterating the Markov matrix or, equivalently, by running the random walk forward. The DM allows us to relate the spectral properties. Briefly, we present the simplified DM framework as follows (Van Der Maaten et al. 2009):

1. The graph of the data is constructed, and the weights W of the edges in the graph are calculated using the Gaussian kernel function. Hence, the entries of W can be defined as follows,

$$w_{ij} = e^{-\frac{\|x_i - x_j\|}{2\sigma^2}} \tag{8.23}$$

   where σ represents the variance of the Gaussian.

2. The normalisation of the matrix W is achieved so that its rows add up to 1. In this way, we define a matrix $P^{(1)}$, which is considered a Markov matrix that describes the forward transition probability of a dynamical process, designed with entries w_{ij} such that,

$$p_{ij}^{(1)} = \frac{w_{ij}}{\sum_k w_{ik}} \tag{8.24}$$

   Here, the matrix $P^{(1)}$ represents the probability of a transition from one point to another point in a single timestep.

3. The random walk forward probabilities can be defined as the probability for t timesteps $P^{(t)}$, i.e. can be given by $(P^{(1)})^t$. Hence, the diffusion distance can be defined using the following equation:

$$D^{(t)}(x_i, x_j) = \sqrt{\sum_k \frac{(p_{ik}^{(t)} - p_{jk}^{(t)})^2}{\psi(x_k)^{(0)}}} \tag{8.25}$$

Here, $\psi(x_k)^{(0)} = \frac{g_i}{\sum_j g_j}$ is a term that assigns more weight to parts of the graph with high density, and $g_i = \sum_j p_{ij}$ is the degree of the node x_i. It can be observed from Eq. (8.24) that pairs of data points with a high transition probability $P^{(t)}$ have a small diffusion distance.

4. To retain the diffusion distances $D^{(t)}(x_i, x_j)$ in the low-dimensional representation of the data Y, DMs use spectral theory on the random walk. Here, the low-dimensional representation Y is formed by the d nontrivial principal eigenvectors of the eigenproblem such that,

$$P^{(t)}v = \lambda v \tag{8.26}$$

As the graph is fully connected, the largest eigenvalue is trivial, and its eigenvector v_1 is thus discarded.

5. Finally, the low-dimensional representation Y can be defined using the following equation:

$$Y = \{\lambda_2 v_2, \lambda_3 v_3, \lambda_4 v_4, \dots, \lambda_{d+1} v_{d+1}\} \tag{8.27}$$

8.4.1 Application of DMs in Machine Fault Diagnosis

A number of studies have begun to examine the application of DMs in machine fault diagnosis. For example, Huang presented a dimensionality-reduction technique called discriminant diffusion maps analysis (DDMA) for machine condition monitoring and fault diagnosis (Huang et al. 2013). This method integrates a discriminant kernel approach into the framework of DM. Three different experiments are used to validate the DDMA technique, including a pneumatic pressure regulator experiment, a vibration-based rolling element bearing test, and an artificial noisy nonlinear test system. The results showed that the DDMA is able to represent the high-dimensional data in a lower-dimensional space effectively. Also, empirical comparisons with PCA, ICA, LDA, LE, LLE, HLLE, and LTSA showed that the low-dimensional features generated by DDMA are much better in different conditions.

Chinde et al. (2015) proposed a data-driven method for damage diagnosis in mechanical structures based on comparing the intrinsic geometry of datasets corresponding to the undamaged and damaged condition of the system. In this method, the spectral DM technique is used to identify the intrinsic geometry of time series data from distributed sensors. The low-dimensional embedding of the dataset corresponding to different damage severity is performed using singular value decomposition (SVD) of the DM. Damage diagnosis of wind turbine blades is used to validate the proposed method. The simulation result demonstrated the effectiveness of the proposed DM-based method in fault diagnosis.

Sipola et al. (2015) introduced a method for detecting and classifying gear faults based on DMs. In this method, the DM is used to produce a low-dimensional representation of the data. The new measurements that are not part of the training are extended to the model with the Nystrom technique. Real gear data is used to train and test the proposed method. The results showed the effectiveness of DM-based dimensionality reduction in gear fault detection.

8.5 Laplacian Eigenmap (LE)

The LE is a geometrically motivated technique for nonlinear dimensionality reduction (Belkin and Niyogi 2002, 2003). The LE algorithm computes a low-dimensional representation of given data in which the distances between the connected points are as close together as possible. For a given n data points $\{x_i\}_{i=1}^n$ where each point lies in a high-dimensional space $x_i \in R^D$, the following procedure of the LE algorithm finds a mapping $\emptyset$, which maps x_i to y_i in a low-dimensional space $\emptyset(x_i) = y_i \in R^d$ where $d \ll D$:

1. The LE constructs a graph $G = (V, E)$ using n points $\{x_i\}_{i=1}^n$, where each point corresponds to one vertex $v_i \in V$ such that

$$|V| = n \tag{8.28}$$

2. Put an edge between v_i and v_j if x_i and x_j are close to each other. Here, two common ways to compute the closeness can be used. These are ε-neighbourhood and k-nearest neighbour, which can be defined using Eqs. (8.29) and (8.30):

$$\|x_i - x_j\|^2 < \varepsilon \tag{8.29}$$

$$x_j \in N_i \; or \; x_i \in N_j \tag{8.30}$$

3. Assign a weight to each edge using a binary technique such that,

$$W_{ij} = \begin{cases} 1, v_i \text{ and } v_j \text{ are connected.} \\ 0, otherwise \end{cases} \tag{8.31}$$

or using a heat kernel such that,

$$W_{ij} = e^{-\frac{\|x_i - x_j\|^2}{t}} \tag{8.32}$$

4. Compute the low-dimensional representation y_i by solving the generalised eigenvalue problem

$$Lv = \lambda Dv \tag{8.33}$$

where $D = W1, L = D - W$.

8.5.1 Application of the LE in Machine Fault Diagnosis

So far, very little attention has been paid to the application of LEs in machine fault diagnosis. Sakthivel et al. (2014) compared the performance of dimensionality-reduction methods in fault diagnosis of a monoblock centrifugal pump using vibration signals. In this study, statistical features extracted from the collected vibration signals are used as the features. Then, dimensionality reduction is performed using a linear dimensionality-reduction technique (PCA) and nonlinear dimensionality-reduction techniques (KPCA, ISOMAP, MVU, DM, LLE, LE, HLLE, and LTSA). For classification purpose, naïve Bayes, Bayes net, and KNN are used to deal with the classification problem. Vibration signals acquired from a centrifugal pump are used to examine the efficiency of these different dimensionality-reduction techniques in machine fault diagnosis. The effectiveness of each method is verified using visual analysis. The results showed that amongst the various dimensionality-reduction techniques, PCA achieved the highest classification of all the classifiers considered in this study.

Recently, Yuan et al. (2018) proposed a method for machine condition monitoring using LE feature conversion and a PSO-based deep neural network (DNN). In this method, first, features are extracted from the collected vibration signals acquired from a machine in the original high-dimensional space using time and frequency domain analysis as well as WPT. Then, the LE is used to transform the data from the original high-dimensional space to a lower-dimensional space. The PSO-based optimised DNN is employed to assess health conditions. Vibration signals acquired from roller bearings are used to validate the proposed method. The results showed the efficiency of the proposed method in roller bearing fault diagnosis.

8.6 Local Linear Embedding (LLE)

LLE is a learning algorithm that computes low-dimensional, neighbourhood-preserving embeddings of high-dimensional inputs (Roweis and Saul 2000; De Ridder and Duin 2002). For a given n data points $\{x_i\}_{i=1}^{n}$ where each point lies in a high-dimensional space $x_i \in R^D$, the procedure of the LLE algorithm can find a mapping $\emptyset$, which maps x_i to y_i in a low-dimensional space $\emptyset(x_i) = y_i \in R^d$ where $d \ll D$, as described by the following three main steps:

1. For each x_i, find its K nearest neighbours $x_{i1}, x_{i2}, \ldots, x_{iK}$.
2. Assume that each data point and its neighbours lie on or close to a locally linear patch of the manifold. The local geometry of these patches can be characterised by linear coefficients that reconstruct each data point from its neighbours. The reconstruction error ε is calculated by the following equation cost function,

$$\varepsilon = \sum_{i=1}^{n} \left| x_i - \sum_{j=1}^{n} w_{ij} x_j \right|^2 \tag{8.34}$$

Here, the weights w_{ij}, which summarise the contribution of the jth data point to the ith reconstruction, can be computed by minimising the cost function subject to $\sum_{j=1}^{n} w_{ij} = 1$ and $w_{ij} = 0$ for all points that are not neighbours of x_i.

3. The reconstruction weights w_{ij} reflect intrinsic geometric properties of the data that are invariant to exactly such transformations. Hence, w_{ij} characterisation of local geometry in the original data space is expected to be equal for local patches on the manifold. Therefore, the same w_{ij} that reconstructs the ith data point in the D-dimension should also reconstruct its embedded manifold coordinates in the low-dimension d-dimension. Consequently, in the final step of the LLE algorithm, each high-dimensional x_i is mapped into the data representation y_i in the low-dimension d-dimension. This can be achieved by selecting y_i to minimise the cost function such that,

$$\Phi = \sum_{i=1}^{n} \left| y_i - \sum_{j=1}^{n} w_{ij} y_j \right|^2 \tag{8.35}$$

The LLE algorithm has only one parameter, which is the number of the nearest neighbours that need to be set. To automatically select the optimal number of nearest neighbours for the LLE algorithm, Kouropteva et al. (2002) proposed a hierarchical method for automatic selection of an optimal parameter value.

8.6.1 Application of LLE in Machine Fault Diagnosis

A search of the literature revealed few studies that considered the application of LLE in machine fault diagnosis. For instance, Li et al. (2013) introduced a fault-detection method for gearboxes using blind source separation (BSS) and nonlinear feature-extraction techniques. In this method, the KICA algorithm is used as the BSS technique for the mixed example signals of the gearbox vibration to discover the characteristic vibration source associated with the gearbox faults. Then, WPT and empirical mode decomposition (EMD) methods are employed to deal with the nonstationary vibrations to obtain the original fault feature vector. Furthermore, the LLE is used to reduce the feature dimension. Lastly, the fuzzy k-nearest neighbour (FKNN) is applied to classify the gearbox's health condition. Two vibration datasets are used to validate this method. These vibration datasets are acquired from: (i) a gear, under 750 rpm with heavy load, and (ii) a roller bearing, under motor drive speed 800 rpm. The results verified that the LLE algorithm can effectively extract distinct features, and hence the FKNN successfully distinguished different fault conditions. In a different study, Su et al. (2014) proposed a method for machinery fault diagnosis based on incremental enhanced supervised local linear embedding (I-ESLLE) and an adaptive nearest neighbour classifier (ANNC). In this method, first, I-ESLLE is used to reduce the dimensionality of fault samples obtained from the collected vibration signals. Then, the reduced dimensional fault samples are used as inputs to ANNC for fault-type recognition. Vibration signals for five kinds of gearbox operating conditions – normal operating condition, gear surface pitting fault, bearing IR fault, bearing OR fault, and bearing rolling element fault – are acquired to validate the proposed method. The results demonstrated the effectiveness of the proposed method in gear fault diagnosis.

Furthermore, Wang et al. (2015) presented a method for machinery fault diagnosis based on a statistical locally linear embedding (S-LLE) algorithm, which is an extension of LLE. In this method, first, features are extracted from the vibration signals using time-domain, frequency-domain, and EMD techniques. Then, S-LLE is used to obtain a low-dimensional space from the original high-dimensional space. Finally, in the low-dimensional feature space, an SVM classifier based on RBF is applied to classify faults. Vibration signals collected from roller bearings are used to test the proposed method. The results verified that S-LLE outperformed PCA, LDA, and LLE.

8.7 Hessian-Based LLE

HLLE is another version of the LLE algorithm based on Hessian eigenmaps (Donoho and Grimes 2003). Briefly, we present the simplified Hessian locally linear embedding algorithm as follows.

For given data $\{x_i\}_{i=1}^{N} \in R^n$, d is the dimension of the parameter space, and k represents the number of neighbours for fitting, where $\min(k, n) > d$. The aim is to compute

$\{w_i\}_{i=1}^{N} \in R^d$ using HLLE. The procedure of the HLLE algorithm can be described by the following steps:

1. For each data point x_i, $i = 1, 2, \ldots, n$, find its k nearest neighbours in Euclidean distance. Let N_i denote the collection of neighbours. For each neighbourhood N_i, $i = 1, 2, \ldots, N$, a k x n matrix X^i has rows that contain the recentered points $x_j - \bar{x}_i \in N_i$. Here $\bar{x}_i = Average\{x_j : j \in N_i\}$.
2. Apply SVD to obtain the matrices U, D, and V; U is k by $min(k, n)$: the first d columns of U give the tangent coordinates of points in N_i.
3. Develop a Hessian estimator. Develop the infrastructure for least-squares estimation of the Hessian. In principle, this is a matrix H^i with the property that if f is a smooth function $f : X \to R$, and $\mathbf{f}_j = (f(x_i))$, then the vector v^i with entries that are acquired from $\mathbf{f}$ by extracting those entries correspond to points in the neighbourhood N_i; at that point, the matrix vector $H^i v^i$ delivers an $d(d+1)/2$ vector with entries that estimate the entries of the Hessian matrix, $\left(\frac{\partial f}{\partial u_i u_j} \right)$.
4. Develop the quadratic form. Construct a symmetric matrix $\mathcal{H}_{ij}$ having the entry,

$$\mathcal{H}_{ij} = \sum_l \sum_r ((H^l)_{r,i}(H^l)_{r,j}) \tag{8.36}$$

Here, H^l is the $\frac{d(d+1)}{2}$ by k matrix associated with approximating the Hessian over the neighbourhood N_i.
5. Apply an eigen analysis of $\mathcal{H}$, and determine the $(d + 1)$–dimensional subspace corresponding to the $d + 1$ smallest eigenvalues. Here, an eigenvalue 0 will be associated with the subspace of constant functions, and the next d eigenvalues correspond to eigenvectors spanning a d-dimensional space $\hat{V}_d$ in which the embedding coordinates are found.
6. Select a basis for $\hat{V}_d$ that has the property that its restriction to a specific fixed neighbourhood $\mathcal{N}_0$ provides an orthonormal basis. Here, the neighbourhood may be selected arbitrarily from those used in the algorithm. The delivered basis vectors $w^1, w^2, \ldots, w^d$ represent the embedding coordinates.

8.7.1 Application of HLLE in Machine Fault Diagnosis

A search of the literature for the application of HLLE in machine fault diagnosis revealed the study of Tian et al. (2016), who proposed a method for bearing diagnostics based on differential geometry. In this method, HLLE is used to extract manifold features from manifold topological structures, and singular values of eigenmatrices along with several specific frequency amplitudes in a spectrogram are extracted afterward to reduce the complexity of the manifold features. Then information geometry-based SVM is employed to classify the fault states. For the health assessment, the manifold distance is used to represent the health information; and the confidence values, which directly reflect the health status, are computed using the Gaussian mixture model. Case studies on Lorenz signals and vibration datasets of roller bearings verified the effectiveness of the proposed methods. In particular, the study showed that the characteristic frequency amplitudes obtained by HLLE are clearer, and the harmonic frequencies are more prominent compared with the Teager energy operator (TEO) and Hilbert-Huang transform (HHT).

8.8   Local Tangent Space Alignment Analysis (LTSA)

LTSA is a method for manifold learning and nonlinear dimensionality reduction (Zhang and Zha 2004). Briefly, we present the LTSA algorithm as follows.

Given $\{x_i\}_{i=1}^N \in R^n$ samples, possibly noisy, from an underlying $d-$ dimensional manifold, the LTSA algorithm produces N-dimensional coordinates $T \in R^{d \times N}$ for the manifold constructed from k local nearest neighbours. The procedure for the LTSA algorithm can be described in three main steps as follows:

(1) Extract local information: for each x_i, $i = 1, 2, \ldots, N$, first find its k nearest neighbours x_{ij} where $j = 1, 2, \ldots, k$. Then, calculate the d largest eigenvectors $v_1, v_2, \ldots, v_d$ of the correlation matrix $(X_i - \bar{x}_i e^T)^T (X_i - \bar{x}_i e^T)$, and set $G_i = \left[\frac{e}{\sqrt{k}}, v_1, \ldots v_d \right]$.

(2) Construct the alignment matrix. To compute the d smallest eigenvectors that are orthogonal to e, apply an eigensolver with an explicitly formed matrix B, which can be computed using the following equation,

$$B(I_i, I_i) \leftarrow B(I_i, I_i) + I - G_i G_i^T, i = 1, 2, \ldots, N \tag{8.37}$$

with initial $B = 0$, where $I_i = \{i_1, i_2, \ldots, i_k\}$ represent the set of indices for the k nearest neighbours of x_i. If a direct eigensolver is used, form the matrix B by locally summing Eq. (8.37). Otherwise, implement a routine that computes matrix-vector multiplication Bu for an arbitrary vector u.

(3) Compute the eigenvector and eigenvalues. Apply an eigenanalysis of B, and determine the $(d+1)-$dimensional subspace corresponding to the $d+1$ smallest eigenvalues. Here, we will have an eigenvector matrix $[u_2, \ldots, u_{d+1}]$ and $T = [u_2, \ldots, u_{d+1}]^T$.

8.8.1   Application of LTSA in Machine Fault Diagnosis

Several studies in the area of machine fault diagnosis have used LTSA to map high-dimensional original features into a low-dimensional feature space. For instance, Wang et al. introduced improved LTSA: a supervised learning – local tangent space alignment analysis (SLLTA) algorithm and a supervised incremental – Local tangent space alignment analysis (SILTSA) algorithm. Based on these two algorithms and a SVM classifier, the authors proposed two methods, SLLTSA-SVM and SILTSA-SVM, for fault diagnosis of roller bearings (Wang et al. 2012). Vibration signals collected from roller bearings are used to validate the proposed methods. The results showed that the SILTSA-SVM method achieved better diagnosis results compared to related methods. Moreover, Zhang et al. (2014) introduced a method called supervised local tangent space alignment (S-LTSA) for machine fault diagnosis. S-LTSA aims to take full advantage of class information to improve classification performance. Vibration signals with an IR fault, OR fault, and ball fault are used to evaluate the proposed method. The results demonstrated the efficiency of the proposed method in machine fault diagnosis. Moreover, compared to the traditional LTSA and PCA, the results showed that S-LTSA performed better.

Dong et al. (2015) introduced a method for diagnosing rotating machine faults based on a morphological filter optimised by PSO and the nonlinear manifold learning algorithm LTSA. In this method, first, the signals are purified using the morphological filter, where the filter's structure element (SE) is selected by the PSO method. Then, EMD is used to decompose the filtered signals. The extracted features are then mapped into the LTSA algorithm to extract the characteristic features. Lastly, the SVM is employed to deal with the fault-diagnosis problem. Bearing fault signals acquired at speed 1797 rpm, with an IR fault, rolling element fault, and OR fault, are used to validate the proposed method. The results verified the effectiveness of the proposed method in removing noise and diagnosing machine faults.

Su et al. (2015a) introduced a dimensionality-reduction method called supervised extended local tangent space alignment (SE-LTSA) for machinery fault diagnosis. In this study, first, the vibration signals are decomposed into several intrinsic mode functions (IMFs) using EMD. Then, the energy of each IMF is calculated, and the IMFs with more energy are selected. In this study, it is found that the first six IMFs contain more than 90% of the total energy. Hence, the first six IMFs are selected. Then, seven time-domain features and five frequency-domain features are extracted from each IMF. SE-LTSA is utilised to reduce the feature-space dimension. Finally, the obtained low-dimensional fault samples are used as input to a k-nearest neighbour classifier (KNNC) for fault recognition. Vibration signals collected from a gearbox in a test rig are used to validate the proposed method. The results demonstrated the effectiveness of the proposed method in machine fault diagnosis.

In another study, Su et al. (2015b) proposed a method for rotating machinery multifault diagnosis based on orthogonal supervised linear tangent space alignment (OSLLTSA) and a least square support vector machine (LS-SVM). In this method, first, the collected vibration signals are decomposed into IMFs using EMD, and a high-dimensional feature set is created by extracting statistical features, autoregressive (AR) coefficients, and instantaneous amplitude Shannon entropy from the IMFs that contain the most information about faults. Then, the OSLLTSA method is used for dimension reduction to produce more sensitive low-dimensional fault features. With these low-dimensional fault features, the LS-SVM is used to classify machinery faults where the parameters of LS-SVM are selected by enhanced particle swarm optimisation (EPSO). Vibration signals collected from roller bearings are used to validate the proposed method. The results confirmed that the proposed method achieved improved accuracy for fault diagnosis.

Furthermore, Li et al. (2015) proposed a method for life-grade recognition of rotating machinery based on supervised orthogonal linear local tangent space alignment (SOLLTSA) and optimal supervised fuzzy c-means clustering (OSFCM). In this method, first, multiple time-frequency features are extracted. Then, the OSLLTSA algorithm is used to compress the extracted features sets of training and testing samples into low-dimensional eigenvectors, which are used as input to OSFCM to recognise life grades. A life-grade recognition example for deep groove ball bearings is used to examine the efficiency of the proposed method. The results showed the effectiveness of the proposed method for life grade recognition.

8.9 Maximum Variance Unfolding (MVU)

MVU, also called semidefinite embedding (SDE), is a dimensionality-reduction technique that uses semidefinite programming (SDP) to model problems of dimensionality reduction (Vandenberghe and Boyd 1996; Weinberger et al. 2005; Van Der Maaten et al. 2009). To keep the maximum variance in the learned low-dimensional feature space, MVU learns the data based on similarities. Thus, it retains both local distances and angles between the pairs of all neighbours of each point in the dataset (Wang 2011). Briefly, we describe the procedure of MVU for a given $\{x_i\}_{i=1}^{N} \in R^n$ as follows:

1. Construct a neighbourhood graph G, in which each data point x_i is connected to its k nearest neighbours x_{ij} where $j = 1, 2, ..., k$. Then, MVU attempts to maximise the sum of the squared Euclidean distances between all points such that the following optimisation is performed:

$$\textit{Maximise} \sum_{ij} \|y_i - y_j\|^2 \tag{8.38}$$

$$\text{s.t } \|y_i - y_j\|^2 = \|x_i - yx_j\|^2 \forall(i, j) \in G \tag{8.39}$$

2. Reformulate the optimisation problem in Eq. (8.38) as an (SDP) problem by defining the kernel matrix K as the outer product of the low-dimensional data representation Y such that,

$$\textit{Maximise trace}(K) \tag{8.40}$$

$$(1)\, s.t, k_{ij} + k_{jj} - 2k_{ij} = \|x_i - yx_j\|^2 \forall(i, j) \in G \tag{8.41}$$

$$(2)\, s.t, \sum_{ij} k_{ij} = 0 \tag{8.42}$$

$$(3)\, s.t, K \geq 0 \tag{8.43}$$

where the matrix K is constructed by solving the digital signal processing (DSP) problem.
3. Compute the low-dimensional data representation by applying an eigenanalysis of the kernel matrix K.

8.9.1 Application of MVU in Machine Fault Diagnosis

Although some research has been carried out on the application of nonlinear dimensionality-reduction techniques in machine fault diagnosis, to the authors' knowledge only one study has attempted to investigate the application of MVU in machine fault diagnosis. In this study, Zhang and Li (2010) proposed a noise-reduction technique for the nonlinear signal based on MVU. In this method, the noisy vibration signal is first embedded into a high-dimensional phase space based on phase-space reconstruction theory. Then, MVU is employed to perform nonlinear reduction on the data of the phase space in order to separate a low-dimensional manifold representing the attractor from noise subspace. Finally, the noise-reduced signal is obtained through reconstructing the low-dimensional manifold. Vibration signals acquired from a rotor rubbing in an aero engine are used to evaluate the performance of MVU-based noise reduction in fault detection. The results showed that the proposed method can effectively extract slight rubbing features that are overwhelmed by noise.

8.10   Stochastic Proximity Embedding (SPE)

SPE is a nonlinear dimensionality-reduction technique, which has four attractive features (Agrafiotis et al. 2010):

(1) It is simple to implement.
(2) It is very fast.
(3) It scales linearly with the size of the data in both time and memory.
(4) It is relatively insensitive to missing data.

SPE uses a self-organising iterative scheme to embed m-dimensional data into p dimensions, such that the geodesic distances in the original m dimensions are preserved in the embedded d dimension. To compute a reduced dimension from the compressively sampled signals using SPE, the following steps are performed.

1. Initialize the coordinates y_i. Select an initial learning rate β.
2. Select a pair of points, i and j, at random and compute their distance $d_{ij} = \|y_i - y_j\|$. If $d_{ij} \neq r_{ij}$ (r_{ij} is the distance of the corresponding proximity), update the coordinates y_i and y_j by:

$$y_i \leftarrow y_i + \beta \frac{1}{2} \frac{r_{ij} - d_{ij}}{d_{ij} + \upsilon} (y_i - y_j) \tag{8.44}$$

and

$$y_j \leftarrow y_j + \beta \frac{1}{2} \frac{r_{ij} - d_{ij}}{d_{ij} + \upsilon} (y_j - y_i) \tag{8.45}$$

Here, υ is a small number to avoid division by zero. For a given number of iterations, this step will be repeated for a prescribed number of steps, and β will be decreased by a suggested decrement $\delta\beta$.

8.10.1   Application of SPE in Machine Fault Diagnosis

So far, very little attention has been paid to the application of SPE in machine fault diagnosis. For example, Wang et al. (2017) proposed a method for fault diagnosis based on sparse filtering and t-distributed stochastic neighbour embedding (t–SNE). To validate their proposed method, the authors adopted five dimensionality techniques with sparse filtering to process a gearbox dataset. The five techniques are PCA, locality preserving projection (LPP), Sammon mapping (SM), LDA, and SPE. In this study, vibration signals are collected from a four-speed motorcycle gearbox with four types of gearbox health conditions: normal, slightly worn, medium worn, and broken tooth. The results showed that the proposed method, i.e. sparse filtering with t–SNE, achieved the best classification accuracy, with 99.87%; the other five methods – i.e., sparse filtering with PCA, SM, LPP, and SPE – achieved 99.62%, 96.43%, 88.74%, 56.73%, and 86.22% classification accuracy, respectively. In a different study, Ahmed and Nandi proposed a three-stage method for monitoring rotating machine health, called CS-SPE, based on compressive sampling and stochastic proximity embedding. This method receives a large amount of vibration data as input and produces fewer features as output, which can be used for fault classification of rotating machines. Given a dataset of a rotating machine vibration $X \in R^{nxL}$, CS-SPE first uses compressive sampling (CS) based on a multiple measurement vectors (MMV) model to produce compressively sampled signals (i.e. compressed data $Y = \{y_1, y_2, ..., y_L\} \in R^m$ where $1 \leq l \leq L, m < < n$) and lets each of these signals fit in with one of the c classes of machine health conditions. Then, SPE is used to select optimal features of the compressively sampled signals. With these selected features, a classifier can be used to classify the fault in the rotating machine. We will discuss this method in detail in Chapter 16.

8.11 Summary

In practice, collected vibration signals usually contain a large collection of responses from several sources in the rotating machine, including some background noise. This makes the direct usage of the acquired vibration signal in rotating machine fault diagnosis – using either manual inspection or automatic monitoring – challenging, owing to the curse of dimensionality. As an alternative to processing the raw vibration signals, the common method is to compute certain features of the raw vibration signal that can describe the signal in essence. A reasonable approach to tackle the challenges involved in dealing with too many samples is to learn a subspace of the high-dimensional data: i.e. map a high-dimensional dataset into a low-dimensional representation. In Chapter 7, we introduced various linear subspace learning techniques that can be used to learn subspace features from a large amount of vibration signals in rotating machines fault diagnosis. However, linear subspaces may be inefficient for some datasets that have a nonlinear structure. In this chapter, we described various nonlinear subspace learning techniques and their application in machine fault diagnosis. These include KPCA, KLDA, KICA, ISOMAP, DM, LE, LLE, HLLE, LTSA, MVU, and SPE. In addition, the application of these techniques in machine fault diagnosis was presented. Most of the introduced techniques and their publically accessible software are summarised in Table 8.1.

Table 8.1 Summary of some of the introduced techniques and their publically accessible software.

Algorithm name	Platform	Package	Function
Kernel PCA	MATLAB	Steven (2016)	km_pca
Kernel CCA			km_cca
Matlab Toolbox for Dimensionality Reduction	MATLAB	Van der Maaten et al. 2007. The software within this toolbox for some nonlinear methods introduced in this chapter include: 1. LLE 2. LE 3. HLLE 4. LTSA 5. MVU 6. KPCA 7. DM 8. SPE	compute_mapping(data, method, # of dimensions, parameters)

References

Agrafiotis, D.K., Xu, H., Zhu, F. et al. (2010). Stochastic proximity embedding: methods and applications. *Molecular Informatics* 29 (11): 758–770.

Bach, F.R. and Jordan, M.I. (2002). Kernel independent component analysis. *Journal of Machine Learning Research* 3 (Jul): 1–48.

Belkin, M. and Niyogi, P. (2002). Laplacian eigenmaps and spectral techniques for embedding and clustering. In: *Advances in Neural Information Processing Systems*, 585–591. http://papers.nips.cc/paper/1961-laplacian-eigenmaps-and-spectral-techniques-for-embedding-and-clustering.pdf.

Belkin, M. and Niyogi, P. (2003). Laplacian eigenmaps for dimensionality reduction and data representation. *Neural computation* 15 (6): 1373–1396.

Benkedjouh, T., Medjaher, K., Zerhouni, N., and Rechak, S. (2013). Remaining useful life estimation based on nonlinear feature reduction and support vector regression. *Engineering Applications of Artificial Intelligence* 26 (7): 1751–1760.

Chinde, V., Cao, L., Vaidya, U., and Laflamme, S. (2015). Spectral diffusion map approach for structural health monitoring of wind turbine blades. In: *American Control Conference (ACC), 2015*, 5806–5811. IEEE.

Coifman, R.R. and Lafon, S. (2006). Diffusion maps. *Applied and Computational Harmonic Analysis* 21 (1): 5–30.

De Ridder, D. and Duin, R.P. (2002). Locally linear embedding for classification. In: *Pattern Recognition Group, Dept. of Imaging Science & Technology, Delft University of Technology, Delft, The Netherlands, Tech. Rep. PH-2002-01*, 1–12.

Dong, S., Chen, L., Tang, B. et al. (2015). Rotating machine fault diagnosis based on optimal morphological filter and local tangent space alignment. *Shock and Vibration*: 1–9.

Dong, S., Luo, T., Zhong, L. et al. (2017). Fault diagnosis of bearing based on the kernel principal component analysis and optimized k-nearest neighbour model. *Journal of Low Frequency Noise, Vibration and Active Control* 36 (4): 354–365.

Donoho, D.L. and Grimes, C. (2003). Hessian eigenmaps: locally linear embedding techniques for high-dimensional data. *Proceedings of the National Academy of Sciences* 100 (10): 5591–5596.

He, Q., Kong, F., and Yan, R. (2007). Subspace-based gearbox condition monitoring by kernel principal component analysis. *Mechanical Systems and Signal Processing* 21 (4): 1755–1772.

Huang, Y., Zha, X.F., Lee, J., and Liu, C. (2013). Discriminant diffusion maps analysis: a robust manifold learner for dimensionality reduction and its applications in machine condition monitoring and fault diagnosis. *Mechanical Systems and Signal Processing* 34 (1–2): 277–297.

Kouropteva, O., Okun, O., and Pietikäinen, M. (2002). Selection of the optimal parameter value for the locally linear embedding algorithm. In: *FSKD*, 359–363. Citeseerx.

Lai, P.L. and Fyfe, C. (2000). Kernel and nonlinear canonical correlation analysis. *International Journal of Neural Systems* 10 (05): 365–377.

Li, F., Chyu, M.K., Wang, J., and Tang, B. (2015). Life grade recognition of rotating machinery based on supervised orthogonal linear local tangent space alignment and optimal supervised Fuzzy C-Means clustering. *Measurement* 73: 384–400.

Li, Z., Yan, X., Tian, Z. et al. (2013). Blind vibration component separation and nonlinear feature extraction applied to the nonstationary vibration signals for the gearbox multi-fault diagnosis. *Measurement* 46 (1): 259–271.

Mika, S., Ratsch, G., Weston, J. et al. (1999). Fisher discriminant analysis with kernels. In: *Neural Networks for Signal Processing IX, 1999. Proceedings of the 1999 IEEE Signal Processing Society Workshop*, 41–48. IEEE.

Qian, H., Liu, Y.B., and Lv, P. (2006). Kernel principal components analysis for early identification of gear tooth crack. In: *2006. WCICA 2006. The Sixth World Congress on Intelligent Control and Automation*, vol. 2, 5748–5751. IEEE.

Roweis, S.T. and Saul, L.K. (2000). Nonlinear dimensionality reduction by locally linear embedding. *Science* 290 (5500): 2323–2326.

Sakthivel, N.R., Nair, B.B., Elangovan, M. et al. (2014). Comparison of dimensionality reduction techniques for the fault diagnosis of mono block centrifugal pump using vibration signals. *Engineering Science and Technology, an International Journal* 17 (1): 30–38.

Samko, O., Marshall, A.D., and Rosin, P.L. (2006). Selection of the optimal parameter value for the Isomap algorithm. *Pattern Recognition Letters* 27 (9): 968–979.

Schölkopf, B., Smola, A., and Müller, K.R. (1998). Nonlinear component analysis as a kernel eigenvalue problem. *Neural Computation* 10 (5): 1299–1319.

Sipola, T., Ristaniemi, T., and Averbuch, A. (2015). Gear classification and fault detection using a diffusion map framework. *Pattern Recognition Letters* 53: 53–61.

Su, Z., Tang, B., Deng, L., and Liu, Z. (2015a). Fault diagnosis method using supervised extended local tangent space alignment for dimension reduction. *Measurement* 62: 1–14.

Su, Z., Tang, B., Liu, Z., and Qin, Y. (2015b). Multi-fault diagnosis for rotating machinery based on orthogonal supervised linear local tangent space alignment and least square support vector machine. *Neurocomputing* 157: 208–222.

Su, Z., Tang, B., Ma, J., and Deng, L. (2014). Fault diagnosis method based on incremental enhanced supervised locally linear embedding and adaptive nearest neighbor classifier. *Measurement* 48: 136–148.

Tenenbaum, J.B., De Silva, V., and Langford, J.C. (2000). A global geometric framework for nonlinear dimensionality reduction. *Science* 290 (5500): 2319–2323.

Tian, Y., Wang, Z., Lu, C., and Wang, Z. (2016). Bearing diagnostics: a method based on differential geometry. *Mechanical Systems and Signal Processing* 80: 377–391.

Van Der Maaten, L., Postma, E., and Van den Herik, J. (2009). Dimensionality reduction: a comparative. *Journal of Machine Learning Research* 10: 66–71.

Van der Maaten, L., Postma, E.O., and van den Herik, H.J. (2007). *Matlab Toolbox for Dimensionality Reduction. MICC, Maastricht University*.

Van Vaerenbergh, S. (2016). Kernel methods toolbox. Mathworks File Exchange Center. https://uk.mathworks.com/matlabcentral/fileexchange/46748-kernel-methods-toolbox?s_tid=FX_rc3_behav.

Vandenberghe, L. and Boyd, S. (1996). Semidefinite programming. *SIAM Review* 38 (1): 49–95.

Wang, G., He, Y., and He, K. (2012). Multi-layer kernel learning method faced on roller bearing fault diagnosis. *Journal of Social Work* 7 (7): 1531–1538.

Wang, J. (2011). *Geometric Structure of High-Dimensional Data and Dimensionality Reduction*. Springer Berlin Heidelberg.

Wang, J., Li, S., Jiang, X., and Cheng, C. (2017). An automatic feature extraction method and its application in fault diagnosis. *Journal of Vibroengineering* 19 (4).

Wang, X., Zheng, Y., Zhao, Z., and Wang, J. (2015). Bearing fault diagnosis based on statistical locally linear embedding. *Sensors* 15 (7): 16225–16247.

Weinberger, K.Q., Packer, B., and Saul, L.K. (2005). Nonlinear dimensionality reduction by semidefinite programming and kernel matrix factorization. In: *Proc. 30th Int. Workshop Artif. Intell. Statist (AISTATS)*, 381–388.

Widodo, A. and Yang, B.S. (2007). Application of nonlinear feature extraction and support vector machines for fault diagnosis of induction motors. *Expert Systems with Applications* 33 (1): 241–250.

Xu, X.L. and Chen, T. (2009). ISOMAP algorithm-based feature extraction for electromechanical equipment fault prediction. In: *2009. CISP'09. 2nd International Congress on Image and Signal Processing*, 1–4. IEEE.

Yin, A., Lu, J., Dai, Z. et al. (2016). Isomap and deep belief network-based machine health combined assessment model. *Strojniski Vestnik-Journal of Mechanical Engineering* 62 (12): 740–750.

Yuan, N., Yang, W., Kang, B. et al. (2018). Laplacian Eigenmaps feature conversion and particle swarm optimization-based deep neural network for machine condition monitoring. *Applied Sciences* 8 (12): 2611.

Zhang, Y. and Li, B. (2010). Noise reduction method for nonlinear signal based on maximum variance unfolding and its application to fault diagnosis. *Science China Technological Sciences* 53 (8): 2122–2128.

Zhang, Y., Li, B., Wang, W. et al. (2014). Supervised locally tangent space alignment for machine fault diagnosis. *Journal of Mechanical Science and Technology* 28 (8): 2971–2977.

Zhang, Y., Li, B., Wang, Z. et al. (2013). Fault diagnosis of rotating machine by isometric feature mapping. *Journal of Mechanical Science and Technology* 27 (11): 3215–3221.

Zhang, Z. and Zha, H. (2004). Principal manifolds and nonlinear dimensionality reduction via tangent space alignment. *SIAM Journal on Scientific Computing* 26 (1): 313–338.

9

Feature Selection

9.1 Introduction

As mentioned in Chapter 6, the curse of dimensionality and irrelevant features are two of the main problems of learning from vibration data collected from a rotating machine. It is clear that high-dimensional data requires large storage and time for signal processing, and this also may limit the number of machines that can be monitored remotely across wireless sensor networks (WSNs) due to bandwidth and power constraints. Dimensionality reduction is one of the reasonable techniques to address the problems of dealing with noisy and high-dimensional data. Returning briefly to the subject of dimensionality reduction, the two main techniques to reduce data dimensionality are feature extraction and feature selection. Feature-extraction techniques project the high-dimensional data of n instances $\{x_i\}_{i=1}^n$ with D feature space, i.e. $x_i \in R^D$, into a new low-dimensional representation $\{y_i\}_{i=1}^n$ with d feature space, i.e. $y_i \in R^d$ where $d \ll D$. The new low-dimensional feature space is often a linear or nonlinear combination of the original features. Various linear and nonlinear feature-extraction techniques that can be used to map the high-dimensional feature space into a low-dimensional feature space were introduced in Chapters 7 and 8, respectively.

Feature-selection, also called *subset selection*, techniques aim to select a subset of features that can sufficiently represent the characteristic of the original features. In view of that, this will reduce the computational cost and may remove irrelevant and redundant features and consequently improve learning performance. In vibration-based machinery fault diagnosis, the main task is to categorise the acquired vibration signal for the corresponding machine condition correctly, which is generally a multiclass classification problem. Hence, the aim of feature selection in machine condition monitoring is to select a subset of features that are able to discriminate between instances that belong to different classes. In other words, the main task of feature selection is to select a subset of features that are the most relevant to the classification problem. John et al. (1994) introduced definitions for three types of features in subset selections: irrelevant features and two degrees of relevant features (weak and strong). In classification problems, irrelevant features can never contribute to classification accuracy. On the other hand, relevant features, whether strongly or weakly relevant, can contribute to classification accuracy, i.e. removal of a relevant feature can result in a change in the classification results. Feature-selection techniques select a small subset of the relevant features from the original features based on a relevance-evaluation criterion.

Condition Monitoring with Vibration Signals: Compressive Sampling and Learning Algorithms for Rotating Machines,
First Edition. Hosameldin Ahmed and Asoke K. Nandi.

The relevance and irrelevance of features in classification problems have been investigated in several studies: for example, Kohavi and John (1997) investigated the relevance and irrelevance of features for a Bayes classifier and showed that all strongly relevant and some of the weakly relevant features are required in terms of an optimal Bayes classifier. Weston et al. (2001) demonstrated that a support vector machine (SVM) can really suffer in high-dimensional spaces where many features are irrelevant. Yu and Liu (2004) showed that feature relevance alone is not enough for effective feature selection of high-dimensional data, and they developed a correlation-based method for relevance and redundancy analysis. Nilsson et al. (2007) analysed two different feature-selection problems: finding a minimal feature set optimal for classification (MINIMAL-OPTIMAL) and finding all features relevant to the target variable (ALL-RELEVANT). They proved that ALL-RELEVANT is much harder than MINIMAL-OPTIMAL.

Based on the availability of class label information, feature-selection methods can be categorised into three main groups: supervised feature-selection methods, semi-supervised methods, and unsupervised methods (Chandrashekar and Sahin 2014; Huang 2015; Miao and Niu 2016; Kumar et al. 2017; Sheikhpour et al. 2017). Supervised feature-selection techniques use labelled data samples to select a subset of features. In supervised feature selection, the relevance is often evaluated by computing the correlation between the feature and class labels. Supervised feature selection needs adequate labelled data to deliver a discriminant feature space. However, in many real applications, we often have datasets that include a large number of unlabelled samples and a small number of labelled samples. To address this problem, various semi-supervised feature-selection techniques have been introduced (Kalakech et al. 2011). On the other hand, unsupervised feature selection is based on unlabelled samples, and therefore it is more challenging than supervised and semi-supervised feature selection (Ang et al. 2016).

A typical feature-selection procedure can be described by the following four main steps (Dash and Liu 1997; Liu and Yu 2005):

(1) *Subset generation.* In this step, a feature subset is chosen based on a selected search approach that depends on two basic principles: (i) the search direction (or directions), which can be forward, where the search starts with an empty set and sequentially adds features; backward, where the search starts with a full set and sequentially removes features; or bidirectional, where the search starts with both ends and adds and removes features concurrently; and (ii) the search strategy, which can be complete, sequential, or random search.

(2) *Subset evaluation.* Each produced subset is evaluated using an evaluation measure, e.g. independent criteria, distance measures, information measures, dependency measures, and consistency measures.

(3) *Stopping criteria.* This defines when the feature selection should stop. Some commonly used stopping criteria are:
   a. The search is complete.
   b. A given number of features or a maximum number of features or number of iterations is reached.
   c. Subsequent addition or removing of any feature will not provide a better subset.
   d. The classification error rate of a subset is less than the acceptable rate for a given classification task.

(4) *Validation.* Test the validity of the selected subsets by comparing the results with previously well-known results, or with the results of other competing feature-selection techniques using artificial datasets, real datasets, or both.

With regard to feature selection for classification problems, Dash and Liu stated that the majority of real-world classification problems require supervised learning where the underlying class probabilities and class-conditional probabilities are unknown, and each instance is associated with a class label. In vibration-based machine fault diagnosis, to make a sensible deduction using a machine learning–based classifier, we often compute certain features of the raw vibration signal that can describe the signal in essence. Sometimes, multiple features are computed to form a feature set. Depending on the number of features in the set, one may need to perform further filtering of the set using a feature-selection technique (Nandi et al. 2013). In fact, feature-selection techniques are employed to select a subset of features that are expected to result in better machine health prediction accuracy for given vibration data. The problem of feature selection in vibration-based machine fault diagnosis can be briefly defined as follows.

Given vibration data X composed of L instances, each of which is characterised by a row vector $\vec{x}_i = \begin{bmatrix} x_{i1} & x_{i2} & \cdots & x_{iM} \end{bmatrix}$ containing values of M features such that,

$$X = \begin{bmatrix} \vec{x}_1 \\ \vec{x}_2 \\ \cdot \\ \cdot \\ \cdot \\ \vec{x}_L \end{bmatrix} = \begin{bmatrix} x_{11} & \cdots & x_{1M} \\ \vdots & \ddots & \vdots \\ x_{N1} & \cdots & x_{NM} \end{bmatrix} \tag{9.1}$$

assume the original set of features for each vector is $F = \{f_1, f_2, ..., f_M\}$. The main task of the feature-selection technique is to select a subset (Z) of m features from F, where $Z \subset F$. The selected subset Z is expected to build a better classification model.

Based on their relationship with learning algorithms, feature-selection techniques can be further grouped into filter models, wrapper models, and embedded models. The subsequent subsections discuss these types of methods in more detail.

9.2 Filter Model-Based Feature Selection

Filter model–based feature selection uses a preprocessing step independent of any machine learning algorithms. It is based on measures of various characteristics of the training data such as similarity, dependency, information, and correlation. Therefore, it is fast and requires low computational complexity compared to other methods. The filter models procedure can be described by the following two main steps:

(1) The filter model ranks features based on certain measures. In this step, filtering can be performed using univariate feature filters that rank every single feature or using multivariate feature filters that evaluate a feature subset (Mitra et al. 2002; Tang et al. 2014; Chandrashekar and Sahin. 2014).
(2) The features with the highest rankings are selected to train classification algorithms.

This section gives brief descriptions of several feature-ranking (FR) methods that can be used to rank the features of a vibration signal.

9.2.1 Fisher Score (FS)

Fisher score (FS) is a filter-based feature-selection method and one of the commonly used supervised feature-selection methods (Duda et al. 2001; Gu et al. 2012). The main idea of FS is to compute a subset of features with a large distance between data points in different classes and a small distance between data points in the same class. To briefly describe the FS method, assume the input matrix $X \in R^{L \times M}$ reduces to the $Z \in R^{L \times m}$ matrix. The FS can be calculated using the following equation,

$$F(Z) = tr\{(S_B)(S_T + \gamma I)^{-1}\} \tag{9.2}$$

where S_B is the between-class scatter matrix, S_t is the total scatter matrix, and γ is a regularisation parameter. S_B and S_T can be computed using Eqs. (9.3) and (9.4), respectively,

$$S_B = \sum_{i=1}^{c} l_i(\mu_i - \mu)(\mu_i - \mu)^T \tag{9.3}$$

$$S_T = \sum_{i=1}^{L} (z_i - \mu)(z_i - \mu)^T \tag{9.4}$$

where L is the total number of observations and c is the number of classes, z_i represents the dataset belong to the cth class, l_i is the size of the ith class in the reduced data space, μ_i is the mean vector of class i, and μ is the overall mean vector of the reduced data that can be computed using the following equation:

$$\mu = \sum_{i=1}^{c} l_i \mu_i \tag{9.5}$$

Let μ_i^j and σ_i^j be the mean and the standard deviation of the ith class, corresponding to the jth feature; and let μ^j and σ^j be the mean and standard deviation of the entire data corresponding to the jth feature. Formally, the FS feature of the jth can be computed by the following equation:

$$FS(X^j) = \frac{\sum_{i=1}^{C} l_i(\mu_i^j - \mu^i)^2}{(\sigma^j)^2} \tag{9.6}$$

Here, $(\sigma^i)^2$ can be represented as follows,

$$(\sigma^i)^2 = \sum_{i=1}^{c} l_i(\sigma_i^j)^2 \tag{9.7}$$

Usually, the FS of each feature is computed independently. Therefore, Gu et al (2012) proposed a generalised FS technique that selects features jointly. In this technique, the authors introduced an indicator variable $p = (p_1, p_2, ..., p_d)^T$, where $p_i \in \{0, 1\}$, to denote whether a feature is selected or not. Hence, Eq. (9.2) can be rewritten as Eq. (9.8),

$$F(p) = tr\{(diag(p)(S_B)diag(p))(diag(p)(S_T + \gamma I)diag(p))^{-1}\} \tag{9.8}$$

$$s.t \, p \in \{0, 1\}^d, p^T 1 = m$$

where $diag(p)$ is a diagonal matrix of p_i's and m is a number of features to be selected. Then, the optimal value of Eq. (9.8) is lower bounded by the optimal value of the following equation,

$$F(W, p) = tr\{(W^T diag(p)(S_B)diag(p)W)(W^T diag(p)(S_T + \gamma I)diag(p)W)^{-1}\} \tag{9.9}$$

$$s.t\ p \in \{0,1\}^d, p^T 1 = m$$

where $W \in R^{d \times c}$

Based on the theory of the generalised FS, the optimal indicator p that maximises the problem in Eq. (9.9) is the same as the optimal p that minimises the following problem in Eq. (9.10),

$$\min_{p,W} \frac{1}{2} \|X^T diag(p)W - G\|_F^2 + \frac{\gamma}{2}\|W\|_F^2 \tag{9.10}$$

$$s.t\ p \in \{0,1\}^d, p^T 1 = m$$

where $G = [g_1, \ldots, g_c] \in R^{L \times c}$, and g_i is a column vector that can be represented as follows:

$$g_{ji} = \begin{cases} \sqrt{\dfrac{L}{L_C}} - \sqrt{\dfrac{L_C}{L}}, & if\ y_j = i \\ -\sqrt{\dfrac{L_C}{L}}, & otherwise \end{cases} \tag{9.11}$$

9.2.2 Laplacian Score (LS)

The Laplacian score (LS) is an unsupervised filter-based technique that ranks features depending on their locality-preserving power. In fact, LS is mainly based on Laplacian eigenmaps and locality-preserving projection, and can be briefly described as follows (He et al. 2006).

Given a dataset $X = [x_1, x_2, \ldots, x_L]$, where $X \in R^{L \times M}$, suppose the LS of the rth feature is LS_r and f_{ri} represent the ith sample of the rth feature where $i = 1, \ldots, M$ and $r = 1, \ldots, L$. First, the LS algorithm constructs the nearest neighbour graph G with M nodes, where the ith node corresponds to x_i. Next, an edge between nodes i and j is placed; if x_i is among the k nearest neighbours of x_j or vice versa, then i and j are connected. The elements of the weight matrix of graph G is S_{ij} and can be defined as follows:

$$S_{ij} = \begin{cases} e^{-\dfrac{\|x_i - x_j\|^2}{t}}, & i\ and\ j\ are\ connected \\ 0, otherwise \end{cases} \tag{9.12}$$

where t is an appropriate constant. The LS_r for each sample can be computed as follows:

$$LS_r = \frac{\tilde{f}_r^T (LS)\tilde{f}_r}{\tilde{f}_r^T D\tilde{f}_r} \tag{9.13}$$

Here, $D = diag\ (S1)$ is the identity matrix, S is the similarity matrix, $1 = [1, \ldots, 1]^T$, $LS = D - S$ is the graph Laplacian matrix, and $\tilde{f}_r$ can be calculated using the following equation,

$$\tilde{f}_r = f_r - \frac{f_r^T D1}{1^T D1} \tag{9.14}$$

where $f_r = [f_{r1}, f_{r2}, \ldots, f_{rM}]^T$.

9.2.3   Relief and Relief-F Algorithms

Relief methods, i.e. Relief and Relief-F, are supervised feature-ranking methods that commonly used as preprocessing techniques for feature-subset selection for both classification and regression (Kira and Rendell 1992; Robnik-Šikonja and Kononenko 2003; Liu and Motoda 2007). The Relief algorithm is a two-class filtering algorithm, and Relief-F is a multiclass extension of Relief. Briefly, we describe these two techniques as follows.

9.2.3.1   Relief Algorithm

Relief is a two-class algorithm that ranks features according to how well their values discriminate between instances that are near each other (Robnik-Šikonja and Kononenko 2003). Briefly, we describe the process of Relief algorithm as follows.

Given two-class training data $X \in R^{L \times M}$, to begin, Relief sets the quality estimation $W[F]$ for all features F to zero. Then, it selects instances randomly. For each selected training instance (x_i) of feature values and the class label value, Relief searches for its two nearest neighbours. The first neighbour is called the *nearest hit* (H), which is a neighbour from the same class. The second neighbour is called the *nearest miss* (Q), which is a neighbour from a different class. Next, it updates $W[F]$ for all features F based on their values for the randomly selected instance x_i, H, and Q as follows:

(1) If the values of x_i and H are different than feature F, then feature F separates two instances with the same class, which is not needed, so we decrease $W[F]$.
(2) If the values of x_i and M are different than feature F, then feature F separates two instances with different classes, which is needed, so we increase $W[F]$.

The whole procedure is repeated k times and can be described in Algorithm 9.1:

Algorithm 9.1   Relief

Input: L learning instances, M features, and *two* classes;
Output: the vector W of estimations of the qualities of features;
   1. Initialise $W[A] = [0, 0, ..., 0]$;
   2. For i:= 1 to k do
   3. Select an instance x_i randomly;
   4. Find nearest hit H and nearest miss Q;
   5. For F:= 1 to f do
   6. W:= W − diff(A, x_i, H)/k + diff(A, x_i, Q)/k;
   7. end;

The Relief algorithm can be used for both numerical and nominal features. For numerical features, the difference between the values of the feature F for two instances x_1 and x_2 can be computed using the function *diff* (F, x_1, x_2) such that,

$$diff(F, x_1, x_2) = \frac{|value\,(F, x_1) - value\,(F, x_2)|}{\max(F) - \min(F)} \tag{9.15}$$

and for nominal features

$$diff(F, x_1, x_2) = \begin{cases} 0; value(F, x_1) = value(F, x_2) \\ 1; otherwise \end{cases} \tag{9.16}$$

9.2.3.2 Relief-F Algorithm

The Relief-F algorithm is an extension of the traditional Relief algorithm that has the ability to deal with noisy, incomplete, and multiclass datasets. It uses a statistical approach to select the important features based on their weight, i.e. quality estimation, W. The good attribute should have the same value of weights for instances from the same class and discriminate between instances from different classes (Kononenko et al. 1997). The main idea of Relief-F is to randomly select instances from the training data and then calculate their l nearest neighbours from the same class, also called the *nearest hit* (H_j), and the other l nearest neighbours from different classes, also called the *nearest miss* (Q_j). The function *diff* (*Attribute, Instance1, and Instance2*) is used to compute the distance between instances to find the nearest neighbours. Similarly, to Relief, Relief-F updates $W[A]$ for all features F based on the values of the randomly selected instance x_i, H_j, and Q_j. Here, we take the average of all the hits and all the misses to be used in the updated formula of the Relief algorithm. Also, to deal with incomplete data, the missing values are preserved probabilistically. This can be done by computing the probability that two given instances have different values for assumed features trained over class values such that,

$$diff(F, x_1, x_2) = 1 - P(value(F, x_2) \mid class(x_1)) \tag{9.17}$$

where the x_1 value is unknown. In case both instances, i.e. x_1 and x_2, have an unknown value, the difference function can be performed using the following equation:

$$diff(F, x_1, x_2) = 1 - \sum_V (P(V \mid class(x_1)) \times P(V \mid class(x_2))) \tag{9.18}$$

The procedure for the Relief-F algorithm is summarised in Algorithm 9.2.

Algorithm 9.2   Relief-F

Input: L learning instances, M features, and C classes;
Output: the vector W of estimations of the qualities of features;

1. Initialise $W = [0,0,...,0]$;
2. For i:= 1 to k do
3. Select an instance x_i randomly;
4. Find l nearest hits H_j;
5. For each class $C \neq class(x_i)$ do
6. Find l nearest misses Q_j
7. For F:= 1 to f do
8. $W[F] := W[F] - \sum_{j=1}^{l} diff(A, x_i, H_j)/(l.k) + \sum_{C \neq class(x_i)} \frac{P(C)}{1 - P(class(x_i))} \sum_j^l diff(A, x_i, H_j)/(l.k);$
9. end;

9.2.4 Pearson Correlation Coefficient (PCC)

The Pearson correlation coefficient (PCC) (Liu and Motoda 2007) is a supervised filter-based ranking technique that examines the relationship between two variables according to their correlation coefficient (r), $-1 \leq r \leq 1$. Here, negative values indicate inverse relations, positive values indicate a correlated relation, and the value 0 indicates no relation. For classification purposes, PCC can be used to rank features based on the correlation between features and class labels. This can be represented mathematically using the following equation:

$$r(i) = \frac{cov(x_i, y)}{\sqrt{var(x_i) * var(y)}} \tag{9.19}$$

Here, x_i is the ith variable, y is the class labels, $var(.)$ denotes the variance, and $cov(.)$ is the covariance. The value of correlation coefficient $r(i)$ indicates how well feature x_i and the class y are correlated, as follows:

(1) If the feature x_i and the class y are fully correlated, $r(i)$ takes the value 1 or -1.
(2) If the feature x_i and the class y are completely independent, $r(i)$ takes the value zero.
(3) Other values close to 1 or -1 indicate high levels of correlations, and those close to zero indicate low levels of correlations.

The higher the correlation between the feature and the class label, the better the selected feature. However, Yu and Liu (2003) argued that linear correlation measures may not be capable of capturing correlations that are not linear in nature. To overcome this limitation, they proposed a correlation measure based on the information-theoretical idea of entropy.

9.2.5 Information Gain (IG) and Gain Ratio (GR)

As described previously, Yu and Liu (2003) proposed a correlation measure based on the information-theoretical concept of entropy to overcome the limitation of PCC as a linear correlation measure. The entropy of a variable X can be defined using the following equation,

$$H(X) = - \sum_i P(x_i) log_2(P(x_i)) \tag{9.20}$$

and the entropy of X after observing values of another variable Y can be defined as follows,

$$H(X \mid Y) = - \sum_j P(y_j) \sum_i P(x_i|y_i) log_2(P(x_i \mid y_i)) \tag{9.21}$$

where $P(x_i)$ is the prior probabilities for all values of X, and $P(x_i|y_i)$ is the posterior probabilities of X given the values of Y. The difference between Eqs. (9.20) and (9.21) is called information gain (IG) and can be computed using the following equation:

$$IG(X \mid Y) = H(X) - H(X \mid Y) \tag{9.22}$$

Based on the *IG*, feature Y is regarded as more correlated to feature X than feature Z if $IG(X|Y) > IG(Z|Y)$. Moreover, IG is normalised with feature entropy such that,

$$SU(X, Y) = 2 \left[\frac{IG(X|Y)}{H(X) + H(Y)} \right] \tag{9.23}$$

where $SU(.)$ is the symmetrical uncertainty that restricts its values to the range $[0, 1]$. A value of 1 indicates that knowing the values of either feature completely predicts the values of the other; and a value of 0 indicates that X and Y are independent.

Furthermore, a new measure called gain ratio (GR), which is a normalised version of IG, can be defined as follows:

$$GR(X, Y) = \frac{(H(X) - H(X|Y))}{H(Y)} \tag{9.24}$$

9.2.6 Mutual Information (MI)

The mutual information (MI) is a measure of dependence between two variables, i.e. how much information is shared between two variables, and can be defined as follows (Liu et al. 2009; Cang and Yu 2012):

$$MI(X, Y) = \sum_{y \in Y} \sum_{x \in X} p(x, y) log \frac{p(x, y)}{p(x)p(y)} \tag{9.25}$$

From the definition, the value of $MI(X, Y)$ will be vey high if X and Y are closely related to each other; otherwise, if the value of $MI(X, Y)$ is equal to zero, this denotes that X and Y are totally unrelated. In classification problems, MI is used to measure the dependence between features $F = \{f_1, f_2, ..., f_m\}$ and class labels $C = \{c_1, c_2, ..., c_k\}$ such that,

$$MI(C, F) = \sum_{c} P(c) \sum_{f} p(f, c) log \frac{p(f, c)}{p(f)} \tag{9.26}$$

9.2.7 Chi-Squared (Chi-2)

Feature ranking and selection using chi-square (chi-2) is based on the χ^2 test statistic (Yang and Pedersen 1997). Chi-2 evaluate the importance of a feature by calculating the χ^2 test with respect to the class labels. The χ^2 value for each feature f in a class label group c can be computed using the following equation,

$$\chi2(f, c) = \frac{L(E_{c,f}E - E_c E_f)^2}{(E_{c,f} + E_c)(E_f + E)(E_{c,f} + E_f)(E_c + E)} \tag{9.27}$$

where L is the total number of examples, $E_{c,f}$ is the number of times f and c co-occur, E_f is the number of time feature f occurs without c, E_c is the number of times c occurs without f, and E is the number of times neither f nor c occurs. A bigger value of χ^2 indicates that the features are highly related.

9.2.8 Wilcoxon Ranking

As described in Vakharia et al. (2016), Wilcoxon ranking (Wilcoxon 1945) is a non-parametric test utilised to estimate the ranking of features in a feature set. It is used to test the null hypothesis (H_0) that two populations have identical distribution functions against the alternative hypothesis (H_1) that the two distribution functions differ only with respect to the median. Consequently, a large sample ζ test can be computed as in Eq. (9.28),

$$\zeta = \frac{R - \mu_R}{\sigma_R} \tag{9.28}$$

where R represents the sum of ranks of the sample, μ_R is the mean of the sample R values, and σ_R represents the standard deviation of the sample R values. μ_R and σ_R can be calculated using Eqs. (9.29) and (9.30), respectively,

$$\mu_R = \frac{n_1(n_1 + n_2 + 1)}{2} \tag{9.29}$$

$$\sigma_R = \sqrt{\frac{n_1 n_2 (n_1 + n_2 + 1)}{12}} \tag{9.30}$$

where n_1 and n_2 are the size of sample 1 and sample 2, respectively (Vakharia et al. 2016).

9.2.9 Application of Feature Ranking in Machine Fault Diagnosis

The use of filter model–based techniques for feature selection for diagnosing rotating machines faults was investigated by several researchers. For example, Tian et al. (2012) introduced a method based on the wavelet packet transform (WPT) and envelope analysis to extract fault features of a roller bearing from masking faulty gearbox signals. In this method, the wavelet packet is selected by PCC-based correlation analysis, and the fault feature of the bearing is extracted from the selected wavelet packet using envelope analysis. The authors assumed that the impulses generated by the faulty bearing and the faulty gearbox are not the same. The results from the vibration signals case study demonstrated that the proposed method successfully extracted fault features of outer race BPFO (bearing pass frequency of outer race) at 30.48 HZ from a vibration signal mixed with the signal of a faulty gearbox. Moreover, the authors showed that the proposed method does not need fault information for the gearbox. Wu et al. (2013) investigated the feasibility of using the multiscale analysis and SVM classification technique to diagnosis roller bearing faults in rotating machinery. In this technique, multiscale analysis is utilised to extract fault features in different scales, such as multiscale entropy (MSE), multiscale permutation entropy (MPE), multiscale root mean square (MSRMS), and multiband spectrum entropy (MBSE). Some of the features are then selected as inputs for the SVM classifier through the FS as well as Mahalanobis distance (MD) evaluations. The FS and MD are used to enhance both the accuracy of the bearing fault classification and computational efficiency. Experimental vibration signals collected from a roller bearing in a normal condition, with an inner race defect (IRD), with a rolling element defect, and with an outer race defect (ORD) are used to evaluate the effectiveness of the proposed method. The results showed that an accurate bearing fault diagnosis can be achieved using the extracted features in different scales.

Moreover, Zheng et al. (2014) introduced a technique for diagnosing roller bearing faults based on multiscale fuzzy entropy (MFE), LS, and variable predictive model-based class discrimination (VPMCD). In this technique, MFE is used to describe the complexity and irregularity of roller vibration signals. Then, LS is employed to the feature vector by sorting the features based on their importance and correlation with the fault information and selecting the most important features to avoid a high dimension for the feature vector. With these selected features, VPMCD is used to deal with the classification problem. Vibration signals collected from roller bearings with eight health conditions are used to validate the proposed method. These are normal condition and three fault conditions: rolling element fault (REF) with two levels of severity (slight and very severe); outer race fault (ORF) with two levels of severity (slight and very severe); and inner race fault (IRF) with three levels of severity (slight, medium, and very severe). The results verified the effectiveness of the proposed method in diagnosing roller bearing faults.

Furthermore, Vakharia et al. (2016) presented a method to detect various bearing faults from acquired vibration signals. In this method, first, 40 statistical features are computed from the time domain, the frequency domain, and the discrete wavelet transform. Then, to select the most informative features and reduce the size of the feature vector, Chi-squared and Relief-F feature ranking techniques are used. Finally, the effect of a ranked feature on the performance of an artificial neural network (ANN) and random forest (RF) are investigated. Vibration signals acquired from a roller bearing with healthy bearing (HB), IRD, ORD, and ball defect (BD) are used to validate the proposed method. The results demonstrated that 93.54% ten-fold cross-validation accuracy is obtained when the Chi-squared feature ranking method is used along with the RF classifier.

Also, Saucedo Dorantes et al. (2016) presented a multifault diagnosis method for electric machines based on high-dimensional feature reduction. In this method, (i) empirical mode decomposition (EMD) is employed to decompose the signal. (ii) Statistical features are extracted from the obtained decompositions. (iii) To preserve the data variance, a combination of a genetic algorithm (GA) and principal component analysis (PCA) is used to perform a feature optimisation. (iv) FS is performed to select features from the extracted features. Finally, (v) with these selected features, ANN is employed to deal with the classification problem. Vibration signals with six different experimental conditions are used to validate the proposed method. The results showed diagnosis accuracy improvement of almost 12% and 20% compared to PCA and linear discriminant analysis (LDA), respectively.

Moreover, Haroun and colleagues investigated the application of the Self-Organising Maps (SOM) for the detection of rolling element bearing faults in one three-phase induction motor. In this study, multiple features extraction techniques from time, frequency, and time-frequency domains were used. Then, Relief-F and minimum redundancy max Relevance (mRMR) features selection techniques were employed to select the optimal features and reduce the dimension of the acquired vibration data. Finally, the SOM was utilised for classification of the different bearing health conditions. The results showed that the combination of feature selection techniques and SOM classifier improved the classification performances of the fault diagnosis process (Haroun et al. 2016).

Li et al. (2017) introduced a rolling bearing fault diagnosis strategy based on improved multiscale permutation entropy (IMPE), LS, and quantum behaved particle swarm optimisation-least square SVM (QPSO-LSSVM). Under this strategy, (i) the IMPE is used to obtain precise and reliable values of the roller bearing vibration signals. (ii) The LS algorithm is used to refine the extracted features to form a new feature vector comprising main unique information. Finally, (iii) the QPSO-LSSVM classifier is utilised to classify the health status of the roller bearings. Experimental vibration signals acquired from a roller bearing under various operating conditions are used to examine the effectiveness of the proposed method. The results verified the effectiveness of the proposed method in diagnosing roller bearing faults. Furthermore, compared with MPE, IMPE achieved higher classification accuracy.

Additionally, Ahmed and Nandi (2017) introduced a method for fault classification of roller bearings using multiple measurement vector compressive sampling (MMV-CS), FS, and SVM. In this method, the combination of MMV-CS and FS is employed to compress the data and then select the most important features of the compressed samples to reduce the computational cost and remove irrelevant and redundant features. Then, the compressed samples with the selected features enter the SVM classifier for classification of the roller bearing's health condition. Vibration signals acquired from roller bearings under six different conditions, including two healthy conditions and four faulty conditions, are used to examine the effectiveness of the proposed method. The results showed high classification accuracy for roller bearing health. In the same vein, Ahmed et al. (2017) presented a method for roller bearing fault classification based on compressive sampling, LS, and a multiclass support vector machine (MSVM). In this method, compressive sampling (CS) is employed to get compressed samples from the raw vibration signals, and LS is used to rank the features of the obtained compressed samples with respect to their importance and correlations with the core fault characteristics. Afterward, based on LS ranking, a small number of the most significant compressed samples are selected to produce the features vector. Finally, with these selected features, the MSVM classifier is used to classify the roller bearing's health condition. Vibration signals collected from roller bearings under six health conditions are used to validate the proposed method. The experimental results showed that the proposed method achieved a high classification accuracy with significantly reduced feature sets.

Recently, Ahmed and Nandi (2018) proposed a combined CS based on a multiple measurement vector (MMV) and feature ranking framework to learn optimally fewer features from a large amount of vibration data from which bearing health conditions are classified. In this framework, CS is used to reduce the amount of original signals by obtaining compressively sampled signals that possess the quality of the original signal. Then, a feature-ranking technique is employed to further filter the obtained compressively sampled signals by ranking their features and selecting a subset of fewer, most-significant features. Based on this framework, the authors considered two techniques for feature selection to select fewer features from the compressively sampled signals. These are: (i) similarity-based methods that assign similar values to compressively sampled signals that are close to each other; three algorithms (LS, FS, and Relief-F) were investigated to select fewer features based on similarity; and (ii) statistical-based methods that measure the importance of features of the compressively sampled signals using different statistical measures; two algorithms (PCC and Chi-2) were investigated to select fewer features based on correlation and an independence test, respectively. These selected features, in combination with three of the popular classifiers – multinomial logistic regression classifier (MLRC), ANN, and SVM – are evaluated for the classification of bearing faults. Two case studies of vibration signals generated by different health conditions of roller bearings are used to verify the efficiency of the proposed framework. The results showed that the proposed framework is able to achieve high classification accuracy in roller bearing fault diagnosis. Also, the various combinations of CS and feature-ranking techniques investigated in this study offer high classification accuracies with logistic regression classifier (LRC), ANN, and SVM, even with a limited amount of data. These results will be described in detail in Chapter 16.

9.3 Wrapper Model–Based Feature Subset Selection

Wrapper-based feature selection is based on the predictive performance of a predefined predictor. It is usually expensive compared to filter-based feature selection. The wrapper model procedure can be described by the following two main steps:

(1) Search for a subset of features using a predefined search strategy.
(2) Evaluate the selected subset of features using a predefined predictor.

This procedure is repeated until there is no improvement in the prediction. Wrapper methods utilise the predictor performance as the objective function to evaluate the selected subset of features. Various types of search techniques can be used to find a subset of features that maximises prediction performance. Wrapper methods can be categorised into sequential selection algorithms and heuristic search algorithms (Chandrashekar and Sahin, 2014). The following subsections describe these algorithms in detail.

9.3.1 Sequential Selection Algorithms

There are two types of sequential selection algorithms: (i) sequential forward selection (SFS) starts with an empty set and sequentially adds features, evaluates the selected features, and repeats until there is no improvement in the prediction; (ii) sequential backward selection (SBS) starts with the complete set of features and sequentially removes features, evaluates the selected features, and repeats until there is no improvement in the prediction (Devijver and Kittler 1982). Moreover, Pudil et al. (1994) suggested floating search methods in feature selection, where a dynamically changing number of features are added or removed at each step of the sequential search methods. These are sequential forward floating selection (SFFS) and sequential backward floating selection (SBFS).

9.3.2 Heuristic-Based Selection Algorithms

In the wrapper approach, the size of the search space for n features is $O(2^n)$, which makes it impractical for high-dimensional feature spaces. The heuristic and meta-heuristic search-based feature-selection algorithms were proposed to improve search performance (Kohavi and John 1997; Bozorg-Haddad et al. 2017). Of these algorithms, meta-heuristic-based algorithms are amongst the most common algorithms used for feature selection in bearing fault diagnosis. These include ant colony optimisation (ACO), GA, and particle swarm optimisation (PSO). In the following subsections, we introduce these methods.

9.3.2.1 Ant Colony Optimisation (ACO)

ACO is a meta-heuristic technique proposed by Dorigo et al. (1996). As described by Blum and Roli (2003), the ACO algorithm is inspired by the foraging behaviour of real ants, which enables them to find the shortest paths between food sources and their nest. Ants deposit a substance called *pheromone* on the ground on their way from food sources to their nest or vice versa. When they decide what direction to go, they select paths with higher probability, which are marked by stronger pheromone concentrations. ACO algorithms are based on a parametrised probabilistic model – the pheromone model, which is used to model chemical pheromone tracks. Artificial ants perform randomised walks on a graph $G(A, B)$, called a *construction graph*, whose vertices are the solution components A; set b are the connections. Components $a_i \in A$ have a related pheromone trail parameter $\mathcal{T}_i$, connections $b_{ij} \in B$ have a related pheromone trail parameter $\mathcal{T}_{ij}$, and the values of these parameters are denoted by t_i and t_{ij}. Also, $a_i \in A$ and $b_{ij} \in B$ have associated heuristic values $\hbar_i$ and $\hbar_{ij}$, respectively. The set of all pheromone parameters is denoted by $\mathcal{T}$, and the set of all heuristic values is denoted by $\mathcal{H}$. These values are utilised by the ants to make probabilistic decisions, also called *transition probabilities*, about how to walk on graph G.

The ant system (AS), which was proposed by Dorigo et al. (1996), is the first ACO algorithm in the literature. The basic idea of *AS* is that at each iteration, each ant $m \in M$ creates a solution s_m, where M denotes the set of ants. Then, pheromone values are updated by these solutions. In this algorithm, first, the pheromone values are initialised to the same small value ($v_0 > 0$). Then, an ant incrementally constructs solution components to the incomplete solution constructed up to now. Afterward, transition probabilities are used to choose the next solution component based on the following transition rule,

$$p(a_r|s_m[a_l]) = \begin{cases} \dfrac{[\hbar_r]^\alpha [t_r]^\beta}{\displaystyle\sum_{a_u \in J(s_m[a_l])} [\hbar_u]^\alpha [t_u]^\beta}, & \text{if } a_r \in J(s_m[a_l]) \\ 0, & otherwise \end{cases} \tag{9.31}$$

where $J(s_m[a_l])$ is the set of solution components that are allowed to be added to the partial solution $s_m[a_l]$; here, a_l denotes the last component that was added, and α and β are parameters to adjust the relative importance of heuristic information and pheromone values. After all the ants build a solution, the AS algorithm applies the online delayed pheromone update as follows,

$$t_j \leftarrow (1-\rho).t_j + \sum_{m \in M} \Delta t_j^{s_m}, \forall \mathcal{T}_i \in \mathcal{T} \tag{9.32}$$

where $0 \le \rho \le 1$ is the pheromone evaporation rate and $\Delta t_j^{s_m}$ can be defined as follows,

$$\Delta t_j^{s_m} = \begin{cases} F(s_m) \text{ if } a_j \text{ is a component of } s_m \\ 0, \text{ otherwise} \end{cases} \tag{9.33}$$

where F is the quality function. The procedure for the *AS* algorithm is summarised in Algorithm 9.3.

Algorithm 9.3   Ant System (AS)

Initialisation: Initialise pheromone values ($\mathcal{T}$), ant number (M), and number of iterations ($\mathcal{K}$);
While $k < \mathcal{K}$ do
for all ants $m \in M$ do
$s_m \leftarrow$ construct a solution ($\mathcal{T}, \mathcal{H}$)
end
Apply Online Delayed Pheromone Update ($\mathcal{T}, \{s_m| \, m \in M\}$)
end

Based on AS, several ACO algorithms have been proposed: for example, a rank-based version of the AS (Bullnheimer et al. 1997); the ant colony system (ACS) (Dorigo and Gambardella 1997); MAX-MIN AS (MMAS) (Stützle and Hoos 2000); a population-based approach for ACO (Guntsch and Middendorf 2002); a hyper-cube framework for ACO (Blum and Dorigo 2004); and beam-ACO (Blum 2005).

9.3.2.2 Genetic Algorithms (GAs) and Genetic Programming

The genetic algorithm (GA) is a heuristic search-based optimisation algorithm inspired by Darwin's theory of survival of the fittest. It was first developed by Holland (1975). The GA algorithm simulates the process of evolution using the principles of natural selection to find the optimal solution to a problem. In 1989, Holland and Goldberg proposed the simple GA that starts with an initial randomly generated set of individuals, also called the *population*. Then, this population is evaluated using a fitness function, where each individual of the population is tested to decide whether it fulfils the stopping criteria. In case none of the individuals satisfies the stopping criteria, the best individuals are selected. Afterward, the selected individuals are utilised as the parents of the new population. The selected parents perform a process called *crossover*, which is the exchange of sub-individuals of two parents to produce two new individuals. Finally, individuals perform a process of mutation, and the resulting new population is utilised for the following generation (Holland and Goldberg 1989).

As described by Wong and Nandi (2004), to solve the optimisation problem, a GA needs the following four items:

1. *Coding*. A way of representing the individuals to solve the optimisation problem. An individual is often characterised by a set of parameters (variables) that are joined into a string to form a chromosome, i.e. solution. Each individual solution in GA is represented by a genome string. This string comprises specific parameters to solve the problem, e.g. the binary genome coding and the list genome coding. In the binary genome coding, each string has a length N, where N is the total number of input features available. A binary '1' represents the presence of the feature at the corresponding index number. Similarly, a binary '0' represents the feature's absence. The advantage of this coding is that it searches through the feature subspace dynamically without a user-defined number of subset features. No constraint is needed with this coding method. The second coding used is the real numbered list genome string. Each genome in this category is of length M, where M is the anticipated number of

features in a feature subset. To initialise, the GA chooses randomly M numbers from a list of integers ranging from 1 to N. However, we do not desire any repetition of the integers, as this means that the same feature is selected more than once. Therefore, a constraint, $1 < f_i < N$ is applied, where f_i represents the *ith* input feature.

2. *Initialisation.* The initial population, which is often generated randomly, represents the initial solutions.

3. *Fitness function.* Also called the *evaluation function*, used to estimate how good an individual is at solving the optimisation problem. For example, in function minimisation problems, the individual that provides the smallest output will be given the highest score.

4. *Genetic operators.* The GA is composed of three main operators, as follows:
   a. *Parent selection.* There are several ways to select a new intermediate population. Based on the performance of individual strings, roulette wheel selection assigns a probability to each string according to their performance. Therefore, poor genome strings will have a slight chance of survival. Unlike a roulette wheel, selection by rank just orders the individuals according to their performance and selects copies of the best individuals for reproduction.
   b. *Crossover.* Crossover occurs with a probability of P_c. A point is chosen for two strings where their genetic information is exchanged. There are also variations: two-point or multipoint crossover.
   c. *Mutation.* Mutation is used to avoid local convergence of the GA. In binary coding, it just means the particular bit chosen for mutation is inverted to its complement. In list coding, the chosen index is replaced with a new index without breaking the constraint. The mutation occurs with a typical mutation probability of 0.05. This probability is kept at such a low value to prevent unnecessary oscillation.

   Other genetic operators like elitism, niche, and diploidy are often classified as advanced genetic operators.

Genetic programming (GP) is a technique for automatically generating computer programs. It is a branch of GAs with the main difference that GP creates computer programs as the solution, while GA creates a string of numbers or parameters that influence the performance of a fixed solution (Zhang et al. 2005).

9.3.2.3 Particle Swarm Optimisation (PSO)

The particle swarm optimisation (PSO) technique is one of the swarm intelligence methods used for solving global optimisation problems. It is inspired by the social behaviour of some biological organisms (Kennedy and Eberhart, 1995). The PSO algorithm utilises a set of particles to explore the search space. These particles are defined by their positions and velocities. Each particle position (x) represents a potential solution. During the optimisation procedure, the algorithm memorises the best position obtained by each particle. Afterward, the swarm memorises the best position obtained by any of its particles. Both the position (x) and the velocity of each particle are updated in each iteration. For instance, the velocity (v) and the position (x) of each particle in iteration k can be updated using Eqs. (9.34) and (9.35), respectively,

$$v(k + 1) = w.v[k] + C_l.r_l[k].(p[k] - x[k] + C_g.r_g[k].(g[k] - x[k]) \tag{9.34}$$

$$x(k + 1) = x[k] + v[k + 1]) \tag{9.35}$$

where w represents an inertia factor that keeps the swarm together and preventing it from extreme expansion and therefore diminishing PSO into a pure random search; C_l and C_g are acceleration factors that control the local and global knowledge of the movement of each particle; and r_l and r_g are independent random numbers. Although Eqs. (9.34) and (9.35) are adopted in implementations, theoretical analyses are conducting using a second-order equivalent equation, as follows:

$$x(k+1)-(1+w-C_l.r_l[k]-C_g.r_g[k])x[k]+w.x[k-1]$$
$$= C_l.r_l[k].(p[k]+C_g.r_g[k].(g[k]-x[k]) \tag{9.36}$$

Rapaic et al. proposed a generalised PSO method. In this method, Eq. (9.36) can be taken as a difference equation describing the motion of a stochastic, second order, discrete-time, linear system with multiple inputs (Rapaic et al. 2008). The basic idea is to replace Eq. (9.36) with a general second-order model such that,

$$x(k+1)+a_1x[k]+a_0x[k-1]=b_l\,p[k]+b_g\,g[k] \tag{9.37}$$

where a_0, a_1, b_l, and b_g are random numbers with appropriate distributions. In order to make Eq. (9.35) a successful optimiser, the following restrictions are imposed:

1. Equation (9.37) should be stable, and the stability margin should grow during the optimisation procedure.
2. The response of the system to perturbed initial conditions should be oscillatory.
3. $p[k] \to g[k] \to g$ as k grows, where g is an arbitrary point of the search space.

All the requirements can be satisfied using the following equation,

$$x[k+1]-2\xi\rho x[k]+\rho^2 x[k-1]=(1-2\xi\rho+\rho^2)(c.p[k]+(1-c).g[k] \tag{9.38}$$

where the stability condition is $0<\rho<1$ and $|\xi|\leq 1$; ρ should be gradually decreased during the optimisation process. The algorithm is oscillatory for $1\leq\xi<1$, and the system becomes more oscillatory as ξ deceases. Parameter c takes values in the range [0, 1].

9.3.3 Application of Wrapper Model–Based Feature Subset Selection in Machine Fault Diagnosis

Many researchers have utilised sequential-selection algorithms to select features from the high-dimensional feature space of vibration data for roller bearing fault diagnosis. For instance, Zhang et al. (2011) proposed a hybrid model for machinery fault diagnosis. The proposed model combines multiple feature-selection models including eight filter models and two wrapper models to select the most significant features from all potentially relevant features. These are data variance, PCC, Relief, FS, class separability, Chi-2, IG, GR, a binary search (BS) model, and an SBS model, respectively. Two case studies are used to demonstrate the effectiveness of the proposed method in machinery fault diagnosis. The results showed that the proposed method is useful in revealing fault-related frequency features. Moreover, Rauber et al. (2015) proposed different feature models that are utilised in a single pool. These models rely on statistical time parameters, a complex envelope spectrum, and wavelet packet analysis. Furthermore, the most significant features are selected using PCA, SFS, SBS, SFFS, and SBFS. Vibration signals collected from roller bearings with several different health conditions are used to validate the proposed method. Furthermore, Islam and colleagues (Islam et al. 2016) proposed a hybrid fault-diagnosis model for bearings, which extracts features from the acoustic emission signal. In this study, several statistical features are extracted the time domain and frequency domain. Then, the authors investigated feature-selection approaches including SFS, SFFS, and GA for selecting optimal features that can be used for fault diagnosis. Acoustic emission (AE) signals collected from a normal healthy bearing (NB), a bearing with a crack in the inner raceway (BCI), a bearing with a crack in the outer raceway (BCO), and a bearing with a crack on the roller element (BCR) under various rotational speeds are used to validate the proposed method. The results showed that by embedding the feature selection in the fault-diagnosis process, diagnosis performance is improved.

Moving on now to consider the application of the ACO technique in machine fault diagnosis, ACO is used by many researchers to select SVM parameters to improve classification performance for roller bearing faults. For instance, Li et al. (2013) proposed a method for detecting roller bearing faults called IACO-SVM, based on improved ACO and SVM. In this method, both the optimal and the worst solutions obtained by the ants are allowed to update the pheromone trail density, and the mesh is applied in the ACO to adjust the range of the optimised parameters. Vibration signals acquired from roller bearings with four health conditions – normal condition, IRF, ORF, and REF – are used to validate the proposed method. First, a wide set of features are computed from the vibration signals by using statistical techniques. Then, these extracted features are used as input for SVM. The experimental results showed that IACO is feasible to optimise the parameters for SVM, and the improved algorithm is effective, with high classification accuracy and fast speed.

Moreover, Zhang et al. (2015) presented ACO for synchronous feature selection and parameter optimisation for SVM in fault diagnosis of rotating machinery. In this method, the feature-extraction step is performed by extracting statistical features from raw vibration signals and the corresponding FFT spectrums. In order to verify the effectiveness of the proposed method, it is compared with two methods based on SVM to diagnosis faults in locomotive roller bearings. These include SVM with the ACO algorithm for parameter optimisation and SVM with the ACO algorithm for feature selection. The results showed the effectiveness of the proposed method in diagnosing machine faults compared with other methods.

With regard to the application of GA in machine fault diagnosis, Jack and Nandi examined the use of GA to select the most significant input features for ANNs from a large set of possible features in machine condition monitoring contexts (Jack and Nandi 1999, 2000). In this study, a number of different statistical features are taken based on moments and cumulants of the vibration data. GA feature selection is controlled through the values contained with the genome generated by the GA. On being passed a genome with $N + 1$ values to be tested, the first N values are used to determine which rows are selected as a subset from the input feature set matrix. Rows corresponding to the numbers contained within the genome are copied into a new matrix containing N rows. The last value of the genome determines the number of neurons present in the hidden layer. The mutation operator used is real Gaussian mutation with a probability of mutation equal to 0.2, and the crossover operator used is the uniform crossover with a probability of 0.75. The fitness function used in the GA is simply the number of correct classifications made over the whole dataset. Real vibration data acquired from bearings with a total of six different conditions – two normal and four fault conditions – are used to validate the proposed method. The experimental results showed that GA is able to select a subset of 6 inputs from a set of 156 features that allowed the ANN to perform with 100% accuracy. Moreover, on a smaller feature set of 66 spectral inputs, the GA is able to select a subset of 8 inputs from the 66 spectral features that allowed the ANN to achieve 99.8% accuracy. In the same vein, Jack and Nandi (2002) examined the performance of SVM and ANN in two-class fault/no-fault recognition examples and attempted to improve the overall generalisation performance of both techniques through the use of a GA-based feature-selection process. In this study, vibration data taken from two machines are used. The first set of raw vibration data with six bearing health conditions is acquired from machine 1, where accelerometers are mounted vertically and horizontally on the bearing housing. The second set of raw vibration data with five health conditions is acquired from an accelerometer on one axis only. The results showed that using three feature sets (statistical feature set, spectral feature set, and combined feature set) extracted from the raw vibration data of machine 1, ANN generalises well and achieves a high success rate. Using spectral features extracted from the raw vibration data of machine 2, both classifiers achieved a 100% success rate.

Moreover, Samanta et al. (2001) presented a procedure for diagnosing gear conditions using GA and ANN. In this study, the time-domain vibration signals of a rotating machine with normal and faulty gears are processed for feature-extraction using the mean, root mean square, variance, skewness, and kurtosis. The selection of input features and the number of nodes in the hidden layer of ANN are optimised using a GA-based technique in combination with ANN. The output layer comprises two binary nodes indicating the condition of the machine, i.e. normal or faulty gears. The experimental results showed the effectiveness of the proposed method in machine fault detection. Also, Saxena and Saad (2007) presented the results of their investigation into the use of GA and ANN for condition monitoring of mechanical systems. In this study, first the raw vibration signals are normalised, and then five feature sets are considered: (i) statistical features from the raw vibration signal; (ii) statistical features from the sum signal; (iii) statistical features from the difference signals; (iv) spectral features; and (v) all the features together. Then, GA is used to select the best features of all the considered features. With these selected features, ANN is used to deal with the fault-classification problem. Vibration signals acquired from bearings are used to examine the effectiveness of the proposed method. The results showed that GA-evolved ANNs clearly outperform the standalone ANNs. Additionally, GP was demonstrated to be effective in selecting the best features in machine fault classification (Zhang et al. 2005; Zhang and Nandi 2007).

We turn now to the experimental evidence for the application of the PSO technique in machine fault diagnosis. Yuan and Chu (2007) proposed a method that jointly optimises feature selection and the SVM parameters with a modified discrete PSO for fault diagnosis. In this method, a correct ratio based on an evaluation method is used to estimate the performance of SVM and serves as the target function in the optimisation problem. A hybrid vector that describes both the features and the SVM parameters is taken as the constraint condition. Application of the proposed method in fault diagnosis of a turbopump rotor showed the effectiveness of the proposed method in fault diagnosis.

Furthermore, Kanović et al. (2011) presented a detailed theoretical and empirical analysis of PSO and generalised PSO with applications in fault diagnosis. Vibration signals from horizontal and vertical vibration sensors mounted on 10 induction motors were acquired and used to validate the proposed method. The results demonstrated the effectiveness of the proposed method in induction motor fault detection.

Additionally, Liu et al. (2013) presented a multifault classification model, called WSVM, based on the kernel method of SVM and wavelet frame. To find the optimal parameters for the proposed method, PSO is used to optimise unknown parameters of WSVM. In this study, EMD is first applied to decompose the vibration signals measured from rolling element bearings. Then, a distance-evaluation technique is applied to remove redundant and irrelevant information and select the salient features to be used for the classification of roller bearing faults. The results showed the effectiveness of the proposed method in roller bearing fault diagnosis. In the same vein, Van and Kang (2015) proposed a method called wavelet kernel function and local Fisher discriminant analysis (WKLFDA), based on a wavelet kernel function and linear local Fisher discriminant analysis (LFDA). PSO is used to seek the optimal parameters of the WKLFDA method. The experimental results for synthetic data and real vibration data measured from roller bearings showed the effectiveness of the proposed method in roller bearing defect classification.

Furthermore, Zhu et al. (2014) proposed a fault feature-extraction method based on hierarchical entropy (HE) and SVM with PSO. In this method, the SampEns of eight hierarchical decomposition nodes are computed to serve as fault feature vectors that consider both the low-frequency components and the high-frequency components of the roller bearing vibration signals. With the extracted HE feature vectors, multiclass SVM with PSO is used to deal with the classification problem. The experimental results verified that HE can depict the features of roller bearing vibration signals more accurately than MSE.

9.4   Embedded Model–Based Feature Selection

Embedded model–based feature-selection methods are built in the classification algorithm to accomplish the feature selection. LASSO, elastic net, the classification and regression tree (CART), C4.5, and SVM–recursive feature elimination (SVM-RFE) are amongst the commonly used embedded methods. Several researchers have utilised embedded model-based feature selection for bearing fault diagnosis. For example, Rajeswari and colleagues (Rajeswari et al. 2015) examined the performance of MSVMs for bearing fault classification using different feature-selection techniques. In the data preprocessing step, the wavelet transform is used to extract the features from the bearing vibration signal. To reduce the dimensionality of the feature SVM-RFE, the wrapper subset method, Relief-F method, and PCA feature-selection techniques are used. Then, two classification algorithms, MSVM and C4.5, are used to deal with the classification problem. Vibration signals collected from roller bearings with four health conditions are used to validate the proposed method. The results showed that the combination of the wrapper feature-selection method and an SVM classifier with 14 selected features gives the maximum accuracy of 96%.

Moreover, Seera and colleagues proposed a hybrid online learning model that combines the fuzzy min-max (FMM) neural network and CART for motor fault diagnosis (Seera et al. 2016). In this study, several time- and frequency-domain features under different conditions are first extracted from the vibration signals. Then, the extracted features are used as the input features for FMM-CART for detection and diagnosis of motor bearing faults. The results demonstrated the effectiveness of the proposed method in motor bearing fault detection and diagnosis. To validate the

Table 9.1 Summary of some of the introduced techniques and their publically accessible software.

Algorithm name	Platform	Package	Function
Sequential feature selection	MATLAB	Statistics and Machine Learning Toolbox	sequentialfs
Importance of attributes (predictors) using Relief-F algorithm			relieff
Wilcoxon rank sum test			ranksum
Rank key features by class separability criteria		Bioinformatics Toolbox	rankfeatures
Feature selection library		Giorgio (2018)	relieff laplacian fisher lasso
Find minimum of function using genetic algorithm		Global Optimisation Toolbox	ga

proposed method, benchmark data samples pertaining to different motor bearing fault conditions are used. Duque-Perez and colleagues used the LASSO technique to improve the performance of a LRC to diagnosis bearing health conditions. A case study of roller bearing data with five health operating conditions is used to validate the proposed method. The results showed the effectiveness of the proposed method in roller bearing fault diagnosis (Duque-Perez et al. 2017).

9.5 Summary

In vibration-based machine fault diagnosis, to make a sensible deduction using a machine learning–based classifier, we often compute certain features of the raw vibration signal that can describe the signal in essence. Sometimes, multiple features are computed to form a feature set. Depending on the number of features in the set, one may need to perform further filtering using a feature-selection technique. The aim of feature selection in vibration-based machine fault diagnosis is to select a subset of features that are able to discriminate between instances that belong to different classes. In other words, the main task of feature selection is to select a subset of features that are the most relevant to the classification problem. Based on the availability of class label information, feature-selection methods can be categorised into three main groups: supervised, semi-supervised, and unsupervised. Moreover, depending on their relationship with learning algorithms, feature-selection techniques can be further grouped into filter models, wrapper models, and embedded models.

This chapter has introduced generally applicable methods that can be used to select the most important features to effectively represent the original features. These include various algorithms of feature ranking, sequential selection algorithms, heuristic-based selection algorithms, and embedded model–based feature-selection algorithms. Most of the introduced techniques and their publicly accessible software are summarised in Table 9.1.

References

Ahmed, H. and Nandi, A.K. (2018). Compressive sampling and feature ranking framework for bearing fault classification with vibration signals. *IEEE Access* 6: 44731–44746.

Ahmed, H.O.A. and Nandi, A.K. (2017). Multiple measurement vector compressive sampling and fisher score feature selection for fault classification of roller bearings. In: *2017 22nd International Conference on Digital Signal Processing (DSP)*, 1–5. IEEE.

Ahmed, H.O.A., Wong, M.D., and Nandi, A.K. (2017). Classification of bearing faults combining compressive sampling, laplacian score, and support vector machine. In: *Industrial Electronics Society, IECON 2017-43rd Annual Conference of the IEEE*, 8053–8058. IEEE.

Ang, J.C., Mirzal, A., Haron, H., and Hamed, H.N.A. (2016). Supervised, unsupervised, and semi-supervised feature selection: a review on gene selection. *IEEE/ACM Transactions on Computational Biology and Bioinformatics* 13 (5): 971–989.

Blum, C. (2005). Beam-ACO—hybridizing ant colony optimization with beam search: an application to open shop scheduling. *Computers & Operations Research* 32 (6): 1565–1591.

Blum, C. and Dorigo, M. (2004). The hyper-cube framework for ant colony optimization. *IEEE Transactions on Systems, Man, and Cybernetics, Part B (Cybernetics)* 34 (2): 1161–1172.

Blum, C. and Roli, A. (2003). Metaheuristics in combinatorial optimization: overview and conceptual comparison. *ACM Computing Surveys (CSUR)* 35 (3): 268–308.

Bozorg-Haddad, O., Solgi, M., and LoÃ, H.A. (2017). *Meta-Heuristic and Evolutionary Algorithms for Engineering Optimization*, vol. 294. Wiley.

Bullnheimer, B., Hartl, R.F., and Strauss, C. (1997). A new rank based version of the Ant System. A computational study. Technical report, Institute of Management Science, University of Vienna.

Cang, S. and Yu, H. (2012). Mutual information based input feature selection for classification problems. *Decision Support Systems* 54 (1): 691–698.

Chandrashekar, G. and Sahin, F. (2014). A survey on feature selection methods. *Computers & Electrical Engineering* 40 (1): 16–28.

Dash, M. and Liu, H. (1997). Feature selection for classification. *Intelligent Data Analysis* 1 (3): 131–156.

Devijver, P.A. and Kittler, J. (1982). *Pattern Recognition: A Statistical Approach*. Prentice Hall.

Dorigo, M. and Gambardella, L.M. (1997). Ant colony system: a cooperative learning approach to the traveling salesman problem. *IEEE Transactions on Evolutionary Computation* 1 (1): 53–66.

Dorigo, M., Maniezzo, V., and Colorni, A. (1996). Ant system: optimization by a colony of cooperating agents. *IEEE Transactions on Systems, Man, and Cybernetics, Part B (Cybernetics)* 26 (1): 29–41.

Duda, R.O., Hart, P.E., and Stork, D.G. (2001). *Pattern Classification Second Edition*, vol. 58, 16. New York: Wiley.

Duque-Perez, O., Del Pozo-Gallego, C., Morinigo-Sotelo, D., and Godoy, W.F. (2017). Bearing fault diagnosis based on Lasso regularization method. In: *2017 IEEE 11th International Symposium on Diagnostics for Electrical Machines, Power Electronics and Drives (SDEMPED)*, 331–337. IEEE.

Giorgio. (2018). Feature selection library. Mathworks File Exchange Center. https://uk
.mathworks.com/matlabcentral/fileexchange/56937-feature-selection-library.

Gu, Q., Li, Z., and Han, J. (2012). Generalized Fisher score for feature selection. arXiv
preprint arXiv: 1202.3725.

Guntsch, M. and Middendorf, M. (2002). A population based approach for ACO. In:
Workshops on Applications of Evolutionary Computation, 72–81. Berlin, Heidelberg:
Springer.

Haroun, S., Seghir, A.N., and Touati, S. (2016). Feature selection for enhancement of
bearing fault detection and diagnosis based on self-organizing map. In: *International
Conference on Electrical Engineering and Control Applications*, 233–246. Cham:
Springer.

He, X., Cai, D., and Niyogi, P. (2006). Laplacian score for feature selection. In: *Advances in
Neural Information Processing Systems*, 507–514.

Holland, J.H. (1975). Adaptation in Natural and Artificial Systems: an Introductory
Analysis With Applications to Biology, Control, and Artificial Intelligence, 4th ed.
Boston, MA: MIT Press.

Holland, J.H. and Goldberg, D. (1989). *Genetic Algorithms in Search, Optimization and
Machine Learning*. Massachusetts: Addison-Wesley.

Huang, S.H. (2015). Supervised feature selection: a tutorial. *Artificial Intelligence Research* 4
(2): 22.

Islam, M.R., Islam, M.M., and Kim, J.M. (2016). Feature selection techniques for increasing
reliability of fault diagnosis of bearings. In: *Proceedings of the 9th International
Conference on Electrical and Computer Engineering (ICECE)*, 396–399. IEEE.

Jack, L.B. and Nandi, A.K. (1999). Feature selection for ANNs using genetic algorithms in
condition monitoring. In: *Proc. Eur. Symp. Artif. Neural Netw (ESANN)*, 313–318.

Jack, L.B. and Nandi, A.K. (2000). Genetic algorithms for feature selection in machine
condition monitoring with vibration signals. *IEE Proceedings-Vision, Image and Signal
Processing* 147 (3): 205–212.

Jack, L.B. and Nandi, A.K. (2002). Fault detection using support vector machines and
artificial neural networks, augmented by genetic algorithms. *Mechanical Systems and
Signal Processing* 16 (2–3): 373–390.

John, G.H., Kohavi, R., and Pfleger, K. (1994). Irrelevant features and the subset selection
problem. In: *Machine Learning Proceedings 1994*, 121–129. Elsevier.

Kalakech, M., Biela, P., Macaire, L., and Hamad, D. (2011). Constraint scores for
semi-supervised feature selection: a comparative study. *Pattern Recognition Letters* 32
(5): 656–665.

Kanović, Ž., Rapaić, M.R., and Jeličić, Z.D. (2011). Generalized particle swarm optimization
algorithm-theoretical and empirical analysis with application in fault detection. *Applied
Mathematics and Computation* 217 (24): 10175–10186.

Kennedy, J. and Eberhart, R. (1995). Particle swarm optimization. In: *Proceedings of the
IEEE International Conference on Neural Networks IV*, 1942–1948.

Kira, K. and Rendell, L.A. (1992). A practical approach to feature selection. In: *Machine
Learning Proceedings 1992*, 249–256. Elsevier.

Kohavi, R. and John, G.H. (1997). Wrappers for feature subset selection. *Artificial
Intelligence* 97 (1–2): 273–324.

Kononenko, I., Šimec, E., and Robnik-Šikonja, M. (1997). Overcoming the myopia of
inductive learning algorithms with RELIEFF. *Applied Intelligence* 7 (1): 39–55.

Kumar, C.A., Sooraj, M.P., and Ramakrishnan, S. (2017). A comparative performance evaluation of supervised feature selection algorithms on microarray datasets. *Procedia Computer Science* 115: 209–217.

Li, X., Zhang, X., Li, C., and Zhang, L. (2013). Rolling element bearing fault detection using support vector machine with improved ant colony optimization. *Measurement* 46 (8): 2726–2734.

Li, Y., Zhang, W., Xiong, Q. et al. (2017). A rolling bearing fault diagnosis strategy based on improved multiscale permutation entropy and least squares SVM. *Journal of Mechanical Science and Technology* 31 (6): 2711–2722.

Liu, H. and Motoda, H. (eds.) (2007). *Computational Methods of Feature Selection*. CRC Press.

Liu, H., Sun, J., Liu, L., and Zhang, H. (2009). Feature selection with dynamic mutual information. *Pattern Recognition* 42 (7): 1330–1339.

Liu, H. and Yu, L. (2005). Toward integrating feature selection algorithms for classification and clustering. *IEEE Transactions on Knowledge and Data Engineering* 17 (4): 491–502.

Liu, Z., Cao, H., Chen, X. et al. (2013). Multi-fault classification based on wavelet SVM with PSO algorithm to analyze vibration signals from rolling element bearings. *Neurocomputing* 99: 399–410.

Miao, J. and Niu, L. (2016). A survey on feature selection. *Procedia Computer Science* 91: 919–926.

Mitra, P., Murthy, C.A., and Pal, S.K. (2002). Unsupervised feature selection using feature similarity. *IEEE Transactions on Pattern Analysis and Machine Intelligence* 24 (3): 301–312.

Nandi, A.K., Liu, C., and Wong, M.D. (2013). Intelligent vibration signal processing for condition monitoring. In: *Proceedings of the International Conference Surveillance*, vol. 7, 1–15.

Nilsson, R., Peña, J.M., Björkegren, J., and Tegnér, J. (2007). Consistent feature selection for pattern recognition in polynomial time. *Journal of Machine Learning Research* 8 (Mar): 589–612.

Pudil, P., Novovičová, J., and Kittler, J. (1994). Floating search methods in feature selection. *Pattern Recognition Letters* 15 (11): 1119–1125.

Rajeswari, C., Sathiyabhama, B., Devendiran, S., and Manivannan, K. (2015). Bearing fault diagnosis using multiclass support vector machine with efficient feature selection methods. *International Journal of Mechanical and Mechatronics Engineering* 15 (1): 1–12.

Rapaic, M.R., Kanovic, Z., Jelicic, Z.D., and Petrovacki, D. (2008). Generalized PSO algorithm—an application to Lorenz system identification by means of neural-networks. In: *9th Symposium on Neural Network Applications in Electrical Engineering, 2008. NEUREL 2008*, 31–35. IEEE.

Rauber, T.W., de Assis Boldt, F., and Varejão, F.M. (2015). Heterogeneous feature models and feature selection applied to bearing fault diagnosis. *IEEE Transactions on Industrial Electronics* 62 (1): 637–646.

Robnik-Šikonja, M. and Kononenko, I. (2003). Theoretical and empirical analysis of ReliefF and RReliefF. *Machine Learning* 53 (1–2): 23–69.

Samanta, B., Al-Balushi, K.R., and Al-Araimi, S.A. (2001). Use of genetic algorithm and artificial neural network for gear condition diagnostics. In: *Proceedings of COMADEM*, 449–456. Elsevier Science Ltd.

Saucedo Dorantes, J.J., Delgado Prieto, M., Osornio Rios, R.A., and Romero Troncoso, R.D.J. (2016). Multifault diagnosis method applied to an electric machine based on high-dimensional feature reduction. *IEEE Transactions on Industry Applications* 53 (3): 3086–3097.

Saxena, A. and Saad, A. (2007). Evolving an artificial neural network classifier for condition monitoring of rotating mechanical systems. *Applied Soft Computing* 7 (1): 441–454.

Seera, M., Lim, C.P., and Loo, C.K. (2016). Motor fault detection and diagnosis using a hybrid FMM-CART model with online learning. *Journal of intelligent manufacturing* 27 (6): 1273–1285.

Sheikhpour, R., Sarram, M.A., Gharaghani, S., and Chahooki, M.A.Z. (2017). A survey on semi-supervised feature selection methods. *Pattern Recognition* 64: 141–158.

Stützle, T. and Hoos, H.H. (2000). MAX–MIN ant system. *Future Generation Computer Systems* 16 (8): 889–914.

Tang, J., Alelyani, S., and Liu, H. (2014). Feature selection for classification: a review. In: *Data Classification: Algorithms and Applications*, 37–46. CRC Press.

Tian, J., Pecht, M., and Li, C. (2012). *Diagnosis of Rolling Element Bearing Fault in Bearing-Gearbox Union System Using Wavelet Packet Correlation Analysis*, 24–26. Dayton, OH.

Vakharia, V., Gupta, V.K., and Kankar, P.K. (2016). A comparison of feature ranking techniques for fault diagnosis of ball bearing. *Soft Computing* 20 (4): 1601–1619.

Van, M. and Kang, H.J. (2015). Wavelet kernel local fisher discriminant analysis with particle swarm optimization algorithm for bearing defect classification. *IEEE Transactions on Instrumentation and Measurement* 64 (12): 3588–3600.

Weston, J., Mukherjee, S., Chapelle, O. et al. (2001). Feature selection for SVMs. In: *Advances in Neural Information Processing Systems*, 668–674. http://papers.nips.cc/paper/1850-feature-selection-for-svms.pdf.

Wilcoxon, F. (1945). Individual comparisons by ranking methods. *Biometrics Bulletin* 1 (6): 80–83.

Wong, M.D. and Nandi, A.K. (2004). Automatic digital modulation recognition using artificial neural network and genetic algorithm. *Signal Processing* 84 (2): 351–365.

Wu, S.D., Wu, C.W., Wu, T.Y., and Wang, C.C. (2013). Multi-scale analysis based ball bearing defect diagnostics using Mahalanobis distance and support vector machine. *Entropy* 15 (2): 416–433.

Yang, Y. and Pedersen, J.O. (1997). A comparative study on feature selection in text categorization. In: *Proc. 14th Int'l Conf. Machine Learning*, vol. 97, 412–420.

Yu, L. and Liu, H. (2003). Feature selection for high-dimensional data: a fast correlation-based filter solution. In: *Proceedings of the 20th International Conference on Machine Learning (ICML-03)*, 856–863.

Yu, L. and Liu, H. (2004). Efficient feature selection via analysis of relevance and redundancy. *Journal of Machine Learning Research* 5 (Oct): 1205–1224.

Yuan, S.F. and Chu, F.L. (2007). Fault diagnostics based on particle swarm optimisation and support vector machines. *Mechanical Systems and Signal Processing* 21 (4): 1787–1798.

Zhang, K., Li, Y., Scarf, P., and Ball, A. (2011). Feature selection for high-dimensional machinery fault diagnosis data using multiple models and radial basis function networks. *Neurocomputing* 74 (17): 2941–2952.

Zhang, L., Jack, L.B., and Nandi, A.K. (2005). Fault detection using genetic programming. *Mechanical Systems and Signal Processing* 19 (2): 271–289.

Zhang, L. and Nandi, A.K. (2007). Fault classification using genetic programming. *Mechanical Systems and Signal Processing* 21 (3): 1273–1284.

Zhang, X., Chen, W., Wang, B., and Chen, X. (2015). Intelligent fault diagnosis of rotating machinery using support vector machine with ant colony algorithm for synchronous feature selection and parameter optimization. *Neurocomputing* 167: 260–279.

Zheng, J., Cheng, J., Yang, Y., and Luo, S. (2014). A rolling bearing fault diagnosis method based on multi-scale fuzzy entropy and variable predictive model-based class discrimination. *Mechanism and Machine Theory* 78: 187–200.

Zhu, K., Song, X., and Xue, D. (2014). A roller bearing fault diagnosis method based on hierarchical entropy and support vector machine with particle swarm optimization algorithm. *Measurement* 47: 669–675.

Part IV

Classification Algorithms

10

Decision Trees and Random Forests

10.1 Introduction

In Part III of this book, we introduced feature learning and selection techniques for vibration data. In the fault-detection and -diagnosis problem framework using these vibration data, the next stage is classification. Classification is a typical supervised learning task: it categorises the acquired vibration signal correctly into the corresponding machine condition, which is generally a multiclass classification problem. As described in Chapter 6, supervised learning can be categorised into batch learning (i.e. offline learning) or online learning (i.e. incremental learning). In batch learning, the data points along with their corresponding labels are used together to learn and optimise the parameters of the model of interest. A simple example of classification is to assign a given vibration signal a 'normal' or 'fault' category. This can be seen in Figure 10.1, where the classifier is trained with several vibration signal examples, x_1, x_2, ..., x_L along with their predefined labels, i.e. classes, (normal condition [NO], fault condition [FA]). The trained classifier then learns to classify new vibration examples, $y_1, y_2, ..., y_k$ to their corresponding classes, i.e. NO or FA.

The literature on classification algorithms has highlighted various techniques to deal with classification problems, e.g. k-nearest neighbours (k-NN) (Duda and Hart 1973); hierarchical-based models, decision trees (DTs) (Quinlan 1986), and random forests (RFs) (Breiman 2001); probability-based models, including naïve Bayes classification (Rish 2001) and logistic regression classification (Hosmer et al. 2013); support vector machines (SVMs) (Cortes and Vapnik 1995); layered models, e.g. artificial neural networks (ANNs) (Jain et al. 1996, 2014) and deep neural networks (DNNs) (Schmidhuber 2015), which can be used for both batch learning and online learning.

Various classification techniques can be used to classify different vibration types based on the features provided. If the vibration signals' features are carefully devised and the parameters of the classifiers are carefully tuned, it is possible to achieve high classification accuracies. This part of the book introduces some widely used state-of-the-art classifiers – decision trees/forests, multinomial logistic regression, SVMs, and ANNs – that have already been used for classification of vibration signal status. In addition to these classifiers, this part also describes recent trends of deep learning in the field of machine condition monitoring and provides an explanation of commonly used techniques and examples of their application in machine fault diagnosis.

To begin with, this chapter introduces DT and RF classifiers by giving the basic theory of the DT diagnosis tool, its data structure, the ensemble model that combines DTs into

Condition Monitoring with Vibration Signals: Compressive Sampling and Learning Algorithms for Rotating Machines,
First Edition. Hosameldin Ahmed and Asoke K. Nandi.
© 2020 John Wiley & Sons Ltd. Published 2020 by John Wiley & Sons Ltd.

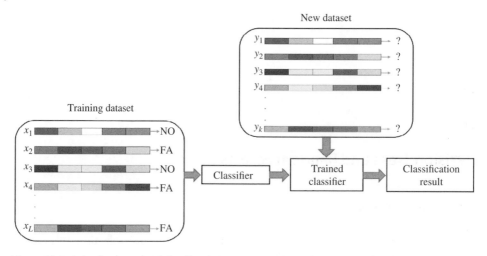

Figure 10.1 A simple example of classification to assign a given vibration signal to the 'normal' or 'fault' category. (See insert for a colour representation of this figure.)

a decision Forest model, and their applications in machine fault diagnosis. The other types of techniques, i.e. Multinomial logistic regression, SVM, ANN, and deep learning, will be covered in details in Chapters 11–14, respectively.

10.2 Decision Trees

The DT is one of the most popular tools in machine learning. There are two main types of DTs: classification trees used when the dependent variable is categorical, and regression trees used when the dependent variable is continuous (Loh 2011; Breiman 2017). Classification trees approximate the discrete classification function by using the tree-based representation. DTs are often constructed recursively using a top-down approach, by partitioning input training data into smaller subspaces until it reaches a subspace representing the most appropriate class label. In fact, most DT algorithms consist of two main phases: a building phase followed by a pruning phase. The stage of constructing a tree from training data is often called *tree induction, tree building,* or *tree growing*. Figure 10.2 shows a simple example of two classes, NO and FA, and two X variables, x_i^1 and x_i^2, where Figure 10.2a presents the data points and the partitions, and Figure 10.2b shows the corresponding DT. Using this classifier, we can easily predict the status of a signal of interest by passing it down the tree.

To deal with multiclass classification, we can use one tree and assign each leaf node a class where each leaf is selected by a class label; there also may be two or more leaves with the same class. For instance, Figure 10.3 shows a typical example of a six-class tree for rolling bearing health conditions. These include two normal conditions – brand new (NO), and worn but undamaged (NW) – and four fault conditions – inner race (IR), outer race (OR), rolling element (RE), and cage (CA).

DTs are well suited for those situations where the instances, i.e. the input, are represented by attribute pair values, i.e. data vectors. They can even tolerate some missing feature values, which happens when it comes to real-time data collection and feature

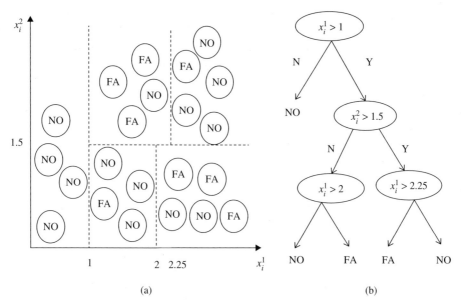

Figure 10.2 A simple example of two classes – normal condition (NO) and fault condition (FA) – and two *X* variables.

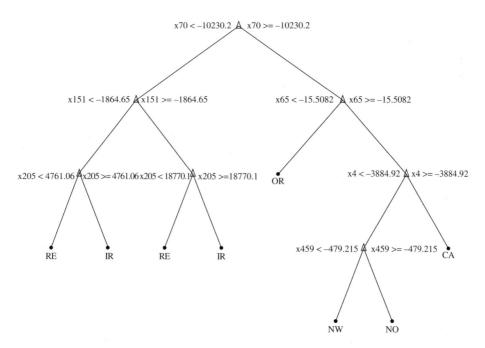

Figure 10.3 Atypical example of a six-class tree for rolling bearing health conditions.

extraction. The structure of a DT is composed of a root, several nodes that are also known as *indicators*, a number of branches, and a number of leaves that also known as *terminals* or *decision nodes*, which include class labels; branches represent the paths from the root to leaves, passing through nodes. The depth of a DT is the longest path from a root to a leaf (Rivest 1987). In the DT algorithm, the tree-based representation is learned from the input data by recognising the best splits and learning to classify the label, i.e. defining when a node is terminal. Each node of DT uses a splitting rule, e.g. GINI index of diversity, information gain, the Marshall correction, or a random selection of attributes, for splitting (Buntine and Niblett 1992; Jaworski et al. 2018), to split the instance space into two or more subspaces.

These splitting rules can be univariate (i.e. consider one attribute) or multivariate (consider multiple attributes) (Brodley and Utgoff 1995; Myles et al. 2004; Lee 2005). The main goal of these splitting rules is to find the best attribute that splits the training tuples into subsets. In DT classification, an attribute of a tuple is often categorical or numerical. However, there are various methods that have been proposed for constructing DTs from uncertain data (e.g. Tsang et al. 2011).

Moreover, DTs can be either binary, in which case the splitting rule splits the subspace into two parts, or multiway. In the binary DT induction method, a bi-partition at a node divides values into two subsets. A multiway split DT use as many partitions as possible (Biggs et al. 1991; Fulton et al. 1995; Kim and Loh 2001; Berzal et al. 2004; Oliveira et al. 2017).

In each DT, the instance is passed down from root to a leaf that includes a final decision, i.e. classification result. In each node in the tree, a test is carried out for a certain attribute of the instance, and each branch of this node corresponds to one possible value of the attribute. To describe the DT classification process in a more understandable way, it is a set of if-then rules. In fact, DTs are typically constructed recursively, following a greedy top-down construction approach. The top-down induction on decision tree (TDIDT) (Quinlan 1986) is a framework that refers to this type of algorithm. Here, the greedy approach defines the best available variable of the current split without considering future splits, which often represents the key idea of learning DTs.

The literature on DTs identified various methods for building DT models such as the classification and regression tree (CART) (Breiman et al. 1984), ID3 (Quinlan 1986), C4.5 (Quinlan 1993), and Chi-square automatic integration detector DT (CHAID) (Berry and Linoff 1997). To build a DT, it is essential to discover at each internal node a rule for splitting the data into subsets. In the case of univariate trees, one finds the attribute that is the most useful in discriminating the input data and finding a decision rule using the attribute. In the case of multivariate trees, one finds a combination of existing attributes that has the most useful discriminatory power (Murthy 1998). The subsequent subsections discuss the algorithms most commonly used by DTs to make splitting decisions. These are univariate and multivariate splitting techniques.

10.2.1 Univariate Splitting Criteria

Univariate splitting criteria consider one attribute to perform the splitting test in an internal node. Hence, the DT learning algorithm, i.e. the inducer, looks for the best available attribute that can be used for splitting. The ideal rule of splitting often results in class purity where all the cases in a terminal node come from the same class. For example, only one node in Figure 10.2 is pure with NO status for all the instances; three of the four terminal nodes are impure. In order to optimise the DT, one can search for the splitting rule S_r that minimises the impurity function F_{imp}. An *impurity function*, F_{imp}, is a function defined on the set of all k-tuples of numbers $(p(c_1), p(c_2), \ldots, p(c_k))$ satisfying $p(c_i) \geq 0 \, \forall \, i \in \{1, \ldots, k\}$ and $\sum_{i=1}^{k} p(c_i) = 1$ such that

(a) An impurity function (F_{imp}) achieves its maximum only at the point $(\frac{1}{k}, \frac{1}{k}, \ldots, \frac{1}{k})$.
(b) F_{imp} achieves its minimum at the points $(1, 0, \ldots, 0), (0, 1, \ldots, 0), \ldots, (0, 0, \ldots, 1)$, which represent the purity points.
(c) F_{imp} is a symmetric function of $(p(c_1), p(c_2), \ldots, p(c_k))$.

The impurity measure of any node N_1 can be defined using the following equation:

$$i(N_1) = F_{imp}(p(c_1 \mid N_1), p(c_2 \mid N_1), \ldots, p(c_k \mid N_1)) \tag{10.1}$$

If a splitting rule S_r in node N_1 divides all the instances into two subsets N_{1L} and N_{1R}, the decrease in impurity can be defined using the following equation,

$$\Delta i(S_r, N_1) = i(N_1) - P_L i(N_{1L}) - P_R i(N_{1R}) \tag{10.2}$$

where P_L and P_R are proportions of N_{1L} and N_{1R}, respectively. If a test in a node N_j is based on an attribute having n values, Eq. (10.2) can be generalised as Eq. (10.3) (Raileanu and Stoffel 2004):

$$\Delta i(S_r, N_1) = i(N_i) - \sum_{j=1}^{n} P(N_j) i(N_j) \tag{10.3}$$

The literature on DTs has highlighted several impurity-based measures that can be used for splitting. The subsequent subsections discuss the most commonly used measures in more detail.

10.2.1.1 Gini Index

The Gini index is an impurity-based measure for categorical data. It was originally a measure of the probability of misclassification that was developed by the Italian statistician Corrado Gini in 1912. It was proposed for DTs by Breiman et al. (1984) as a measure of node impurity. The Gini index measures the deviations between the probability distributions of the target attribute's values. Given a node N, an impurity function based on Gini index measure assigns a training instance to a class label c_i with probability $p(c_i \mid N)$. Hence, the probability that the instance is class j can be estimated by $p(c_j \mid N)$. Based on this rule, the estimated probability of misclassification is the Gini index, which can be expressed mathematically as in Eq. (10.4):

$$Gini(c, N) = \sum_{i=1}^{k} \sum_{j=1, j \neq i}^{k} p(c_i \mid N) p(c_j \mid N) = \sum_{i=1}^{k} p(c_i \mid N)(1 - p(c_i \mid N))$$

$$= 1 - \sum_{i=1}^{k} (p(c_i \mid N))^2 \tag{10.4}$$

When a node N is pure, the Gini index value is zero. For example, using the Gini index from Eq. (10.4), the impurity of the pure node in Figure 10.2 described with NO status for all the instances is $(1 - (4/4)^2 = 0)$. For the root node that satisfies $x_i^1 > 2.25$, which has four instances with NO status and one instance with FA status (see Figure 10.2a), the Gini index is $(1 - (4/5)^2 - (1/5)^2 = 0.32)$. If a splitting rule S_r in node N divides all the instances into two subsets N_L and N_R, then the Gini index can be computed using Eq. (10.5),

$$Gini_A(S_r, N) = \frac{|N_1|}{|N|} Gini(N_1) + \frac{|N_2|}{|N|} Gini(N_2) \tag{10.5}$$

The reduction in impurity can be expressed mathematically using Eq. (10.6):

$$\Delta Gini(A) = Gini(S_r, N) - (Gini_A(S_r, N)) \tag{10.6}$$

An alternative to the Gini index is the twoing criterion, which is a binary criterion that can be defined using the following equation:

$$Twoing(N) = 0.25 \left(\frac{|N_0|}{|N|} \right) \left(\frac{|N_1|}{|N|} \right) \left(\sum_{c_i \in C} p_{N_0, c_i} - p_{N_1, c_i} \right)^2 \tag{10.7}$$

10.2.1.2 Information Gain

Information gain (IG) is a measure of how much information a feature provides about the class. It is an impurity-based measure that utilises the entropy as a measure of impurity. The information entropy can be represented mathematically using Eq. (10.8),

$$Info(N) = - \sum_{i=1}^{C} p(N, c_i) log_2(p(N, c_i)) \tag{10.8}$$

where C is the number of classes and $p(N, c_i)$ is the proportion of cases in N that belong to c_i. The corresponding information gained with k outcomes can be defined as the difference between the entropy before the split and the entropy after the split. This can be represented mathematically using Eq. (10.9) (Quinlan 1996a,b):

$$IG(N) = Info(N) - \sum_{i=1}^{k} \frac{|N_i|}{|N|} Info(N_i) \tag{10.9}$$

Even though IG is a good impurity-based measure, it biases toward attributes with many values. To overcome this problem, Quinlan proposed a new measure called the *information gain ratio* that penalises IG for selecting attributes that generate many small subsets (Quinlan 1986). The gain ratio can be expressed mathematically using Eq. (10.10):

$$GR(N) = \frac{IG(N)}{SplitINFO} \tag{10.10}$$

Here, *SplitINFO* represents the split information value that denotes the potential information generated by splitting the training data N into k subsets such that

$$SplitINFO(N, A) = \sum_{i=1}^{k} \frac{|N_i|}{|N|} log \frac{|N_i|}{|N|} \tag{10.11}$$

where A is candidate attribute, i is a possible value of A, N is a set of examples, and N_i is subset when $X_A = i$.

Another technique for measuring the statistical importance of the IG is the likelihood-ratio, which can be defined as (Rokach and Maimon 2005b)

$$G^2(a_i, N) = 2.\ln(2).|N|.IG(a_i, N) \tag{10.12}$$

10.2.1.3 Distance Measure

As an alternative selection criterion, De Mántaras proposed a feature-selection measure for DT induction based on the distance between partitions. Based on this measure, the selected feature in a node induces the partition that is closest (in terms of the distance) to the correct partition of the subset of examples in this node (De Mántaras 1991). The distance measure between partitions generated by feature A and class C can be expressed using Eq. (10.13),

$$(A, C) = 1 - \frac{IG(A, C)}{Info(A, c)} \tag{10.13}$$

where $Info(A, c)$ is a joint entropy of A and C that can be defined using Eq. (10.14):

$$Info(A, C) = \sum_a \sum_c p_{ac} log_2 p_{ac} \tag{10.14}$$

10.2.1.4 Orthogonal Criterion (ORT)

The orthogonal criterion (ORT) is a binary criterion that can be defined using the following equation (Fayyad and Irani 1992):

$$ORT(N) = 1 - COS\theta(p_{c,N_1}, p_{c,N_2}) \tag{10.15}$$

Here, $\theta(p_{c,N_1}, p_{c,N_2})$ is the angle between the probability distribution of the target attribute in partitions N_1 and N_2.

There are also many other univariate splitting measures that have been proposed for DTs. Because of the limitations on space, we cannot detail them here. Readers who are interested are referred to (Rounds 1980; Li and Dubes 1986; Taylor and Silverman 1993; Dietterich et al. 1996; Friedman 1977; Ferri et al. 2002). Moreover, comparative studies of splitting criteria have been conducted (e.g. Safavian and Landgrebe 1991; Buntine and Niblett 1992; Breiman 1996b; Shih 1999; Drummond and Holte 2000; Rokach and Maimon 2005a,b).

10.2.2 Multivariate Splitting Criteria

Unlike univariate splits that consider one attribute to perform the splitting test in an internal node, multivariate splits consider a number of features to be used in a single node split test. Most of the multivariate splits are based on the linear combination of the input features. In fact, finding the best multivariate split is more difficult than finding the best univariate split. In the literature of multivariate splits, various methods have been used for finding the best linear combination. These include greedy search, linear programming, linear discriminant analysis, perceptron training, hill climbing search, etc. Table 10.1 shows a summary of most commonly used methods for multivariate splits in different studies of DTs.

In each case, i.e. univariate or multivariate splitting, the splitting continues until a stopping criterion is satisfied. A simple stopping criterion is the condition of node purity where all the instances in the node belong to only one class. In this case, there

Table 10.1 Summary of the most commonly used methods for multivariate splits in different studies of decision trees.

Method	Studies
Greedy learning	Breiman et al. 1984; Murthy 1998.
Non-greedy learning	Norouzi et al. 2015.
Linear programming	Lin and Vitter 1992; Bennett 1992; Bennett and Mangasarian 1992, 1994; Brown and Pittard 1993; Michel et al. 1998.
Linear discriminant analysis	You and Fu 1976; Friedman 1977; Qing-Yun and Fu 1983; Loh and Vanichsetakul 1988; Likura and Yasuoka 1991; Todeschini and Marengo 1992; John 1996; Li et al. 2003.
Perceptron learning	Utgoff 1989; Heath et al. 1993; Sethi and Yoo 1994; Shah and Sastry 1999; Bennett et al. 2000; Bifet et al. 2010.
Hill-climbing search	Murphy and Pazzani 1991; Brodley and Utgoff 1992; Murthy et al. 1994.

is no need for further splitting. Other stopping rules include: (i) when the number of cases in a node is less than a predefined N_{stop}, the node is declared terminal (Bobrowski and Kretowski 2000); (ii) when the tree depth, i.e. the longest path from a root to a leaf is reached; (iii) when the optimal splitting criterion is not larger than a predefined threshold; and (iv) when the number of nodes, i.e. the predefined maximum number of nodes, has been reached.

10.2.3 Tree-Pruning Methods

As mentioned earlier, the most common DT induction contains two main stages: (i) building a complete tree able to classify all training instances, and (ii) pruning the built tree back. Pruning methods originally proposed in (Breiman et al., 1984) were widely used to avoid overfitting, i.e. obtain the right-sized tree, which is a pruned version of the original built tree, also called a *subtree* (Mingers 1989; Murthy 1998). In reality, using the splitting measures presented, the decision-growing process will continue splitting until a stopping criterion is reached. Nevertheless, the procedure for DT building may lead to a large tree size and/or overfitting. Pruning methods can reduce overfitting and tree size. In fact, pruning methods are basically tree growing in reverse and are also described as *bottom-up pruning* (Kearns and Mansour 1998). Using a tight stopping measure often results in creating small, underfitted DTs, while using a loose stopping measure often result in generating large, overfitted DTs. To solve this problem, the overfitted tree is reduced into a smaller tree (subtree) by removing sub-branches that are not contributing to generalization accuracy (Rokach and Maimon 2005a).

The literature for DTs has defined various methods for pruning DTs. The subsequent subsections detail the most commonly used techniques.

10.2.3.1 Error-Complexity Pruning

Error-complexity pruning, also known as *cost-complexity pruning* (Breiman et al., 1984), is a two-stage error-minimisation pruning method. In the first stage, a series of trees $T_1, T_2, ..., T_m$ is generated on training examples, where T_1 is the original tree before pruning. Then, T_{i+1} is generated by replacing one or more of the subtrees in the previous T_i where $T_{i+1} < T_i$. For example, we can generate $T_2 < T_1$ by pruning away the branch ζ_{T_1} of T_1 such that,

$$T_2 = T_1 - \zeta_{T_1} \qquad (10.16)$$

In general, this can be represented using the following equation:

$$T_{i+1} = T_i - \zeta_{T_i} \qquad (10.17)$$

Continuing this way, we generate a decreasing series of subtrees such that $T_1 > T_2 > \ldots > T_m$. For each internal, the pruned branches are those that achieve the lowest increase in apparent error rate per pruned leaf. The cost-complexity criterion for a subtree $T \subset T_0$ can be computed using the following equation,

$$C_a(T) = C(T) + a \mid \widetilde{\zeta}(T) \mid \qquad (10.18)$$

where $C(T)$ is the misclassification of a DT T, $\widetilde{\zeta}(T)$ is the number of terminal nodes, T_0 is the original tree before pruning, and a is a complexity parameter. The error-complexity, cost-complexity function C_a of pruned subtree $T - T_\zeta$ to that of the branch at node ζ can be represented using Eq. (10.19):

$$g(\zeta) = \frac{C(T) - C(T_\zeta)}{|\widetilde{\zeta}(T)| - 1} \qquad (10.19)$$

In the second stage, for each internal node, the error complexity is computed and the one with the smallest value is converted to a leaf node. The best-pruned tree is then selected.

10.2.3.2 Minimum-Error Pruning

The minimum-error pruning method (Cestnik and Bratko 1991) is a bottom-up technique to select a single tree with the minimum error on the training dataset. With m probability estimates, the expected error in a given node rate can be represented using the following equation,

$$E_\zeta = \frac{N - n_c + (1 - p_{ac})m}{N + m} \qquad (10.20)$$

where N is the total number of instances in the node, n_c is the number of instances in class c that minimise the expected error, p_{ac} is the a prior probability of class c, and m is the parameter of the estimation method.

10.2.3.3 Reduced-Error Pruning

The reduced-error pruning is a simple direct technique for tree pruning, proposed by Quinlan (1987). In this technique, rather than generating a series of trees and then selecting one of them, a more direct process is used. The simple idea is to assess each non-terminal node with respect to the classification error in a separate test set, also called the *pruning set*. The assessment examines the change in the classification error over the pruning set that may happen if the non-terminal node is replaced by the best possible leaf. If the new tree would result in an equal or lower number of errors, then the subtree of this non-terminal node is replaced by the leaf.

10.2.3.4 Critical-Value Pruning

Critical-value pruning (Mingers 1987) sets a threshold, also called the *critical value*, to examine the importance of a non-terminal node in a DT. If the node does not reach the threshold, it will be pruned; and if the node satisfies this condition, it will then be kept. However, if a node meets the pruning condition but the latter nodes, i.e. its children, do not satisfy the pruning condition, this subtree should be kept. Based on this technique, the larger the threshold value selected, the smaller the resulting tree that is formed.

10.2.3.5 Pessimistic Pruning

Pessimistic pruning (Quinlan 1993) uses the pessimistic statistical correlation test instead of the pruning set or cross-validation. It uses a measure called *continuity correction* for a binomial distribution to find a realistic estimate of the misclassification rate $r(\zeta)$ such that,

$$r(\zeta) = \frac{e(\zeta)}{N(\zeta)} \tag{10.21}$$

where $e(\zeta)$ is the number of misclassified instances at node ζ and $N(\zeta)$ is the number of training set instances at node ζ. The misclassification rate using the continuity correction can be represented using the following equation:

$$\widetilde{r}(\zeta) = \frac{e(\zeta) + 1/2}{N(\zeta)} \tag{10.22}$$

The misclassification rate for a subtree T_ζ can be represented using Eq. (10.23):

$$\widetilde{r}(T_\zeta) = \frac{\sum e(i) + \widetilde{\zeta}(T)/2}{\sum N(i)} \tag{10.23}$$

From Eqs. (10.22) and (10.23), $N(\zeta) = \sum N(i)$, as they define the same set of instances. Hence, the misclassification rates in Eqs. (10.22) and (10.23)v can be represented as the number of misclassifications for a node and for a subtree using Eqs. (10.24) and (10.25), respectively:

$$\widetilde{n}(\zeta) = e(\zeta) + 1/2 \tag{10.24}$$

$$\widetilde{n}(T_\zeta) = \frac{\sum e(i) + \widetilde{\zeta}(T)/2}{\sum N(i)} \tag{10.25}$$

Based on this technique, a subtree is kept if its corrected number of misclassifications is less than that of a node by at least one standard error; otherwise it is pruned. The standard error for the number of misclassifications can be computed using the following equation:

$$STE(\widetilde{n}(T_\zeta)) = \sqrt{\frac{\widetilde{n}(T_\zeta)(N(\zeta)) - \widetilde{n}(T_\zeta)}{N(\zeta)}} \tag{10.26}$$

10.2.3.6 Minimum Description Length (MDL) Pruning

The minimum description length (MDL)-based DT pruning method (Mehta et al. 1995) uses the MDL criterion to find a model within a class that allows the shortest encoding of the class series in the training instances. Given a DT node ζ containing N training examples, belonging to class labels $c_1, c_2, \ldots, c_k$, the encoding length of the classification of all instances of ζ can be represented mathematically using the following equation:

$$L_c(\zeta) = \log\left(\frac{N}{N_{c_1}, N_{c_2}, \ldots, N_{c_k}}\right) + \log\left(\frac{N+k-1}{k-1}\right) \tag{10.27}$$

The MDL-based method chooses DTs that can be encoded with fewer bits, i.e. the shortest encoding of the class labels.

There are also many other pruning methods reported in the literature. Readers who are interested are referred to (Wallace and Patrick 1993; Bohanec and Bratko 1994; Almuallim 1996; Fournier and Crémilleux 2002). Moreover, several studies have been conducted to compare the performance of various pruning methods (Quinlan 1987; Mingers 1989; Esposito et al. 1997; Knoll et al. 1994; Patil et al. 2010).

10.2.4 Decision Tree Inducers

DT inducers are methods that can be used to build a DT from a given training dataset. The literature on DTs has highlighted various methods for DT induction. We briefly describe some of the commonly used DT induction methods, including CART (Breiman et al. 1984), ID3 (Quinlan 1986), C4.5 (Quinlan 1993), and CHAID (Berry and Linoff 1997). The procedures for these methods include the application of the splitting criterion and pruning techniques that have been described.

10.2.4.1 CART
CART (Breiman et al. 1984) is one of the most commonly used methods for DT induction. It creates binary trees from training data: i.e. the splitting rule splits each internal node ζ into two parts ζ_L and ζ_R that can be used for both classification and regression problems. Both the Gini index and twoing criteria have been implemented in CART. To handle missing measurement values, CART uses a technique of surrogate splits, i.e. if a case has x_i missing in its measurement, one decides whether it goes to ζ_L and ζ_R by using the best surrogate split. The built tree is then pruned using cost-complexity pruning.

10.2.4.2 ID3
The iterative Dichotomiser 3 (ID3) (Quinlan 1986) is a simple DT induction method that uses IG as splitting criteria, which uses the greatest attribute of IG as a root node of a DT, but does not apply any pruning processes and does not handle missing values. In this method, the DT stops growing when all the instances in the node belong to only one class or when the IG is not bigger than zero. The main disadvantage of the ID3 method is that it requires the data description to include only discrete features.

10.2.4.3 C4.5
C4.5 is a DT induction method proposed by Quinlan in 1993. It is an extension of the DT inducer ID3 (Quinlan 1986) that deals with both continuous and discrete features. Also, unlike ID3, C4.5 handle missing values and apply the pruning process after DT construction. It uses the gain ratio as a splitting criterion and performs error-based pruning after the DT growing phase. In C4.5, the DT stops growing when the number of examples to be split is less than a predefined threshold (Quinlan 1993).

The commercial version of C4.5 is C5.0, which is a developed version that is more efficient than C4.5. Also, J48 is an open source Java implementation of the C4.5 method in the Weka data-mining tool (Xiaoliang et al. 2009; Moore et al. 2009). A good comparative study of ID3, CART, and C4.5 can be found in (Singh and Gupta 2014).

10.2.4.4 CHAID

CHAID, which is a modification of the automatic interaction detector (AID), is one of the earliest tree classification methods (Kass 1980). The basic idea of this method is to find the pair of values with the least significant difference for each input feature x_i with respect to the target attribute using the p value of the statistical test. Pearson's Chi-squared test is used to measure the significant difference. For each selected pair, the obtained p value is compared to a certain merge threshold, i.e. a predefined level of significance. If the p value is larger than the merge threshold, CHAID merges the values and searches for a further possible pair to be merged. This procedure is repeated until the pairs for which the p value is smaller than the defined level of significance are not identified. CHAID does not apply any pruning processes, but it handles missing values by considering them a single category.

Several advantages and disadvantages of DTs have been reported in the literature. The following is a summary of these advantages and disadvantages:

(a) Advantages of DTs:
   1. Easy to build, easy to use, and easy to interpret.
   2. Have the ability to handle both numerical and categorical features, e.g. CART.
   3. Have the ability to handle missing measurement values.
   4. Have the ability to handle outliers.
(b) Disadvantages of DTs:
   1. Some of the DT algorithms, e.g. ID3 and C4.5, deal only with target attributes that have only discrete values.
   2. Cause overfitting.
   3. The prediction model becomes unstable with data that has a small variance.

There are also many other DT inducers reported in the literature, including the fast algorithm for classification trees (Loh and Vanichsetakul., 1988); multivariate adaptive regression splines (Friedman 1991); CAL5 (Muller and Wysotzki 1994); quick, unbiased, efficient, statistical tree (QUEST) (Loh and Shih 1997); PUBLIC (Rastogi and Shim 1998); classification rule with unbiased interaction selection and estimation (CRUISE) (Kim and Loh 2001); and conditional inference trees (CTree) (Hothorn et al. 2015).

10.3 Decision Forests

As described, DTs are easy to build, easy to use, and easy to interpret. However, in practice, they have two main issues: (i) using one tree easily causes overfitting, and (ii) they are unstable. Significant improvements in classification accuracy have resulted from building an ensemble of trees and allowing them to vote for the most popular class (Breiman 2001). A number of independent methods have been proposed to build ensembles of trees. In these methods, (i) the original training dataset is randomly divided into several subsets that can be used to induce multiple classifiers. Then, (ii) a combination process is used to produce a single classification for a given example. Various techniques

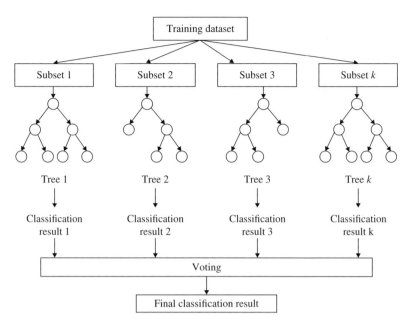

Figure 10.4 Atypical example of a random forest classifier.

have been used for building ensembles, e.g. bagging (Breiman 1996a); wagging, which is a variant of bagging (Bauer and Kohavi 1999); and boosting-based methods, including AdaBoost and Arc-x4 (Quinlan 1996a,b; Freund and Schapire 1997). An empirical comparison of these methods, i.e. bagging, boosting, and their variants, has been conducted by Freund and Schapire (Freund and Schapire 1997). The basic idea of all these methods is that for the kth tree, a random vector v_k is produced, independent of the past random vectors $v_1, v_2, v_3, \ldots, v_{k-1}$ but with the same distribution. With these generated random vectors, the generated k trees are used to vote for the most popular class label (Breiman 2001). These processes are called random forests (RFs). In other words, an RF (see Figure 10.4) is an ensemble learning method that constructs multiple DTs during the training process; the final output is the mode of the classes output by each individual DT (Nandi et al. 2013).

Breiman showed that RFs are an effective tool in prediction. Also, he explained why RFs do not overfit as more trees are added, but generate a limiting value of the generalization error (Breiman 2001).

Due to the compressibility and ease of interpreting their results, DTs are being considered in a large diversity of applications, e.g. automatic modulation recognition (Nandi and Azzouz 1995, 1998; Azzouz and Nandi 1996); bioinformatics (Che et al. 2011); image processing (Lu and Yang 2009); and medicine (Podgorelec et al. 2002). The following subsection presents their application in machine fault diagnosis.

10.4 Application of Decision Trees/Forests in Machine Fault Diagnosis

DTs and RFs are popular techniques that have been applied to machine fault diagnosis. Many studies related to machine fault diagnosis have been published. In this section, a brief discussion of the application of DTs and RFs in machine fault diagnosis is presented.

The early application of DTs for fault diagnosis was proposed by Patel and Kamrani in 1996 (Patel and Kamrani 1996). The study presented an intelligent decision support system for diagnosis and maintenance of automated systems called ROBODOC, which is a DT-based system that uses the shallow knowledge of a maintenance expert in the form of impulse factor (IF) … THEN rules. Yang et al. (2005) proposed an expert system called VIBEX for vibration fault diagnosis of rotating machinery using DT and decision tables. In this method, a decision table based on the cause-symptom matrix is used as the probabilistic method for diagnosing abnormal vibrations; and a DT is introduced to build a knowledge base, which is essential for vibration expert systems. VIBEX embeds a cause-result matrix containing 1800 confidence factors, which makes it suitable to monitor and diagnosis rotating machinery.

Tran et al. (2006) proposed the application of the DT method to classify the faults of induction motors. In this method, feature extraction is applied beforehand to extract useful information from the raw data using statistical feature parameters from the time domain and frequency domain. Then, a DT method is applied as a classification model for fault diagnosis of induction motors with vibration signals and current signals. The experimental results showed that the DT achieved high fault-classification accuracies for induction motors. Sugumaran and Ramachandran proposed automatic rule learning using ID3 DTs for a fuzzy classifier in fault diagnosis of roller bearings (Sugumaran and Ramachandran 2007). In this method, a DT is used to generate rules automatically from the feature set. Statistical features are extracted from the collected vibration data, and good features that discriminate the different fault conditions of the bearing are selected using a DT. The rule set for the fuzzy classifier is obtained once again using the DT. The results are found to be encouraging. However, this method requires a large number of data points in the dataset to achieve good results.

Sun et al. (2007) proposed a method based on C4.5 DTs and principal component analysis (PCA) for rotating machinery fault diagnosis. In this method, PCA is used to extract the features of the acquired data. Then C4.5 is trained by using the samples to generate a DT with diagnosis knowledge. Finally, the tree model is used to make a diagnosis. Six kinds of running conditions including normal, unbalance, rotor radial rub, oil whirl, shaft crack, and a simultaneous state of unbalance and radial rub, are simulated to validate the proposed method. The result showed the efficiency of the proposed method for rotating machinery fault diagnosis.

Yang et al. (Yang et al. 2008) investigated the possibilities of applying the RF algorithm in machine fault diagnosis and proposed a hybrid method combined with the genetic algorithm (GA) to improve classification accuracy. The proposed method is demonstrated by a case study of induction motor fault diagnosis. (i) The experimental results showed that the normal RF achieved satisfactory fault diagnosis accuracy. (ii) By applying GA to do the parameter optimisation of the number of trees and random split number of the RF, the experimental results showed that classification accuracy achieved 98.89%, which is 3.33% higher than the best classification accuracy achieved by the normal RF.

Yang and colleagues (Yang et al. 2009) proposed a fault-diagnosis method based on an adaptive neuro-fussy inference system (ANFIS) in combination with DTs. In this method, CART, which is one of the DT methods, is used as a feature-selection procedure to select pertinent features from the dataset.

Saravanan and Ramachandran (2009) proposed a method for fault diagnosis of a spur bevel gearbox using a discrete wavelet and DT. In this method, the discrete wavelet is used for feature extraction and a J48 DT is used for feature selection as well as for classification. The analysis of vibration signals produced by a bevel gearbox in various conditions and faults is used to validate this method. (i) Statistical features are extracted for all the wavelet coefficients of the vibration signals using Daubechies wavelet db1–db15. (ii) J48 is used for feature selection and classification of various conditions of the gearbox. The experimental results showed that the maximum average classification accuracy achieved was 98.57%.

Sakthivel et al. (2010a,b) proposed the use of a C4.5 DT algorithm for fault diagnosis of monoblock centrifugal pumps through statistical features extracted from vibration signals under good and faulty conditions. The experimental results showed that C4.5 and the vibration signals are good candidates for practical applications of fault diagnosis of monoblock centrifugal pumps. Sakthivel et al. (2010b) also presented the utilization of DTs and rough sets to generate the rules from statistical features extracted from vibration signals under normal and faulty conditions of a monoblock centrifugal pump. In this study, a fuzzy classifier is built using a DT and rough set rules and tested using test data. The experimental results showed that overall classification accuracy obtained using the DT-fuzzy hybrid system is better than the overall classification accuracy obtained using the rough set-fuzzy hybrid system.

Saimurugan et al. (2011) presented the use of c-support vector classification (c-SVC) and nu-SVC models of SVM with four kernel functions for classification of faults using statistical features extracted from vibration signals under good and faulty conditions of the rotational mechanical systems. The DT algorithm (C.4.5 algorithm) was used to select the prominent features. These features were given as input for training and testing the c-SVC and nu-SVC models of SVM, and their fault classification accuracies were compared. Seera et al. (2012) proposed a method called FMM-CART to detect and classify comprehensive fault conditions of induction motors using a hybrid fuzzy min-max (FMM) neural network and CART. In this method, the motor current harmonics are employed to form the input features to FMM-CART for detection and diagnosis of individual and multiple motor fault conditions. The process of FMM-CART starts with FMM, which is formed using hyper-box fuzzy sets; then the hyper boxes generated from FMM training serve as input to the CART DT. The final tree is then used for fault classification. The experimental results showed that FMM-CART is able to achieve 98.25% accuracy for five motor conditions in noise-free conditions. Moreover, Seera et al. (2017) proposed a hybrid model, FMM-RF, for classification of ball bearing faults with vibration signals. This method is based on the FMM neural network and the RF model. In this method, power spectrum and sample entropy features are used during feature extraction, where important features are extracted. With these features, the resulting hyper boxes from FMM are used as input to CART, where the centroid point and the confidence factor are used for tree building. Then, bagging is conducted for RF, and the majority voting scheme is used to combine the predictions from an ensemble of trees. Experiments of benchmark and real-world dataset showed accurate performance using the FMM-RF model. The best classification accuracies obtained from benchmark and real-world datasets were 99.9% and 99.8%, respectively.

Karabadji et al. (2012) developed a GA-based technique for DT selection in industrial machine fault diagnosis. In this technique, for DT selection, GAs are used to validate DT performance on training and testing sets to achieve the most robustness. Several trees are generated, and then the GA is used to find the best representative tree. The experimental results of vibration signals from an industrial fan using RF for fault diagnosis demonstrated robustness and good performance. An enhanced version has the DT evolve using a genetic programming paradigm, which has been used for fault classification (Guo et al. 2005). Cerrada et al. (2016) proposed a two-step fault-diagnosis method for spur gears. In this method, the diagnostic system is performed using a GA and a classifier based on the RF.

Muralidharan and Sugumaran (2013) proposed a fault-diagnosis algorithm for monoblock centrifugal pumps using wavelets and DT. In this method, the continuous wavelet transform (CWT) is calculated from the acquired vibration signals using different families and different levels, which form the feature set. These features are then fed as input to the classifier J48, and the classification accuracies are calculated. The experimental results showed the efficiency of the proposed method in monoblock fault diagnosis. Jegadeeshwaran and Sugumaran proposed a vibration-based monitoring technique for automobile hydraulic brake systems. In this technique, the C4.5 DT algorithm is used to extract statistical features from vibration signals; feature selection is also carried out. The selected features are classified using the C4.5 DT algorithm and best-first DT algorithm, with pre-pruning and post-pruning techniques. The experimental results showed that the best-first DT classifier with pre-pruning is more accurate compared to the best-first DT classifier with post-pruning and C4.5 DT classifier (Jegadeeshwaran and Sugumaran 2013). Saimurugan et al. (2015) conducted a study of the classification ability of the DT and SVM in gearbox fault detection. This study includes two fault classes, two gear speeds (first and fourth gear), and three loading conditions. Feature extraction was performed using statistical features, and classification accuracies were obtained using both the SVM and J48 DT methods.

Li et al. (2016) presented a method called deep RF fusion (DRFF) for gearbox fault diagnosis using acoustic and vibration signals simultaneously. In this method, the statistical parameters of the wavelet packet transform (WPT) are first obtained from the acoustic and vibration signals respectively. Then, two deep Boltzmann machines (DBMs) are developed for deep representations of the WPT statistical parameters. Finally, the RF is used to fuse the outputs of the two DBMs. Experimental results of gearbox fault diagnosis under different operating conditions showed that the DRFF has the ability to improve the performance of gearbox fault diagnosis.

Zhang et al. (2017) proposed a fault-diagnosis method based on the fast-clustering algorithm and DT for rotating machinery with imbalanced data. In this method, the fast-clustering algorithm is used to search for essential samples from the majority data of the imbalanced fault sample set. Hence, the balanced fault sample set comprises the clustered data, and the minority data is built. Then, a DT is trained using the balanced fault sample set to obtain the fault diagnosis model. The experimental results of the gearbox fault dataset and rolling bearing fault dataset showed that the proposed fault-diagnosis model has the ability to diagnose rotating machinery faults for imbalanced data.

Table 10.2 Summary of some of the introduced techniques and their publically accessible software.

Algorithm name	Platform	Package	Function
Fit binary classification decision tree for multiclass classification	MATLAB	Statistics and Machine Learning Tool- box – Classification – Classification trees	fitctree
Produce sequence of subtrees by pruning			prune
Compact tree			compact
Mean predictive measure of association for surrogate splits in decision tree			surrogateAssociation
Decision tree and decision forest	MATLAB	Wang (2014)	RunDecisionForest TrainDecisionForest

10.5 Summary

This chapter has provided a review of DTs and RFs and their application in machine fault diagnosis, including several DT analysis techniques. The algorithms most commonly used by DTs to make splitting decisions were described. These include univariate splitting criteria, e.g. Gini index, information gain, distance measure, and orthogonal criterion; and multivariate splitting criteria, e.g. greedy learning, linear programming, linear discriminant analysis, perceptron learning, and hill-climbing search. Moreover, DT pruning methods including error-complexity pruning, minimum-error pruning, critical-value pruning, etc., were presented. In addition, the most commonly used DT inducers, i.e. methods that can be used to build a DT from a given training dataset, e.g. CART, the iterative Dichotomiser 3 (ID3), C4.5, and CHAID were briefly described. The procedures for these methods include the application of the splitting criteria and pruning techniques that have been described.

Also, a brief discussion of the application of DTs and RFs in machine fault diagnosis was provided in this chapter. Some of the introduced techniques and their publicly accessible software are summarised in Table 10.2.

References

Almuallim, H. (1996). An efficient algorithm for optimal pruning of decision trees. *Artificial Intelligence* 83 (2): 347–362. Elsevier.

Azzouz, E.E. and Nandi, A.K. (1996). Modulation recognition using artificial neural networks. In: *Automatic Modulation Recognition of Communication Signals*, 132–176. Boston, MA: Springer.

Bauer, E. and Kohavi, R. (1999). An empirical comparison of voting classification algorithms: bagging, boosting, and variants. *Machine Learning* 36 (1–2): 105–139. Springer.

Bennett, K.P. (1992). *Decision Tree Construction Via Linear Programming*, 97–101. Center for Parallel Optimization, Computer Sciences Department, University of Wisconsin.

Bennett, K.P. and Mangasarian, O.L. (1992). Robust linear programming discrimination of two linearly inseparable sets. *Optimization Methods and Software* 1 (1): 23–34. Taylor & Francis.

Bennett, K.P. and Mangasarian, O.L. (1994). Multicategory discrimination via linear programming. *Optimization Methods and Software* 3 (1–3): 27–39. Taylor & Francis.

Bennett, K.P., Cristianini, N., Shawe-Taylor, J., and Wu, D. (2000). Enlarging the margins in perceptron decision trees. *Machine Learning* 41 (3): 295–313. Springer.

Berry, M.J. and Linoff, G. (1997). *Data Mining Techniques: For Marketing, Sales, and Customer Support*. Wiley.

Berzal, F., Cubero, J.C., Marın, N., and Sánchez, D. (2004). Building multi-way decision trees with numerical attributes. *Information Sciences* 165 (1–2): 73–90. Elsevier.

Bifet, A., Holmes, G., Pfahringer, B., and Frank, E. (2010). Fast perceptron decision tree learning from evolving data streams. In: *Pacific-Asia Conference on Knowledge Discovery and Data Mining*, 299–310. Berlin, Heidelberg: Springer.

Biggs, D., De Ville, B., and Suen, E. (1991). A method of choosing multiway partitions for classification and decision trees. *Journal of Applied Statistics* 18 (1): 49–62. Taylor & Francis.

Bobrowski, L. and Kretowski, M. (2000). Induction of multivariate decision trees by using dipolar criteria. In: *European Conference on Principles of Data Mining and Knowledge Discovery*, 331–336. Berlin, Heidelberg: Springer.

Bohanec, M. and Bratko, I. (1994). Trading accuracy for simplicity in decision trees. *Machine Learning* 15 (3): 223–250. Springer.

Breiman, L. (1996a). Bagging predictors. *Machine Learning* 24: 123–140. Springer.

Breiman, L. (1996b). Some properties of splitting criteria. *Machine Learning* 24 (1): 41–47. Springer.

Breiman, L. (2001). Random forests. *Machine Learning* 45 (1): 5–32. Springer.

Breiman, L. (2017). *Classification and Regression Trees*. Routledge Taylor & Francis.

Breiman, L., Friedman, J.H., Olshen, R.A. et al. (1984). Classification and regression trees. Wadsworth International Group. LHCb collaboration.

Brodley, C.E. and Utgoff, P.E. (1992). *Multivariate Versus Univariate Decision Trees*. Amherst, MA: University of Massachusetts, Department of Computer and Information Science.

Brodley, C.E. and Utgoff, P.E. (1995). Multivariate decision trees. *Machine Learning* 19 (1): 45–77. Springer.

Brown, D.E. and Pittard, C.L. (1993). Classification trees with optimal multi-variate splits. In: *International Conference on Systems, Man and Cybernetics, 1993.'Systems Engineering in the Service of Humans,' Conference Proceedings*, vol. 3, 475–477. IEEE.

Buntine, W. and Niblett, T. (1992). A further comparison of splitting rules for decision-tree induction. *Machine Learning* 8 (1): 75–85. Springer.

Cerrada, M., Zurita, G., Cabrera, D. et al. (2016). Fault diagnosis in spur gears based on genetic algorithm and random forest. *Mechanical Systems and Signal Processing* 70: 87–103. Elsevier.

Cestnik, B. and Bratko, I. (1991). On estimating probabilities in tree pruning. In: *European Working Session on Learning*, 138–150. Berlin, Heidelberg: Springer.

Che, D., Liu, Q., Rasheed, K., and Tao, X. (2011). Decision tree and ensemble learning algorithms with their applications in bioinformatics. In: *Software Tools and Algorithms for Biological Systems*, 191–199. New York, NY: Springer.

Cortes, C. and Vapnik, V. (1995). Support-vector networks. *Machine Learning* 20 (3): 273–297. Springer.

De Mántaras, R.L. (1991). A distance-based attribute selection measure for decision tree induction. *Machine Learning* 6 (1): 81–92. Springer.

Dietterich, T., Kearns, M., and Mansour, Y. (1996). Applying the weak learning framework to understand and improve C4. 5. In: *International Conference on Machine Learning (ICML)*, 96–104.

Drummond, C. and Holte, R.C. (2000). Exploiting the cost (in) sensitivity of decision tree splitting criteria. *Proceedings of the International Conference on Machine Learning* 1 (1): 239–246.

Duda, R.O. and Hart, P.E. (1973). *Pattern Classification and Scene Analysis. A Wiley-Interscience Publication*. New York: Wiley.

Esposito, F., Malerba, D., Semeraro, G., and Kay, J. (1997). A comparative analysis of methods for pruning decision trees. *IEEE Transactions on Pattern Analysis and Machine Intelligence* 19 (5): 476–491.

Fayyad, U.M. and Irani, K.B. (1992). The attribute selection problem in decision tree generation. In: *American Association for Artificial Intelligence (AAAI)*, 104–110.

Ferri, C., Flach, P., and Hernández-Orallo, J. (2002). Learning decision trees using the area under the ROC curve. In: *International Conference on Machine Learning (ICML)*, vol. 2, 139–146.

Fournier, D. and Crémilleux, B. (2002). A quality index for decision tree pruning. *Knowledge-Based Systems* 15 (1–2): 37–43. Elsevier.

Freund, Y. and Schapire, R.E. (1997). A decision-theoretic generalization of on-line learning and an application to boosting. *Journal of Computer and System Sciences* 55 (1): 119–139. Elsevier.

Friedman, J.H. (1977). A recursive partitioning decision rule for nonparametric classification. *IEEE Transactions on Computers* (4): 404–408.

Friedman, J.H. (1991). Multivariate adaptive regression splines. *The Annals of Statistics*: 1–67.

Fulton, T., Kasif, S., and Salzberg, S. (1995). Efficient algorithms for finding multi-way splits for decision trees. In: *Machine Learning Proceedings 1995*, 244–251.

Guo, H., Jack, L.B., and Nandi, A.K. (2005). Feature generation using genetic programming with application to fault classification. *IEEE Transactions on Systems, Man, and Cybernetics, Part B (Cybernetics)* 35 (1): 89–99.

Heath, D., Kasif, S., and Salzberg, S. (1993). Induction of oblique decision trees. In: *International Joint Conferences on Artificial Intelligence (IJCAL)*, vol. 1993, 1002–1007.

Hosmer, D.W. Jr., Lemeshow, S., and Sturdivant, R.X. (2013). *Applied Logistic Regression*, vol. 398. Wiley.

Hothorn, T., Hornik, K. and Zeileis, A. (2015). ctree: Conditional Inference Trees. The Comprehensive R Archive Network.

Jain, A.K., Mao, J., and Mohiuddin, K.M. (1996). Artificial neural networks: a tutorial. *Computer* 29 (3): 31–44.

Jain, L.C., Seera, M., Lim, C.P., and Balasubramaniam, P. (2014). A review of online learning in supervised neural networks. *Neural Computing and Applications* 25 (3–4): 491–509. Springer.

Jaworski, M., Duda, P., and Rutkowski, L. (2018). New splitting criteria for decision trees in stationary data streams. *IEEE Transactions on Neural Networks and Learning Systems* 29 (6): 2516–2529.

Jegadeeshwaran, R. and Sugumaran, V. (2013). Comparative study of decision tree classifier and best first tree classifier for fault diagnosis of automobile hydraulic brake system using statistical features. *Measurement* 46 (9): 3247–3260. Elsevier.

John, G.H. (1996). Robust linear discriminant trees. In: *Learning from Data*, 375–385. New York, NY: Springer.

Karabadji, N.E.I., Seridi, H., Khelf, I., and Laouar, L. (2012). Decision tree selection in an industrial machine fault diagnostics. In: *International Conference on Model and Data Engineering*, 129–140. Berlin, Heidelberg: Springer.

Kass, G.V. (1980). An exploratory technique for investigating large quantities of categorical data. *Applied statistics*: 119–127. Wiley.

Kearns, M.J. and Mansour, Y. (1998). A fast, bottom-up decision tree pruning algorithm with near-optimal generalization. In: *International Conference on Machine Learning (ICML)*, vol. 98, 269–277.

Kim, H. and Loh, W.Y. (2001). Classification trees with unbiased multiway splits. *Journal of the American Statistical Association* 96 (454): 589–604. Taylor & Francis.

Knoll, U., Nakhaeizadeh, G., and Tausend, B. (1994). Cost-sensitive pruning of decision trees. In: *European Conference on Machine Learning*, 383–386. Berlin, Heidelberg: Springer.

Lee, S.K. (2005). On generalized multivariate decision tree by using GEE. *Computational Statistics & Data Analysis* 49 (4): 1105–1119. Elsevier.

Li, X. and Dubes, R.C. (1986). Tree classifier design with a permutation statistic. *Pattern Recognition* 19 (3): 229–235. Elsevier.

Li, X.B., Sweigart, J.R., Teng, J.T. et al. (2003). Multivariate decision trees using linear discriminants and tabu search. *IEEE Transactions on Systems, Man, and Cybernetics-Part A: Systems and Humans* 33 (2): 194–205.

Li, C., Sanchez, R.V., Zurita, G. et al. (2016). Gearbox fault diagnosis based on deep random forest fusion of acoustic and vibratory signals. *Mechanical Systems and Signal Processing* 76: 283–293. Elsevier.

Likura, Y. and Yasuoka, Y. (1991). Utilization of a best linear discriminant function for designing the binary decision tree. *International Journal of Remote Sensing* 12 (1): 55–67. Taylor & Francis.

Lin, J.H. and Vitter, J.S., 1992. Nearly optimal vector quantization via linear programming.

Loh, W.Y. (2011). Classification and regression trees. *Wiley Interdisciplinary Reviews: Data Mining and Knowledge Discovery* 1 (1): 14–23. Wiley.

Loh, W.Y. and Shih, Y.S. (1997). Split selection methods for classification trees. *Statistica Sinica*: 815–840.

Loh, W.Y. and Vanichsetakul, N. (1988). Tree-structured classification via generalized discriminant analysis. *Journal of the American Statistical Association* 83 (403): 715–725.

Lu, K.C. and Yang, D.L. (2009). Image processing and image mining using decision trees. *Journal of Information Science & Engineering* (4): 25. Citeseer.

Mehta, M., Rissanen, J., and Agrawal, R. (1995). MDL-based decision tree pruning. In: *Proceedings of the First International Conference on Knowledge Discovery and Data Mining (KDD)*, vol. 21, No. 2, 216–221.

Michel, G., Lambert, J.L., Cremilleux, B., and Henry-Amar, M. (1998). A new way to build oblique decision trees using linear programming. In: *Advances in Data Science and Classification*, 303–309. Berlin, Heidelberg: Springer.

Mingers, J. (1987). Expert systems—rule induction with statistical data. *Journal of the Operational Research Society* 38 (1): 39–47. Springer.

Mingers, J. (1989). An empirical comparison of pruning methods for decision tree induction. *Machine Learning* 4 (2): 227–243. Springer.

Moore, S.A., D'addario, D.M., Kurinskas, J., and Weiss, G.M. (2009). Are decision trees always greener on the open (source) side of the fence? In: *Proceedings of DMIN*, 185–188.

Muller, W. and Wysotzki, F. (1994). A splitting algorithm, based on a statistical approach in the decision tree algorithm CAL5. In: *Proceedings of the ECML-94 Workshop on Machine Learning and Statistics* (eds. G. Nakhaeizadeh and C. Taylor).

Muralidharan, V. and Sugumaran, V. (2013). Feature extraction using wavelets and classification through decision tree algorithm for fault diagnosis of mono-block centrifugal pump. *Measurement* 46 (1): 353–359. Elsevier.

Murphy, P.M. and Pazzani, M.J. (1991). ID2-of-3: Constructive induction of M-of-N concepts for discriminators in decision trees. In: *Machine Learning Proceedings 1991*, 183–187. Elsevier.

Murthy, S.K. (1998). Automatic construction of decision trees from data: a multi-disciplinary survey. *Data Mining and Knowledge Discovery* 2 (4): 345–389. Springer.

Murthy, S.K., Kasif, S., and Salzberg, S. (1994). A system for induction of oblique decision trees. *Journal of Artificial Intelligence Research* 2: 1–32.

Myles, A.J., Feudale, R.N., Liu, Y. et al. (2004). An introduction to decision tree modeling. *Journal of Chemometrics: A Journal of the Chemometrics Society* 18 (6): 275–285. Wiley.

Nandi, A.K. and Azzouz, E.E. (1995). Automatic analogue modulation recognition. *Signal Processing* 46 (2): 211–222. Elsevier.

Nandi, A.K. and Azzouz, E.E. (1998). Algorithms for automatic modulation recognition of communication signals. *IEEE Transactions on Communications* 46 (4): 431–436.

Nandi, A.K., Liu, C., and Wong, M.D. (2013). Intelligent vibration signal processing for condition monitoring. In: *Proceedings of the International Conference Surveillance*, vol. 7, 1–15.

Norouzi, M., Collins, M., Johnson, M.A. et al. (2015). Efficient non-greedy optimization of decision trees. In: *NIPS'15 Proceedings of the 28th International Conference on Neural Information Processing Systems - Volume 1*, 1729–1737.

Oliveira, W.D., Vieira, J.P., Bezerra, U.H. et al. (2017). Power system security assessment for multiple contingencies using multiway decision tree. *Electric Power Systems Research* 148: 264–272. Elsevier.

Patel, S.A. and Kamrani, A.K. (1996). Intelligent decision support system for diagnosis and maintenance of automated systems. *Computers & Industrial Engineering* 30 (2): 297–319. Elsevier.

Patil, D.D., Wadhai, V.M., and Gokhale, J.A. (2010). Evaluation of decision tree pruning algorithms for complexity and classification accuracy. *International Journal of Computer Applications* 11 (2).

Podgorelec, V., Kokol, P., Stiglic, B., and Rozman, I. (2002). Decision trees: an overview and their use in medicine. *Journal of Medical Systems* 26 (5): 445–463. Springer.

Qing-Yun, S. and Fu, K.S. (1983). A method for the design of binary tree classifiers. *Pattern Recognition* 16 (6): 593–603. Elsevier.

Quinlan, J.R. (1986). Induction of decision trees. *Machine Learning* 1 (1): 81–106. Springer.

Quinlan, J.R. (1987). Simplifying decision trees. *International Journal of Man-Machine Studies* 27 (3): 221–234. Elsevier.

Quinlan, J.R. (1993). *C4. 5: Programs for Machine Learning*. Elsevier.

Quinlan, J.R. (1996a). Bagging, boosting, and C4. 5. In: *AAAI' 96 Proceedings of the thirteenth national conference on Artificial intelligence* - Volume 1, 725–730.

Quinlan, J.R. (1996b). Improved use of continuous attributes in C4. 5. *Journal of Artificial Intelligence Research* 4: 77–90.

Raileanu, L.E. and Stoffel, K. (2004). Theoretical comparison between the gini index and information gain criteria. *Annals of Mathematics and Artificial Intelligence* 41 (1): 77–93. Springer.

Rastogi, R. and Shim, K. (1998). PUBLIC: a decision tree classifier that integrates building and pruning. In: *Proceedings of the International Conference on Very Large Data Bases (VLDB)* , vol. 98, 24–27.

Rish, I. (2001). An empirical study of the naive Bayes classifier. In: *IJCAI 2001 Workshop on Empirical Methods in Artificial Intelligence*, vol. 3, No. 22, 41–46. New York: IBM.

Rivest, R.L. (1987). Learning decision lists. *Machine Learning* 2 (3): 229–246. Springer.

Rokach, L. and Maimon, O. (2005a). Decision trees. In: *Data Mining and Knowledge Discovery Handbook*, 165–192. Boston, MA: Springer.

Rokach, L. and Maimon, O. (2005b). Top-down induction of decision trees classifiers-a survey. *IEEE Transactions on Systems, Man, and Cybernetics, Part C (Applications and Reviews)* 35 (4): 476–487.

Rounds, E.M. (1980). A combined nonparametric approach to feature selection and binary decision tree design. *Pattern Recognition* 12 (5): 313–317. Elsevier.

Safavian, S.R. and Landgrebe, D. (1991). A survey of decision tree classifier methodology. *IEEE Transactions on Systems, Man, and Cybernetics* 21 (3): 660–674.

Saimurugan, M., Ramachandran, K.I., Sugumaran, V., and Sakthivel, N.R. (2011). Multi component fault diagnosis of rotational mechanical system based on decision tree and support vector machine. *Expert Systems with Applications* 38 (4): 3819–3826. Elsevier.

Saimurugan, M., Praveenkumar, T., Krishnakumar, P., and Ramachandran, K.I. (2015). A study on the classification ability of decision tree and support vector machine in gearbox fault detection. In: *Applied Mechanics and Materials*, vol. 813, 1058–1062. Trans Tech Publications.

Sakthivel, N.R., Sugumaran, V., and Babudevasenapati, S. (2010a). Vibration based fault diagnosis of monoblock centrifugal pump using decision tree. *Expert Systems with Applications* 37 (6): 4040–4049. Elsevier.

Sakthivel, N.R., Sugumaran, V., and Nair, B.B. (2010b). Comparison of decision tree-fuzzy and rough set-fuzzy methods for fault categorization of mono-block centrifugal pump. *Mechanical Systems and Signal Processing* 24 (6): 1887–1906. Elsevier.

Saravanan, N. and Ramachandran, K.I. (2009). Fault diagnosis of spur bevel gear box using discrete wavelet features and decision tree classification. *Expert Systems with Applications* 36 (5): 9564–9573. Elsevier.

Schmidhuber, J. (2015). Deep learning in neural networks: an overview. *Neural Networks* 61: 85–117. Elsevier.

Seera, M., Lim, C.P., Ishak, D., and Singh, H. (2012). Fault detection and diagnosis of induction motors using motor current signature analysis and a hybrid FMM–CART model. *IEEE Transactions on Neural Networks and Learning Systems* 23 (1): 97–108.

Seera, M., Wong, M.D., and Nandi, A.K. (2017). Classification of ball bearing faults using a hybrid intelligent model. *Applied Soft Computing* 57: 427–435. Elsevier.

Sethi, I.K. and Yoo, J.H. (1994). Design of multicategory multifeature split decision trees using perceptron learning. *Pattern Recognition* 27 (7): 939–947. Elsevier.

Shah, S. and Sastry, P.S. (1999). New algorithms for learning and pruning oblique decision trees. *IEEE Transactions on Systems, Man, and Cybernetics, Part C (Applications and Reviews)* 29 (4): 494–505.

Shih, Y.S. (1999). Families of splitting criteria for classification trees. *Statistics and Computing* 9 (4): 309–315. Springer.

Singh, S. and Gupta, P. (2014). Comparative study ID3, cart and C4. 5 decision tree algorithm: a survey. *International Journal of Advanced Information Science and Technology (IJAIST)* 27 (27): 97–103.

Sugumaran, V. and Ramachandran, K.I. (2007). Automatic rule learning using decision tree for fuzzy classifier in fault diagnosis of roller bearing. *Mechanical Systems and Signal Processing* 21 (5): 2237–2247. Elsevier.

Sun, W., Chen, J., and Li, J. (2007). Decision tree and PCA-based fault diagnosis of rotating machinery. *Mechanical Systems and Signal Processing* 21 (3): 1300–1317. Elsevier.

Taylor, P.C. and Silverman, B.W. (1993). Block diagrams and splitting criteria for classification trees. *Statistics and Computing* 3 (4): 147–161. Springer.

Todeschini, R. and Marengo, E. (1992). Linear discriminant classification tree: a user-driven multicriteria classification method. *Chemometrics and Intelligent Laboratory Systems* 16 (1): 25–35. Elsevier.

Tran, V.T., Yang, B.S., and Oh, M.S. (2006). Fault diagnosis of induction motors using decision trees. In: *Proceeding of the KSNVE Annual Autumn Conference*, 1–4.

Tsang, S., Kao, B., Yip, K.Y. et al. (2011). Decision trees for uncertain data. *IEEE Transactions on Knowledge and Data Engineering* 23 (1): 64–78.

Utgoff, P.E. (1989). Perceptron trees: a case study in hybrid concept representations. *Connection Science* 1 (4): 377–391. Taylor & Francis.

Wallace, C.S. and Patrick, J.D. (1993). Coding decision trees. *Machine Learning* 11 (1): 7–22.

Wang, Q. (2014). Decision tree and decision forest. Mathworks File Exchange Center. https://uk.mathworks.com/matlabcentral/fileexchange/39110-decision-tree-and-decision-forest.

Xiaoliang, Z., Hongcan, Y., Jian, W., and Shangzhuo, W. (2009). Research and application of the improved algorithm C4. 5 on decision tree. In: *International Conference on Test and Measurement, 2009. ICTM'09*, vol. 2, 184–187. IEEE.

Yang, B.S., Lim, D.S., and Tan, A.C.C. (2005). VIBEX: an expert system for vibration fault diagnosis of rotating machinery using decision tree and decision table. *Expert Systems with Applications* 28 (4): 735–742. Elsevier.

Yang, B.S., Di, X., and Han, T. (2008). Random forests classifier for machine fault diagnosis. *Journal of Mechanical Science and Technology* 22 (9): 1716–1725. Springer.

Yang, B.S., Oh, M.S., and Tan, A.C.C. (2009). Fault diagnosis of induction motor based on decision trees and adaptive neuro-fuzzy inference. *Expert Systems with Applications* 36 (2): 1840–1849. Elsevier.

You, K.C. and Fu, K.S. (1976). An approach to the design of a linear binary tree classifier. In: *Proc. Symp. Machine Processing of Remotely Sensed Data*, 3A–10A.

Zhang, X., Jiang, D., Long, Q., and Han, T. (2017). Rotating machinery fault diagnosis for imbalanced data based on decision tree and fast clustering algorithm. *Journal of Vibroengineering* 19 (6): 4247–4259.

11

Probabilistic Classification Methods

11.1 Introduction

As described in Chapter 10, classification is a typical supervised learning task, i.e. the training data $X = \{x_1, x_2, \ldots, x_n\}$ along with their corresponding labels $y \in \{c_1, c_2, \ldots, c_k\}$ are used together to perform the learning and optimising parameters of the classification model of interest. The trained classification model is then can be used to classify new examples to their correct classes. A simple example of classification would be assigning a given vibration signal into a 'normal' or 'fault' class. The literature on classification algorithms has highlighted various data classification methods. Of these methods, probabilistic classification methods are amongst the most commonly used data classification methods. The basic idea of the probabilistic classification is to compute the posterior probabilities $p(c_k \mid x)$ of the instances being a member of each of the likely labels. There are two ways to compute the posterior probabilities: (i) Determine the parameters of the modelled class-conditional densities $p(x \mid c_k)$ and class priors $p(c_k)$ using maximum likelihood and then use Bayes' theorem to find $p(c_k \mid x)$, which is known as a *probabilistic generative models*; and (ii) maximising a likelihood function defined through $p(c_k \mid x)$, known as a *probabilistic discriminative model*.

In this chapter, we introduce two probabilistic models for classification: the hidden Markov model (HMM), which is a probabilistic generative model, and the logistic regression model (LR), which is a probabilistic discriminative model. These classifiers, i.e. HMM and LR, have been widely used in machine fault diagnosis.

11.2 Hidden Markov Model

The HMM is an extension of Markov chains (Bishop 2006; Rabiner 1989). The Markov chain is a stochastic model that describe the probabilities of a sequence of possible states where the predicted state only depends on the current state (Gagniuc 2017). For a given N different states, $s_1, s_2, \ldots, s_N$, of a system, where at any time the system can be described as being one of these states, the Markov chain of this system can be represented as follow:

$$P(s_1, s_2, \ldots, s_N) = p(s_1) \prod_{i=2}^{N} p(s_i \mid s_{i-1}) \tag{11.1}$$

Condition Monitoring with Vibration Signals: Compressive Sampling and Learning Algorithms for Rotating Machines, First Edition. Hosameldin Ahmed and Asoke K. Nandi.
© 2020 John Wiley & Sons Ltd. Published 2020 by John Wiley & Sons Ltd.

In HMMs, the observations are probabilistic functions rather of the states. The main target of an HMM is to find the probability of the observation being in a specific state where the observation is a probabilistic function of the state. It can be described as a doubly stochastic process, where the underlying stochastic process consists of a series of states, each of which is related to another stochastic process that emits observable symbols. In other words, it involves the stochastic transition from one state to another state and the stochastic output symbol generated at each state. Therefore, it can be considered as a simple dynamic Bayesian (Murphy and Russell 2002). Each HMM is characterised by a set of the following elements:

a. A set of N hidden states in the model, defined as $\{S = s_1, s_2, ..., s_N\}$, where the state at time t is q_t.
b. A set of M different observation symbols per state, defined as $\{V = v_1, v_2, ..., v_M\}$.
c. The state transition probability distribution $A = \{a_{ij}\}$. Here, a_{ij} denotes the transition probability from state s_i at time t to state s_j at time $t+1$ such that,

$$a_{ij} = p[q_{t+1} = s_j \mid q_t = s_i], 1 \leq i, j \leq N \tag{11.2}$$

where the transition probabilities satisfy the normal stochastic constraints, $a_{ij} \geq 0$, $1 \leq i, j \leq N$, and $\sum_{j=1}^{N} a_{ij} = 1$.

d. The observation symbol probability distribution in each state $B = \{b_j(k)\}$. Here, $b_j(k)$ represents the probability that symbol v_k is emitted in the state s_j such that

$$b_j(k) = p[v_k at\ t \mid q_t = s_j], 1 \leq j \leq N, 1 \leq k \leq M \tag{11.3}$$

where $b_j(k) \geq 0$, and $\sum_{k=1}^{M} b_j(k) = 1$.

If the observation is continuous, we use HMMs with continuous observation densities – in this case, a finite mixture of the form

$$b_j(O) = \sum_{m=1}^{M} \xi_{jm} \aleph[O, \mu_{jm}, U_{jm}], 1 \leq j \leq N \tag{11.4}$$

where a Gaussian density is usually used for $\aleph$, O is the vector being modelled, ξ_{jm} is coefficient, μ_{jm} is a mean vector, and U_{jm} is a covariance matrix.

e. The initial state distribution $\pi = \{\pi_i\}$, where π_i is the probability that the initial state is s_i at t = 0. This can be represented by the following equation:

$$\pi_i = p[q_1 = s_i], 1 \leq j \leq N \tag{11.5}$$

Accordingly, the complete description of an HMM requires the descriptions of two model parameters, N and M, observation symbols, and three probability measures A, B, and π, which can be represented as $\lambda = (A, B, \pi)$ to denote the complete parameter set of the discrete model. In case of a continuous HMM, λ can be represented as $\lambda = (A, \xi_{jm}, \mu_{jm}, U_{jm}, \pi)$. Given a set of output sequences, the main task of the learning algorithm is to find the maximum likelihood estimation of an HMM parameter. Usually, the local maximum likelihood can be obtained using an iterative procedure such as the Baum-Welch method (Welch 2003).

Generally, there are three basic problems that need to be solved for the HMM to be beneficial in practical applications. These problems and their solutions can be summarised as follows:

1. *Compute $P(O \mid \lambda)$, given the observation sequence $O = o_1, o_2, \ldots, o_T$ and $\lambda = (A, B, \pi)$.*
   $P(O \mid \lambda)$ can be computed efficiently using a forward-backward procedure (Rabiner 1989). Consider the forward variable $\vartheta_t(j) = p(o_1, o_2, \ldots, o_t, q_t = s_i \mid \lambda)$. We can solve for $\vartheta_t(i)$ inductively, as follows:

   a. Initialisation: $\vartheta_1(j) = \pi_i b_i(o_1)$ where $1 \le i \le N$.

   b. Induction: $\vartheta_{t+1}(j) = \left[\sum_{i=1}^{N} \vartheta_t(j) a_{ij} \right] b_j(o_{t+1})$ where $1 \le t \le T - 1$ and $1 \le i \le N$.

   c. Termination: $P(O \mid \lambda) = \sum_{i=1}^{N} \vartheta_T(i)$.

2. *Find the most likely state sequence, given the observation sequence $O = o_1, o_2, \ldots, o_T$ and $\lambda = (A, B, \pi)$.*
   To find the single best sequence $Q = q_1, q_2, \ldots, q_T$ for a given observation $O = o_1, o_2, \ldots, o_T$, we can use Viterbi algorithm (Forney 1973) such that,

   $$\delta_t(i) = \max_{q_1, q_2, \ldots, q_T} p[q_1, q_2, \ldots, q_T = s_i, o_1, o_2, \ldots, o_T \mid \lambda] \tag{11.6}$$

   where $\delta_t(i)$ is the best score along a single path at time t, which accounts for the first t observations and ends in the state s_i. Also, by induction, $\delta_{t+1}(i)$ can be given as follows:

   $$\delta_{t+1}(i) = [\max_i \delta_t(i) a_{ji}] . b_i(o_{t+1}) \tag{11.7}$$

   The complete process for finding the most likely state sequence can be summarised as follows:

   a. Initialisation: $\delta_1(i) = \pi_i b_i(o_1)$, where $1 \le i \le N$; $\psi_1(i) = 0$, where ψ is an array used to keep track of the argument for each t and j.

   b. Recursion: $\delta_t(j) = \max_i [\delta_{t-1}(i) a_{ji}] . b_i(o_t)$; $\psi_t(i) = \arg \max_i [\delta_{t-1}(i) a_{ji}]$, where $2 \le t \le T$ and $1 \le i \le N$.

   c. Termination: $p^* = \max_i [\delta_T(i)]$; $q_T^* = \arg \max_i [\delta_T(i)]$.

   d. State sequence, i.e. path, backtracking: $q_t^* = \psi_{t+1}(q_{t+1}^*)$.

3. *Adjust the HMM parameters $\lambda = (A, B, \pi)$ to maximise $P(O \mid \lambda)$.*
   The Baum-Welch method (Welch 2003) can be used to solve this problem. We define Baum's auxiliary function as follows,

   $$Q(\lambda, \bar{\lambda}) = \sum_Q P(q \mid O, \lambda) \log[P(O, Q \mid \bar{\lambda})] \tag{11.8}$$

   where

   $$\max_{\bar{\lambda}} Q(\lambda, \bar{\lambda}) \rightarrow P(O \mid \bar{\lambda}) \ge P(O \mid \lambda) \tag{11.9}$$

   Here, the maximisation of $Q(\lambda, \bar{\lambda})$ leads to an increased likelihood.

HMMs are being considered in a large diversity of applications, e.g. speech recognition (Gales and Young 2008), computational biology (Krogh et al. 1994), stock market forecasting (Hassan and Nath 2005), etc. The following subsection briefly describes their application in machine fault diagnosis.

11.2.1 Application of Hidden Markov Models in Machine Fault Diagnosis

HMMs are used in different studies of machine fault detection and diagnosis. The basic idea of using HMMs in machine fault diagnosis is that first a set of features is extracted in the time domain, frequency domain, and/or time-frequency domain, individually or in a mixture. In order to estimate the parameters of the HMM, these features are then converted into observation sequences. For instance, Ocak and Loparo (2001) introduced a bearing fault detection and diagnosis technique based on a HMM of vibration signals. Under this technique, first, features are extracted from amplitude-demodulated vibration signals acquired from both normal and faulty bearings. The feature extraction is performed by first dividing vibration signals into windows of equal length; then the features for each window are selected to be the reflection coefficients of the polynomial transfer function of the linear autoregressive model. These features are then utilised to train the HMMs. To test the effectiveness of this method in bearing fault diagnosis, the authors used experimental data acquired from drive-end ball bearings from an induction motor. The results showed that the proposed method has the ability to achieve high classification accuracy in diagnosing bearing faults.

Lee et al. (2004) introduced the continuous hidden Markov model (CHMM) to diagnosis mechanical fault signals. In this study, several modifications have been made to the conventional CHMM, including initialisation using a maximum-distance clustering method, using filter banks, scaled forward/backward variables, and diagonal covariance matrix and modification of training equations for multiple observation vector sequences. Sampled data from a rotor simulator are used to validate the proposed method. These include seven machine health conditions: normal, resonance, stable after resonance, bearing housing looseness, misalignment, flexible coupling damage, and unbalance. The experimental results demonstrated the effectiveness of the proposed method in diagnosing rotor faults.

Li et al. (2005) introduced the HMM to detect and identify faults in the speed-up and slow-down processes for the rotating machinery. In this study, several feature vectors obtained from the original signals using the fast Fourier transform (FFT), wavelet transform, bispectrum, etc., are used as fault features, and the HMM is used as the classifier. Four running conditions (unbalance, rubbing, oil whirl, and pedestal looseness) are performed on a Bently rotor kit. Two resonance speeds in the rotor test rig – around 3800 and 6900 rpm, respectively – and five rotating speed regions – 500–3000, 3000–4500, 4500–6700, 6700–7400, and 7400–8000 rpm – are used to validate the proposed method. The experimental results showed the effectiveness of the proposed method in machine fault classification.

Moreover, Nelwamondo et al. (2006) introduced HMM and Gaussian mixture models (GMMs) to classify bearing faults. In this study, linear and nonlinear features are extracted from time-domain vibration signals using multiscale fractal dimension (MFD), mel - frequency cepstral coefficients, and kurtosis. With these features, the HMM and GMM are applied to deal with the classification problem. Real vibration data with four conditions – normal, IR fault, OR fault, and RE bearing fault – are collected from the drive-end-bearing. The experimental results showed that the HMM outperforms the GMM in the classification of bearing health conditions. Miao and Makis (2007) proposed a modelling framework for the classification of machine conditions using the wavelet modulus maxima distribution and HMMs. In this framework, the authors used the modulus maxima distribution as the input observation sequence of the system. Then, the machinery condition is identified by selecting the HMM that maximises the probability of the observation sequence. Two-stage machine fault classification based on the HMM is introduced. The first stage is set up to classify two health conditions (normal and failure); this is a fault-detection stage. The second stage is set up to classify multi-conditions; this is a fault-classification stage. To validate the first stage of the proposed framework, three sets of real gearbox vibration data are used to classify two conditions: normal and failure. In stage two, three HMM models are set up to classify three different conditions: adjacent tooth failure, distributed tooth failure, and normal. The experimental results demonstrated the effectiveness of the proposed framework in machine fault classification.

Furthermore, Miao et al. (2007) investigated the classification performance of HMMs and support vector machines (SVMs) in machinery condition monitoring. In this investigation, a real vibration dataset measured from a gearbox driven by an electrical motor in a laboratory environment is used to compare the classification performance of the HMM and SVM. Two machinery conditions, normal and failure, are considered in this investigation. Based on the results of the conducted experiments, the authors stated that the SVM has better classification performance and generalisation compared to the HMM. Also, Yu J. (2012) proposed a method to assess machine health degradation using the HMM and contribution analysis. In this study, a dynamic principal component analysis (DPCA) is employed to extract features from vibration signals, and the HMM is examined for its applicability to real-world scenarios where no prior knowledge about the fault-severity data is obtainable. Real vibration data from five representative bearings with IR, OR, and ball defects are used to validate the proposed method. The experimental results demonstrated that the proposed method was able to recognise the slight degradation of bearings at an early stage. Also, the proposed method was shown to be able to effectively reduce data dimension and to obtain sensitive features from machine health degradation.

Also, Zhou et al. (2016) presented a method, called the shift-invariant dictionary learning-hidden Markov model (SIDL-HMM), for detecting and diagnosing bearing faults using shift-invariant dictionary learning (SIDL) and HMMs. In this method, SIDL is used to extract features from the raw vibration signals, and a separate HMM is trained and developed for all possible bearing fault types. The trained HMMs are used for faults classification of unlabelled vibration samples where the log likelihood of each trained HMM on the test data is calculated and the maximum value determines the certain state of the test sample. Simulated and experimental data with a specific notch size are used to validate the proposed method. The experimental data are acquired from rolling bearings under different conditions: normal, IR fault, OR fault, and RE fault. The experimental results demonstrated that SIDL-HMM can identify different bearing fault types.

Recently, Sadhu et al. (2017) proposed a technique to enhance the performance of HMMs for fault detection in rotating components using a series of preprocessing steps. In this technique, first, the wavelet packet transform (WPT) is used to extract denoised time-scale signatures from the original vibration signal. Then, the condition indicators are calculated by demodulating the extracted time-scale components using a Teager-Kaiser energy operator. With these condition indicators, the authors used a decision tree to select only relevant features. Finally, a Gaussian mixing model-based HMM is used for fault detection. Two case studies, fault detection of bearings in an induction motor and a low-speed conveyor belt section, are used to validate the proposed method. The experimental results demonstrated that the proposed method was able to achieve improved performance in fault detection of the rotating machine compared to the traditional HMM.

Liu et al. (2017) introduced a hybrid general hidden Markov model–based condition monitoring method (GHMM-CM) for roller bearings. In this method, the original vibration signals are decomposed into multiple mode components using a variational mode decomposition (VMD) method. Then, the authors used the multiscale permutation entropy (MPE) technique to extract the interval-valued features from the decomposed signal. Next, principal component analysis (PCA) is employed to reduce the dimensionality of the extracted features. Finally, the general hidden Markov model (GHMM), based on generalised interval probability, is used to identify and classify the fault types and feature severity levels. To validate the proposed method, experimental vibration data from a roller bearing, provided by the bearing data centre of Case Western Reserve University (CWRU), is used. The experimental results demonstrated the effectiveness of GHMM-CM in roller bearings fault diagnosis.

11.3   Logistic Regression Model

Regression analysis is a predictive modelling technique that describes the relationship between a dependent variable and one or more independent descriptive variables. This technique can be used for time series modelling, as described in Section 3.4; forecasting (Montgomery et al. 1990); and classification (Phillips et al. 2015). There are several types of regression analysis techniques, e.g. linear regression, logistic regression, polynomial regression, stepwise regression, ridge regression (Seber and Lee 2012; Hosmer Jr et al. 2013), and lasso regression (Tibshirani 1996). Of these techniques, the logistic regression (LR) model has become one of the most commonly used techniques in machine learning in many fields. LR is often used where the dependent variable is binary or dichotomous, i.e. the dependent variable c (labels) takes only two possible values, for example $c^{(i)} \in \{0, 1\}$ and $c^{(i)} \in \{Fault, No\ Fault\}$. The LR model is a probabilistic discriminative model that learns $P(y \mid X)$ directly from the training data by maximising a likelihood function defined through $p(y \mid X)$. Considering training data $X = \{x_1, x_2, \ldots, x_n\}$, the logistic regression model with binary labels $c_i \in \{0, 1\}$ can be defined using the following equation,

$$P(y = 1 \mid x) = h_1(x) = g(-\theta^T x) = \frac{1}{1 + e^{-\theta^T x}} \tag{11.10}$$

where $g(-\theta^T x)$ is the logistic function that also goes with the name sigmoid function and can be defined as

$$g(z) = \frac{1}{1 + e^{-z}} \tag{11.11}$$

Figure 11.1 shows a plot of the logistic function. It can be clearly seen that $g(z)$ are bounded between 0 and 1.

Also, due to the fact that $\sum P(y) = 1$, $P(y = 0 \mid x)$ can be computed using the following equation:

$$P(y = 0 \mid x) = h_0(x) = 1 - P(y = 1 \mid x) = 1 - \frac{1}{1 + e^{-\theta^T x}} \tag{11.12}$$

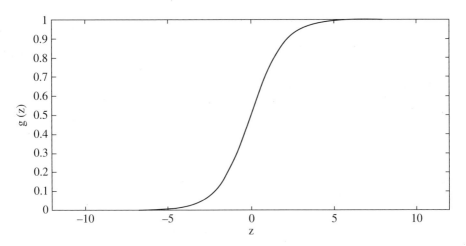

Figure 11.1 Illustration of a logistic function.

Or, more efficiently,

$$P(y = 0 \mid x) = \frac{e^{-\theta^T x}}{1 + e^{-\theta^T x}} \tag{11.13}$$

Combining Eqs. (11.10) and (11.13), we obtain the following:

$$P(y = y_k \mid x; \theta) = (P(y = 0 \mid x))^{y_k}(P(y = 0 \mid x))^{1-y_k} \tag{11.14}$$

The parameters of the logistic regression are often determined using likelihood. The likelihood of the parameters of N training examples can be given using the following equation:

$$L(\theta) = \prod_{n=1}^{N} (g(\theta^T x^n))^{y(n)}((1 - g(\theta^T x^n))^{1-y(n)} \tag{11.15}$$

Here, $\theta = [\theta_0, \theta_1, ..., \theta_n]$ are model parameters that are trained to maximise the likelihood. However, the log likelihood is often used as it is easier mathematically to work with. In this case, the log likelihood can be expressed using the following equation:

$$log\, L(\theta) = \sum_{n=1}^{N} log((g(\theta^T x^n))^{y(n)}((1 - g(\theta^T x^n))^{1-y(n)})$$

$$= \sum_{n=1}^{N} log(\, P(y^{(n)} = y_k \mid x^{(n)}; \theta\,)) \tag{11.16}$$

To find the value of θ that maximises the log likelihood, we differentiate $log\, L(\theta)$ with respect to θ such that,

$$\frac{\partial\, log\, L(\theta)}{\partial \theta} = \frac{\partial \sum_{n=1}^{N} log(\, P(y^{(n)} = y_k \mid x^{(n)}; \theta\,))}{\partial \theta} \tag{11.17}$$

By taking partial derivatives, we obtain the following equation:

$$\frac{\partial \log L(\theta)}{\partial \theta_i} = \sum_{n=1}^{N} (y^{(n)} - g(\theta^T x^{(n)}) x_i^n \tag{11.18}$$

To find the maximum of $\log L(\theta)$, we can solve the problem in Eq. (11.18) using gradient ascent (Bottou 2012), also called gradient descent when a function is minimised, which updates θ_i by multiplying the derivative by a constant learning rate $\mathcal{L}$ such that,

$$\theta_i \leftarrow \theta_i + \mathcal{L} \sum_{n=1}^{N} (y^{(n)} - g(\theta^T x^{(n)}) x_i^n \tag{11.19}$$

Another method that can be used is stochastic gradient ascent, which depends on examples that are randomly picked at each iteration (Bottou 2012; Elkan 2012). This method is commonly used when the training time is important. Also, a different algorithm that can be used to maximise the log-likelihood of the logistic regression model is Newton's method (Lin et al. 2008).

11.3.1 Logistic Regression Regularisation

The learning process of logistic regression can cause θ to become very large, and this may lead to overfitting, which may result in poor prediction performance of the trained model. To avoid the problem of overfitting, one may penalise the objective function, i.e. the log-likelihood function, by adding a regularisation term. One commonly used approach is to add the term $L2$ norm to the objective function such that,

$$\log L(\theta) = \sum_{n=1}^{N} \log(P(y^{(n)} = y_k \mid x^{(n)}; \theta)) - \frac{\lambda}{2} ||\theta||^2 \tag{11.20}$$

where λ is the regularisation parameter. With this regularised log likelihood, the gradient ascent update becomes,

$$\theta_i \leftarrow \theta_i + \mathcal{L} \sum_{n=1}^{N} (y^{(n)} - g(\theta^T x^{(n)}) x_i^n - \mathcal{L} \lambda \theta_i \tag{11.21}$$

11.3.2 Multinomial Logistic Regression Model (MLR)

Multinomial logistic regression (MLR), also called softmax regression in artificial neural networks (ANNs), is a linear supervised regression model that generalises the logistic regression where labels are binary, i.e. $c^{(i)} \in \{0, 1\}$ to multiclassification problems that have labels $\{1 \ldots c\} c^{(i)} \in \{1, \ldots, K\}$, where c is the number of classes. Briefly, we present the simplified MLR model as follows.

In MLR with multi-labels $c^{(i)} \in \{1, \ldots, c\}$ $c^{(i)} \in \{1, \ldots, K\}$, the aim is to estimate the probability $P(c = c^{(i)} \mid x) P(c = k \mid x)$ for each value of $c^{(i)} = 1$ to c, such that

$$h_\theta(x) = \begin{bmatrix} P(c = 1 \mid x; \theta) \\ P(c = 2 \mid x; \theta) \\ \cdot \\ \cdot \\ \cdot \\ P(c = K \mid x; \theta) \end{bmatrix} = \frac{1}{\sum_{j=1}^{K} \exp(\theta^{(j)T} x)} \begin{bmatrix} e^{\theta^{(1)T} x} \\ e^{\theta^{(2)T} x} \\ \cdot \\ \cdot \\ \cdot \\ e^{\theta^{(K)T} x} \end{bmatrix} \tag{11.22}$$

where $\theta^{(1)}, \theta^{(2)}, \ldots, \theta^{(K)} \in R^n$ are the parameters of the MLR model.

The LR and MLR are being considered in a large diversity of applications, e.g. forecasting (Montgomery et al. 1990), cancer classification (Zhou et al. 2004), hazard prediction (Ohlmacher and Davis 2003), etc. The following subsection describes briefly their application in machine fault diagnosis.

11.3.3 Application of Logistic Regression in Machine Fault Diagnosis

LR is used in different studies of machine fault detection and diagnosis. For example, Yan and Lee (2005) proposed a LR-based prognostic method for online performance degradation assessment and failure mode classification. In this method, the WPT is used to extract features from acquired data where wavelet packet energies are used as features and Fisher's criterion is used to select important features. With these selected features, LR is used to examine machine performance and identify failure types. The experimental results showed promising results for the analysis of both stationary and nonstationary signals. Caesarendra et al. (2010) proposed a combination between a relevance vector machine (RVM) and LR to examine machine failure degradation and prediction from incipient failure until final failure happened. In this study, LR is utilised to estimate failure degradation of rolling bearings using run-to-failure datasets, and the results are then considered as target vectors of failure probability. Then, the RVM is trained using run-to-failure bearing data and target vectors of failure probability obtained using LR. The experimental results showed the effectiveness of the proposed method as machine degradation assessment mode.

Moreover, Pandya et al. (2014) proposed a method for fault diagnosis of RE bearings using the MLR classifier and WPT. In this method, the features for classification are extracted through WPT using RIBO 5.5 wavelet. The whole classification is done using energy and kurtosis features. The efficiencies of three classifiers – ANN, SVM, and MLR – are compared. The experimental results showed that the MLR classifier is more effective than the other two classifiers, i.e. ANN and SVM. Phillips et al. (2015) proposed a method for machinery condition classification using oil samples and a binary logistic regression classifier. In this study, the authors argued that LR offers easy interpretability to industry experts, providing understanding about the drivers of the human classification process and the consequences of potential misclassification. A comparative study based on the predictive performance of LR, ANN, and SVM is conducted. The experimental results of a real-world oil analysis dataset from engines on mining trucks demonstrated that LR outperforms ANN and SVM techniques in terms of predicting healthy/not healthy engine conditions of engines.

Furthermore, Ahmed et al. (2017) suggested a strategy based on compressive sampling (CS) for bearing fault classification that uses fewer vibration measurements. In this strategy, compressive sampling is used to compress the acquired vibration signals. Then, three techniques are used to process these compressed data: using the compressed data directly as the input of a classifier, and extracting features from the data using linear discriminant analysis (LDA) and PCA. The MLR is used to deal with the classification problem. The experimental results showed the effectiveness of the proposed strategy in classifying rolling bearing faults.

Habib et al. (2017) proposed a method for mechanical fault detection and classification using pattern recognition based on the bispectrum algorithm. In this method, features extracted from the vibration bispectrum are used for fault classification of critical rotating components in an AH-64D helicopter tail rotor drive train system. The extracted features are reduced using PCA. Then, six classifiers – naïve Bayes, a linear discriminant, quadratic discriminant, SVM, MLR, and ANN – are selected to deal with the classification problem. The experimental results using features extracted from the conventional power spectrum showed that MLR and ANN achieved the highest classification performance, with 94.79% accuracy.

Recently, Li et al. (2018) proposed a method for fault diagnosis of rolling bearings at variable speeds using the Vold-Kalman filter (VKF), refined composite multi-scale fuzzy entropy (RCMFE), and Laplacian score (LS). In this method, (i) the VKF is used to remove the fault-unrelated components from the fault representation. (ii) The RCMFE is employed to extract features from the denoised vibration signal. With these extracted features, LS is used to improve the fault features by sorting the scale factors. Finally, (iii) with these selected features, LR is employed to deal with the classification problem. The experimental results demonstrated the efficiency of the proposed method in diagnosing rolling bearing faults under speed fluctuation.

To avoid the burden of storage requirements and processing time of a tremendously large amount of vibration data, Ahmed and Nandi (2018) proposed a combined CS-based, multiple measurement vector (MMV), and feature-ranking framework to learn optimally fewer features from a large amount of vibration data from which bearing health conditions can be classified. Based on this framework, the authors investigated different combinations of MMV-based CS and feature-ranking techniques to learn features from vibration signals acquired from rolling bearings. Three classification algorithms (MLR, ANN, and SVM) are tested to evaluate the proposed framework for the classification of bearing faults. The experimental results showed that the proposed framework is able to achieve high classification accuracy in all the faults considered in this study. In particular, results from this proposed framework with MLR and ANN showed the efficiency of this framework with high classification accuracies using different values of the sampling rate (α) and a number of selected features (k) for all the considered CS and feature-selection technique combinations.

11.4 Summary

This chapter has introduced two probabilistic models for classification: HMM, which is a probabilistic generative model, and logistic regression, which is a probabilistic

Table 11.1 Summary of some of the introduced techniques and their publically accessible software.

Algorithm name	Platform	Package	Function
Hidden Markov model posterior state probabilities	MATLAB	Statistics and Machine Learning Toolbox – Hidden Markov Models	hmmdecode
Multinomial logistic regression		Statistics and Machine Learning Toolbox – Regression	mnrfit
Maximum likelihood estimates		Statistics and Machine Learning Toolbox – Probability Distributions	mle
Gradient descent optimisation		Allison J. (2018)	grad_descent

discriminative model. These classifiers, i.e. HMM and LR, have been widely used in machine fault diagnosis. The first part of this chapter presented the HMM and different techniques for training this model and their application in machine fault diagnosis. The second part was devoted to a description of the LR model and a generalised LR model, which goes with the name MLR or multiple logistic regression, and its applications in machine fault diagnosis. Most of the introduced techniques and their publicly accessible software are summarised in Table 11.1.

References

Ahmed, H. and Nandi, A.K. (2018). Compressive sampling and feature ranking framework for bearing fault classification with vibration signals. *IEEE Access* 6: 44731–44746.

Ahmed, H.O.A., Wong, M.D., and Nandi, A.K. (2017). Compressive sensing strategy for classification of bearing faults. In: *2017 IEEE International Conference on Acoustics, Speech and Signal Processing (ICASSP)*, 2182–2186. IEEE.

Allison, J. (2018). Simplified gradient descent optimization. Mathworks File Exchange Center. https://uk.mathworks.com/matlabcentral/fileexchange/35535-simplified-gradient-descent-optimization.

Bishop, C. (2006). *Pattern Recognition and Machine Learning*, vol. 4. New York: Springer.

Bottou, L. (2012). Stochastic gradient descent tricks. In: *Neural Networks: Tricks of the Trade*, 421–436. Berlin, Heidelberg: Springer.

Caesarendra, W., Widodo, A., and Yang, B.S. (2010). Application of relevance vector machine and logistic regression for machine degradation assessment. *Mechanical Systems and Signal Processing* 24 (4): 1161–1171.

Elkan, C. (2012). Maximum likelihood, logistic regression, and stochastic gradient training. In: *Tutorial Notes at CIKM*, 11. http://cseweb.ucsd.edu/~elkan/250Bwinter2012/logreg.pdf.

Forney, G.D. (1973). The viterbi algorithm. *Proceedings of the IEEE* 61 (3): 268–278.

Gagniuc, P.A. (2017). *Markov Chains: From Theory to Implementation and Experimentation*. Wiley.

Gales, M. and Young, S. (2008). The application of hidden Markov models in speech recognition. *Foundations and Trends in Signal Processing* 1 (3): 195–304.

Habib, M.R., Hassan, M.A., Seoud, R.A.A., and Bayoumi, A.M. (2017). Mechanical fault detection and classification using pattern recognition based on bispectrum algorithm. In: *Advanced Technologies for Sustainable Systems*, 147–165. Cham: Springer.

Hassan, M.R. and Nath, B. (2005). Stock market forecasting using hidden Markov model: a new approach. In: *Proceedings - 5th International Conference on Intelligent Systems Design and Applications, 2005, ISDA'05*, 192–196. IEEE.

Hosmer, D.W. Jr., Lemeshow, S., and Sturdivant, R.X. (2013). *Applied Logistic Regression*, vol. 398. Wiley.

Krogh, A., Brown, M., Mian, I.S. et al. (1994). Hidden Markov models in computational biology: applications to protein modeling. *Journal of Molecular Biology* 235 (5): 1501–1531.

Lee, J.M., Kim, S.J., Hwang, Y., and Song, C.S. (2004). Diagnosis of mechanical fault signals using continuous hidden Markov model. *Journal of Sound and Vibration* 276 (3–5): 1065–1080.

Li, Z., Wu, Z., He, Y., and Fulei, C. (2005). Hidden Markov model-based fault diagnostics method in speed-up and speed-down process for rotating machinery. *Mechanical Systems and Signal Processing* 19 (2): 329–339.

Li, Y., Wei, Y., Feng, K. et al. (2018). Fault diagnosis of rolling bearing under speed fluctuation condition based on Vold-Kalman filter and RCMFE. *IEEE Access* 6: 37349–37360.

Lin, C.J., Weng, R.C., and Keerthi, S.S. (2008). Trust region newton method for logistic regression. *Journal of Machine Learning Research* 9 (Apr): 627–650.

Liu, J., Hu, Y., Wu, B. et al. (2017). A hybrid generalized hidden Markov model-based condition monitoring approach for rolling bearings. *Sensors* 17 (5): 1143.

Miao, Q. and Makis, V. (2007). Condition monitoring and classification of rotating machinery using wavelets and hidden Markov models. *Mechanical Systems and Signal Processing* 21 (2): 840–855.

Miao, Q., Huang, H.Z., and Fan, X. (2007). A comparison study of support vector machines and hidden Markov models in machinery condition monitoring. *Journal of Mechanical Science and Technology* 21 (4): 607–615.

Montgomery, D.C., Johnson, L.A., and Gardiner, J.S. (1990). *Forecasting and Time Series Analysis*, 151. New York etc.: McGraw-Hill.

Murphy, K.P. and Russell, S. (2002). Dynamic bayesian networks: representation, inference and learning. PhD thesis. University of California at Berkeley, Computer Science Division.

Nelwamondo, F.V., Marwala, T., and Mahola, U. (2006). Early classifications of bearing faults using hidden Markov models, Gaussian mixture models, mel-frequency cepstral coefficients and fractals. *International Journal of Innovative Computing, Information and Control* 2 (6): 1281–1299.

Ocak, H. and Loparo, K.A. (2001). A new bearing fault detection and diagnosis scheme based on hidden Markov modeling of vibration signals. In: *2001 IEEE International Conference on Acoustics, Speech, and Signal Processing, 2001. Proceedings. (ICASSP'01)*, vol. 5, 3141–3144. IEEE.

Ohlmacher, G.C. and Davis, J.C. (2003). Using multiple logistic regression and GIS technology to predict landslide hazard in Northeast Kansas, USA. *Engineering Geology* 69 (3–4): 331–343.

Pandya, D.H., Upadhyay, S.H., and Harsha, S.P. (2014). Fault diagnosis of rolling element bearing by using multinomial logistic regression and wavelet packet transform. *Soft Computing* 18 (2): 255–266.

Phillips, J., Cripps, E., Lau, J.W., and Hodkiewicz, M.R. (2015). Classifying machinery condition using oil samples and binary logistic regression. *Mechanical Systems and Signal Processing* 60: 316–325.

Rabiner, L.R. (1989). A tutorial on hidden Markov models and selected applications in speech recognition. *Proceedings of the IEEE* 77 (2): 257–286.

Sadhu, A., Prakash, G., and Narasimhan, S. (2017). A hybrid hidden Markov model towards fault detection of rotating components. *Journal of Vibration and Control* 23 (19): 3175–3195.

Seber, G.A. and Lee, A.J. (2012). *Linear Regression Analysis*, vol. 329. Wiley.

Tibshirani, R. (1996). Regression shrinkage and selection via the lasso. *Journal of the Royal Statistical Society. Series B (Methodological)* 58: 267–288.

Welch, L.R. (2003). Hidden Markov models and the Baum-Welch algorithm. *IEEE Information Theory Society Newsletter* 53 (4): 10–13.

Yan, J. and Lee, J. (2005). Degradation assessment and fault modes classification using logistic regression. *Journal of Manufacturing Science and Engineering* 127 (4): 912–914.

Yu, J. (2012). Health condition monitoring of machines based on hidden Markov model and contribution analysis. *IEEE Transactions on Instrumentation and Measurement* 61 (8): 2200–2211.

Zhou, X., Liu, K.Y., and Wong, S.T. (2004). Cancer classification and prediction using logistic regression with Bayesian gene selection. *Journal of Biomedical Informatics* 37 (4): 249–259.

Zhou, H., Chen, J., Dong, G., and Wang, R. (2016). Detection and diagnosis of bearing faults using shift-invariant dictionary learning and hidden Markov model. *Mechanical Systems and Signal Processing* 72: 65–79.

12

Artificial Neural Networks (ANNs)

12.1 Introduction

An artificial neural network (ANN) is a mathematical model that mimics biological neurons. A *neuron* is a special biological cell that process information. It receives signals using dendrites, processes the received signals using a cell body, and send signals to other neurons using an axon (Jain et al. 1996; Graupe 2013). An ANN is a group of connected nodes called *artificial neurons*, which mimic the neurons in a biological neural network (NN). In fact, an ANN often consists of a series of algorithms that work together to recognise underlying relationships in a set of data. The first model of neurons was formulated by McCulloch and Pitts (1943). It was a binary threshold unit computational model that computes a weighted sum of the input signals, $x_1, x_2, ..., x_n$, and produces an output of 1 if the weighted sum is above a given threshold. Otherwise, it produces an output of 0 (Jain et al. 1996). This model can be represented mathematically using the following equation,

$$y = f\left(\sum_{i=1}^{n}(w_i x_i - \tau) \right) \tag{12.1}$$

where $f(.)$ is a unit step function at 0, w_i is the weight of x_i, and τ is a threshold. This model has been developed in many ways. For example, Rosenblatt argued that the McCulloch and Pitts neuron model is not capable of learning because its parameters, i.e. the weight and the threshold coefficients, are fixed. Hence, the perceptron learning algorithm for the McCulloch and Pitts neuron model is introduced in (Rosenblatt 1958). Widrow and Hoff (1960) introduced the delta rule, also known as adaline, which is an adaptive linear neuron learning algorithm. Moreover, Cowan (1990) presented a short account of various investigations of NN properties in the period from 1943–1968. In 1982, Hopfield presented a model called neural networks based on the McCulloch and Pitts neuron and some characteristics of neurobiology and adapted to integrated circuits (Hopfield 1982). As described by Chua and Yang (1988), the basic characteristics of NNs are synchronous parallel processing, continuous-time dynamics, and global interaction of network elements.

The literature on NNs has highlighted various types of ANNs, e.g. multilayer perceptron (MLP), radial basis function (RBF), probabilistic neural network (PNN), etc. This chapter introduces some widely used ANN algorithms that have already been used for machine fault diagnosis using vibration signals. To begin with, this chapter presents

Condition Monitoring with Vibration Signals: Compressive Sampling and Learning Algorithms for Rotating Machines,
First Edition. Hosameldin Ahmed and Asoke K. Nandi.

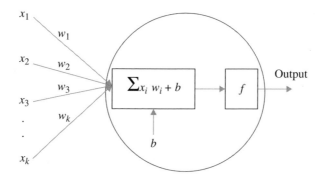

Figure 12.1 Model of an artificial neuron (Ahmed and Nandi 2018).

essential concepts of ANNs; then the chapter describes three different types of ANN (i.e. MLP, RBF, and Kohonen) that can be used for fault classification. In addition, the applications of these methods in machine fault diagnosis are described.

12.2 Neural Network Basic Principles

An ANN is a group of connected nodes called artificial neurons, which mimic the neurons in a biological NN. It is a supervised learning algorithm that has the ability to learn real, discrete, and vector-valued target functions (Nandi et al. 2013). In fact, an ANN often consists of a series of algorithms that work together to recognise underlying relationships in a set of data. The neuron receive inputs, multiplies it by the weights of each input, and combines the results of the multiplication. Then, the combined multiplications of the signals and weights are passed to a transfer function to generate the output of the neuron (Figure 12.1).

As illustrated in Figure 12.1, the artificial neuron is a computational model that transform a set of input signals $X = x_1, x_2, \ldots, x_k$ into a single output using a structure of two functions as follows:

- A weighted sum function, which is a net value function that uses the inputs and their corresponding weights to produce a value (v) that summarises the input data such that,

$$v = \sum_{i=1}^{k} (w_i x_i + b) \tag{12.2}$$

- An activation function, which transfers v into the output of the neuron such that

$$output = f(v) \tag{12.3}$$

The literature on ANNs has highlighted several types of artificial neurons based on the types of net value functions and activation functions. For example, a linear neuron that uses a weighted sum function as a net function and linear or piecewise linear function as an activation function, a sigmoid-based neuron that uses a weighted sum as a net function and a sigmoid function as an activation function, and a distance-based neuron that uses a distance function as a net function and a linear or piecewise linear as an activation function.

An ANN is a group of connected artificial neurons. Based on the type of connectivity between these neurons, various types of ANN architectures can be defined, as follows:

(1) *Layered network.* A layered network organises its neurons into hierarchical layers, which involves an input layer, hidden layer(s), and output layer. Each layer consists of a number of neurons that perform specific functions. Examples of layered networks include the feedforward neural network (FFNN) and multilayer perceptron (MLP), which involve an input layer, one to several hidden layers, and an output layer, and which often use a weighted sum function as a net function and a linear or sigmoid function as an activation function; RBF networks, which involve an input layer, a hidden radial basis layer, and an output linear layer that uses a Gaussian function as an activation function (Lei et al. 2009); and learning vector quantisation (LVQ) networks, which have a feedforward structure with a single computational layer of neurons where input neurons are connected directly to output neurons (Kohonen 1995).

(2) *Feedback network.* The feedback network, also called a recurrent or interactive network, often involves loops in the network. Typical examples of feedback networks include recurrent neural networks that consist of both feedforward and feedback connections between layers and neurons (Chow and Fang 1998).

The following subsections describe three different types of ANN (i.e. MLP, RBF, and Kohonen) that can be used for fault classification. In addition, the applications of these methods in machine fault diagnosis are described.

12.2.1 The Multilayer Perceptron

The MLP, also called a multilayer FFNN, involves an input layer, one to several hidden layers, and an output layer, and is one of the most commonly used NNs. It often uses a weighted sum function as a net function and a linear or sigmoid function as the activation function. An example of the MLP model, with an input layer, one hidden layer, and output layer, is illustrated in Figure 12.2. As shown in the figure, the layers are organised successively following the flow of data from the input layer that receives the input data $X = x_1, x_2, \ldots, x_k$ using its neurons and passes it to each neuron in the hidden layer, then processed using the hidden layer neurons and ending at the output layer. The process in

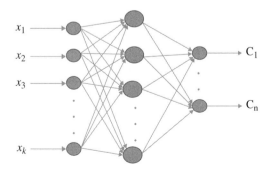

Figure 12.2 A multilayer perceptron model for ANN (Ahmed and Nandi 2018).

each neuron of MLP can be represented mathematically using the following equation,

$$y = f(v(x)) = \sum_{i=1}^{k}(w_i x_i + b) \tag{12.4}$$

where y is the output, the neuron has k inputs, w_i is a weight corresponding to the ith input x_i, and b is a bias term. Then, the produced value of each neuron is sent to each neuron in the output layer, which acts as a classification step. In order to minimise the error between the MLP's output and the target output, many error metrics can be used to train the MLP, e.g. minimum classification error (MCE), mean square error (MSE), and least mean log square (Gish 1992; Liano 1996). The optimisation of the MLP training is based on minimising the following objective function:

$$J(w) = \int J(x; w)p(x)dx \tag{12.5}$$

Most techniques include selecting some initial value w_0 for the weight vector and then moving through the weight space in a sequence of steps that can be represented using Eq. (12.6) (Bishop 2006),

$$w_{\eta+1} = w_\eta + \Delta w_\eta \tag{12.6}$$

where η is the iteration step. Different techniques use a different selection for the update of Δw_η. Most techniques utilise the gradient information. As described in Chapter 11, gradient descent optimisation can be used when the objective function is minimised. This can be done by selecting Δw_η in Eq. (12.6) to encompass a small step in the direction of the negative gradient such that,

$$w_{\eta+1} = w_\eta - \mathcal{L}\nabla w_\eta \tag{12.7}$$

where $\mathcal{L} > 0$ is the learning rate. Here, after each update, the gradient is re-examined for the updated weight, and this process is repeated. Also, there are other methods such as conjugate gradients and quasi-Newton methods that can be used to optimise the objective function. Unlike gradient decent, these methods, i.e. conjugate gradient and quasi-Newton, decreases the error function at each iteration unless the weight vector has reached a local or global minimum.

To evaluate the gradient of error function $E(w)$, the backpropagation, which describes the evaluation of derivatives, is used. It is used to find a local minimum of the error function $E(w)$. Let $E_n(w)$ denote the error function for a specific input pattern n such that

$$E_n(w) = \frac{1}{2}\sum_{i=1}^{k}(y_{nk} - t_{nk})^2 \tag{12.8}$$

where $y_{nk} = y_k(x_n, w)$. The gradient of this error function with respect to a weight w_{ji} can be defined using the following equation:

$$\frac{\partial E_n(w)}{\partial w_{ji}} = \frac{1}{2}\sum_{i=1}^{k}(y_{nk} - t_{nk})^2 \tag{12.9}$$

Generally, each neuron computes a weighted sum of its inputs such that

$$a_j = \sum_i w_{ji} z_i \tag{12.10}$$

Here, z_i is the activation of a neuron that sends a connection to unit j, and w_{ji} is the weight corresponding to that connection. Also, we can define Eq. (12.9) as follows,

$$\frac{\partial E_n(w)}{\partial w_{ji}} = \frac{\partial E_n(w)}{\partial a_j} \frac{\partial a_j}{\partial w_{ji}} \tag{12.11}$$

where

$$\delta_j = \frac{\partial E_n(w)}{\partial a_j} \tag{12.12}$$

$$z_i = \frac{\partial a_j}{\partial w_{ji}} \tag{12.13}$$

Then,

$$\frac{\partial E_n(w)}{\partial w_{ji}} = \delta_j z_i \tag{12.14}$$

Similarly, we can define δ_j as

$$\delta_j = \frac{\partial E_n(w)}{\partial a_j} = \sum_k \frac{\partial E_n(w)}{\partial a_k} \frac{\partial a_k}{\partial a_j} \tag{12.15}$$

More efficiently,

$$\delta_j = h'(a_j) \sum_k w_{kj} \delta_k \tag{12.16}$$

Here, $h(a_j) = z_j$. To evaluate the derivatives, we first compute δ_j and then apply it to Eq. (12.14).

12.2.2 The Radial Basis Function Network

The RBF network is a layered network that has the same structure as MLP. It uses a Gaussian kernel function as the activation function. Assume a c-class classification problem with K hidden neurons in the RBF and c output nodes. The output function, i.e. the Gaussian function, of the neuron can be computed using the following equation,

$$f(x) = e^{\left(-\frac{\|x - \mu_k\|^2}{2\sigma^2}\right)} \tag{12.17}$$

where $\|x - \mu_k\|$ is the activity of the kth neuron, μ_k is the centre of the NN in the kth hidden neuron, and σ is the width of the Gaussians. In the RBF NN, the input nodes pass the input signals to the connections arcs, and the first layer connections are not weighted, i.e. each hidden node (RBF unit) receives all the input values unchanged. The output can be calculated using Eq. (12.18):

$$y_c = \sum_{k=1}^{K} w_{k,c} f(x) \tag{12.18}$$

The training methodology of the RBF network is based on minimising the MSE between the output vector and the target vector. The RBF training process often consists of two steps: (i) define the hidden layer parameters, i.e. the set of centres and the number of hidden nodes, and (ii) determine the connection weights between the hidden layer and the output layer (Maglogiannis et al. 2008).

12.2.3 The Kohonen Network

The Kohonen neural network (KNN), also called a self-organising map (SOM), is a NN model and algorithm that implements characteristic nonlinear projections from high-dimensional space onto a low-dimensional array of neurons (Kohonen et al. 1996). In engineering, the most straightforward applications of the SOM are in the identification and monitoring of complex machine and process conditions. The SOM can be defined as a nonlinear, ordered, smooth mapping of a high-dimensional input data space onto the low-dimensional array. Let $X = \{x_1, x_2, ..., x_N\}^T$ be a set of input signals and $r_i = \{v_{i,1}, v_{i,2}, ..., v_{i,n}\}^T$ be a parametric real vector related to each element in the SOM array. We define the 'image' index b of input vector x on the SOM array using the following equation,

$$b = arg\,\min_i\{d(x, r_i)\} \tag{12.19}$$

where $d(x, r_i)$ is the distance between x and r_i. Our main task here is to determine r_i such that the mapping is ordered and descriptive of the x distribution. One way to determine r_i is by using an optimisation process following the idea of vector quantisation (VQ). The basic idea is to place a codebook vector r_c into the space of x signals where r_c is closest to x in the signal space. VQ then minimises the expected quantisation error E such that,

$$E = \int f[d(x, r_c)]p(x)dx \tag{12.20}$$

where $d(x, r_c)$ is the distance between the codebook vector and the signal, x, and $p(x)$ is the probability density function of x. The parameter r_i that minimises E is the solution of the VQ problem. An alternative technique to estimate r_i that minimises E is by using stochastic approximation such that,

$$E'(t) = \sum_i h_{ci}\,f[d(x(t), r_i(t))] \tag{12.21}$$

The approximate optimisation algorithm can be represented as

$$r_i(t + 1) = r_i(t) - \frac{1}{2\lambda(t)\frac{\partial E'}{\partial r_i(t)}} \tag{12.22}$$

Various SOM algorithms can be represented using Eq. 12.20.

There are also many other methods, including fuzzy-ANNs (Filippetti et al. 2000), learning vector quantization (LVQ) (Wang and Too 2002), etc. Because of limitations on space, we cannot detail them all here. Readers who are interested are referred to the more comprehensive review (Meireles et al. 2003).

ANNs are being considered in a large diversity of applications, e.g. image processing (Egmont-Petersen et al. 2002), medical diagnosis (Amato et al. 2013), stock market index prediction (Guresen et al. 2011), etc. The following subsection briefly describes their application in machine fault diagnosis.

12.3   Application of Artificial Neural Networks in Machine Fault Diagnosis

For the application of ANN algorithms in machine fault diagnosis problems, the following selections concerning ANN system parameters have to be considered first. These include,

- Type of ANN to be used (distance-based or weighted sum–based NN)
- Number of layers and number of nodes in each layer
- Activation function to be used
- Training technique to be used
- Number of epochs to be used
- Validation technique to be used (different validation techniques will be described in Chapter 15)

Many research efforts have been dedicated to the application of NNs in condition monitoring and fault diagnosis. In an early example of research into machine diagnosis using ANNs, Knapp and Wang proposed a back-propagation NN for machine fault diagnosis (Knapp and Wang 1992). D'Antone (1994) implemented the MLP in a parallel architecture based on a transputer-based system to speed up the computation time of the network in an expert system used for fault diagnosis. Peck and Burrows (1994) presented an approach based on ANNs to predict and identify machine component conditions in advance of complete failure. In 1966, McCormick and Nandi described the use of ANNs as a method for automatically classifying the machine condition from the vibration time series. For example, a method for extracting features to use as ANN inputs is described in McCormick and Nandi (1996a). In this method, a combination of the horizontal and vertical vibration time series is used to produce a complex time series. Then, the zero-lag higher-order moments of the magnitude of this time series, its derivative, and its integral are estimated. Moreover, in (McCormick and Nandi 1996a,b, 1997b), statistical estimates of vibration signals such as the mean and variance are also used as indications of faults in rotating machinery. The authors stated that using these estimates jointly can give a more robust classification than using them individually as an input to ANNs.

Furthermore, McCormick and Nandi (1997a, b) described the usage of ANNs as a classifier in machine condition monitoring using vibration time series. In these studies, several methods for the extraction of features that used as inputs to ANNs are described and compared. Real vibration signals of a rotating shaft with four different machine condition – no faults (NO), only rub applied (N-R), only weight added, and both rub and weight applied (W-R) – are used to validate the proposed method. Based on their experimental results, the authors compared the ANN as a classification system with other methods – the thresholding classification and the nearest centroid classification – and found that thresholding was not an appropriate method as it could only use one feature to make a decision. The nearest-neighbour classification as a simple multi-feature technique performed slightly better; however, it requires large storage space and is very computationally expensive and is therefore not suitable for real-time implementation. Therefore, for a continuous monitoring system, the authors recommended the use of ANNs.

Subrahmanyam and Sujatha (1997) developed two NN techniques based on the supervised error back propagation (EBP) learning algorithm and unsupervised adaptive resonance theory-2 (ART2) based training paradigm, for recognizing bearing conditions. Real vibration signals acquired from a normal bearing and two different faulty bearings under several load and speed conditions are used to validate the proposed methods. A number of statistical parameters are obtained from the original vibration signals. The trained NNs are used for the identification of bearing conditions. The experimental results demonstrated the effectiveness of the proposed methods in roller bearing fault diagnosis. Li et al. (1998) presented an approach of using frequency-domain vibration signals and NN to detect common bearing faults from motor vibration data. In this study, vibration signals with varying severity caused by bearing looseness, IR fault, RE fault, and combinations of them are used to validate the proposed approach. The fast Fourier transform (FFT) is employed to extract features from these vibration signals. With these extracted features, a three-layer NN with 10 hidden nodes is used to deal with the fault-detection problem. The experimental results showed that NNs can be used effectively in the detection of various bearing faults using vibration signals from motor bearings.

Jack and Nandi (1999, 2000) examined the use of a genetic algorithm (GA) to select the most-significant input features for ANNs from a large set of possible features in machine condition monitoring contexts. Real vibration data acquired from bearings with a total of six different conditions – two normal conditions and four fault conditions – are used to validate their proposed method. The experimental results showed that the GA is able to selecting a subset of 6 inputs from a set of 156 features that allow the ANN to perform with 100% accuracy. In the same vein, Samanta et al. (2001) presented a procedure for diagnosing gear conditions using GAs and ANNs. In this study, the time domain vibration signals of a rotating machine with normal and faulty gears are processed for feature extraction using the mean, the root mean square, the variance, skewness, and kurtosis. The selection of input features and the number of nodes in the hidden layer of the ANN are optimised using a GA-based technique in combination with the ANN. The output layer consists of two binary nodes indicating the condition of the machine, i.e. normal or faulty gears. The experimental results showed the effectiveness of the proposed method in machine fault detection. Also, Saxena and Saad (2007) presented the results of their investigation into the use of GAs and ANNs for condition monitoring of mechanical systems. In this study, the raw vibration signals are first normalised, and then five feature sets are considered: (i) statistical features from the raw vibration signals; (ii) statistical features from the sum signal; (iii) statistical features from the difference signals; (iv) spectral features; and (v) all the features considered together. Then, a GA is used to select the best features of all the considered features. With these selected features, the ANN is used to deal with the fault classification problem. Vibration signals acquired from bearings are used to examine the effectiveness of the proposed method. The results showed that GA-evolved ANNs clearly outperform the stand-alone ANNs.

An alternative method to optimise ANN-based fault diagnosis techniques is to use the UTA feature-selection algorithm. To investigate the effectiveness of UTA compared to a GA, Hajnayeb et al. (2011) designed an ANN-based fault-diagnosis system for a gearbox using a number of vibration features. In this study, several statistical features – maximum value, root mean square, kurtosis, crest factor, etc. – extracted from the time domain vibration signals are used as characteristics features. Then, UTA and a GA are used to select the best subset features from the extracted features. The results showed that the UTA method is as accurate as the GA, despite its simple algorithm.

Filippetti et al. (2000) presented a review of artificial intelligence (AI) based methods for diagnosing faults in induction motor drives. This review covers the application of expert systems, ANNs, and fuzzy logic systems and combined systems, e.g. (fuzzy-ANNs). Moreover, Li et al. (Li et al. 2000) presented a method for motor bearing fault diagnosis using NNs and time/frequency domain bearing vibration analysis. In this study, the authors used real and simulated vibration signals with looseness, IR, and ball faults to validate their proposed method. The experimental results demonstrated that ANNs can be effectively used in the fault diagnosis of motor bearings through appropriate measurement and interpretation of motor bearing vibration signals.

Vyas and Satishkumar (2001) presented an ANN-based design for fault recognition in a rotor-bearing system. In this study, a back-propagation learning algorithm (BPA) and a multilayer network, which contained layers created of nonlinear neurons and a normalisation technique, are used. Real vibration signals with several fault conditions – rotor with no fault, rotor with mass unbalance, rotor with bearing cap loose, rotor with misalignment, play in spider coupling, and rotor with both mass unbalance and misalignment – collected from a laboratory rotor rig are used to validate the proposed method. Statistical moments of the collected vibration signals are employed to train the ANNs. Several experiments are conducted to investigate the adaptability of different network architectures, and an overall success rate up to 90% is achieved.

Kaewkongka et al. (2001) described a method for rotor dynamic machine condition monitoring using the continuous wavelet transform (CWT) and ANNs. In this method, the CWT is used to extract features from collected vibration signals. The extracted features are then converted into a greyscale image that is used as a characteristic feature of each signal. Four types of machine operating conditions are investigated: balanced shaft, unbalanced shaft, misalignment shaft, and faulty bearing. The backpropagation neural network (BPNN) is used as a classification tool. The experimental results showed that the proposed method achieved a classification accuracy of 90%. Yang et al. (2002b) introduced third-order spectral techniques for the diagnosis of motor bearing conditions using ANNs. In this study, seven techniques based on the power spectrum, bispectrum, bicoherence spectrum, bispectrum diagonal slice, bicoherence diagonal slice, summed bispectrum, and summed bicoherence are examined as signal preprocessing techniques in fault diagnosis of induction motor roller bearings. Four health conditions of the bearings – normal, cage fault, IR fault, and OR fault – are considered in this study. The obtained features are used as inputs to the ANN to identify the bearing conditions. The experimental results demonstrated that the method based on the summed bispectrum achieved better results than the other six methods introduced in this study.

Hoffman and Van Der Merwe evaluated the performance of three different neural classification techniques – Kohonen SOMs, nearest neighbour rule (NNR), and RBF networks – on multiple fault-diagnosis problems. In this study, the authors used real vibration measurements acquired from roller bearings with six classes: three classes with no bearing fault and imbalance masses of 0, 12, and 24 g; and three classes with bearing OR faults and imbalance masses of 0, 12, 24 g. Then, several features are extracted from the collected vibration signals, including six time domain features – mean, root mean square, crest factor, variance, skewness, and kurtosis – and four frequency domain features – amplitude at rotational frequency in horizontal and vertical directions, ball pass frequency of the OR (BPFO) of the faulty bearing, higher-frequency domain components (HFDs), and BPFO obtained from the envelope spectra. These features are then normalised and used as input to the SOM, NNR, and RBF. Based on the experimental results, the authors demonstrated that the SOM can be used for multiclass problems if the class labels are known and the features are chosen and normalised correctly. NNR and RBF can accurately discriminate between different combinations of multiple fault conditions as well as identify the severity of fault conditions. Also, RBF classifiers have a speed advantage over NNR (Hoffman and Van Der Merwe 2002).

Wang and Too (2002) proposed a method for machine fault detection using higher-order statistics (HOS) and ANNs. In this method, HOS is used to extract features from vibration signals. With these extracted features, two types of NN (SOM and LVQ) are used to deal with the classification problem. SOM is used for collecting data at the initial stage, and LVQ is used at the recognition stage. Vibration signals acquired from a rotating pump with eight different health conditions are used to validate the proposed method. The experimental results demonstrated the effectiveness of the proposed method in recognizing rotating machine faults. Yang et al. (2002a) presented a review of a variety of AI-based diagnosis methods used for rotating machinery condition monitoring. Of these methods, the authors presented a review of eight types of NNs applied in diagnosing rotating machinery faults: backpropagation for feed-forward networks (BPFFN), FFNN, recurrent neural network (RNN), RBF, back propagation (BP), MLP, Kohonen SOM, and LVQ.

Shuting et al. introduced an adaptive RBF network called a two-level cluster algorithm for fault diagnosis of a turbine-generator using vibration signals (Shuting et al. 2002). This method can automatically calculate the number, centre, and width of hidden layer neurons in the RBF network. In this study, practical acquired vibration data with three conditions – normal operation, rotor excitation winding short circuit, and stator winding fault – are used to validate the introduced method. The results verified that the proposed method achieved high diagnosis precision in turbo-generator vibration fault diagnosis. Samanta et al. (2003) presented a study that compared the performance of bearing fault detection using ANNs and support vector machines (SVMs) with all the considered signal features and fixed parameters of the two classifiers. In this study, several statistical features of both the original vibration signals and some with preprocessing – such as differentiation and integration, low- and high-pass filtering, and spectral data of the signals – are used. GA is used for selection of input features and classifier parameters. Experimental vibration signals collected from roller bearings are used to validate the proposed methods. The results demonstrated that ANNs and SVMs achieved 100% classification accuracy using only six GA-based features.

Wu and Chow (2004) developed an RBF NN-based fault-detection technique for induction machine fault detection. In this study, after the collected vibration signals are transformed into the frequency domain using FFT, four feature vectors are extracted from the power spectra of the machine vibration signal. Then, the extracted features are used as inputs for the RBF NN for fault detection and classification. The authors proposed a cell-splitting grid algorithm to automatically determine the optimal network architecture of the RBF network. Unbalanced electrical faults and mechanical faults operating at different rotating speeds are used to validate the proposed method. The results showed the effectiveness of the proposed method for induction machine fault detection.

Yang et al. (2004) proposed a method for diagnosing faults in rotating machinery based on NN, which synthesises ART and the learning strategy of KNN. In this study, first, the discrete wavelet transform (DWT) is used to decompose the time domain signal into three levels. Then the transformed signal and the original signal are estimated using eight parameters: mean, standard deviation, root mean square, shape factor, skewness, kurtosis, crest factor, and entropy. Vibration signals acquired from a machinery fault simulator are used to test the proposed method. The experimental results showed the proposed method's effectiveness.

Guo et al. (2005) proposed a genetic programming (GP) based method for feature extraction from raw vibration data recorded from a rotating machine with six different conditions. Then, the extracted features are used as the inputs to ANN and SVM classifiers for the recognition of the six bearing conditions. The experimental results demonstrated that the proposed method achieved improved classification results compared with those using extracted features by classical methods. Castro et al. (2006) presented a method for diagnosing bearing faults based on ANNs and DWT. In this study, the authors used DWT to extract features from the original signals. Then, the extracted features are used as inputs to three different NNs – MLP, RBF, and PNN – to identify the bearing conditions. Vibration signals collected from motor bearings with a normal condition, IR fault, OR fault, and ball fault are used to validate the proposed method. The results showed that the PNN achieved better classification results than MLP and RBF.

Rafiee et al. (2007), presented an ANN-based process for fault detection and identification in gearboxes using a feature vector extracted from the standard deviation of wavelet packet coefficients of vibration signals. In this process, first, the collected vibration signals are preprocessed using the following steps: (i) synchronisation of vibration signals, achieved by applying interpolation using piecewise cubic Hermite interpolation (PCHI), and (ii) feature extraction using the standard deviation of wavelet packet coefficients. Then, a two-layer MLP made up of an input layer, a hidden layer, and an output layer is used to deal with the fault-diagnosis problem. Experimental vibration signals collected from experimental testing of the gearbox are used to test the proposed method. The results showed that an MLP network with a $16:20:5$ structure achieved 100% accuracy for identifying gear failures.

Sanz et al. (2007) presented a technique for rotating machine condition monitoring, combining auto-associative neural networks (AANNs) and a wavelet transform (WT) using vibration analysis. In this study, DWT is used to extract features from the collected vibration signals; then, from these features, the identification of significant changes is performed using AANNs. Real vibration signals obtained under two different operational conditions corresponding to high and low loads are used to test the proposed method. The results demonstrated the effectiveness of the proposed method in online fault diagnosis for rotating machinery.

Yang et al. (2008), presented a fault-diagnosis method for wind turbine gearboxes, based on ANN. In this method, to separate noise signals from tested vibration signals, the collected vibration signals are decomposed into four-level details by using wavelet decomposition and then reconstructing the real signals. Then, a three-layer BPNN is used to diagnosis the health condition of the gearbox. Experimental vibration signals with four kinds of typical patterns of gearbox faults, collected from a gearbox diagnostic testbed, are used to examine the proposed method. The results indicated that the BPNN is an efficient tool to solve complicated state-identification problems in gearbox fault diagnosis.

Al-Raheem et al. (2008) introduced a technique for detecting and diagnosing faults in rolling bearings using the Laplace-wavelet transform and ANNs. In this method, the Laplace-wavelet transform is used to extract features from the time domain vibration signals of rolling bearings. Then, the extracted features are used as input to ANNs for rolling bearing fault classification. The parameters of the Laplace-wavelet shape and the ANN classifier are optimised using a GA algorithm. Real and simulated bearing vibration data are used to test the effectiveness of the proposed method. The results showed the effectiveness of the proposed method in diagnosing rolling bearing faults.

Tyagi (2008) presented a comparative study of SVM classifiers and ANNs' application for rolling bearing fault diagnosis using WT as a preprocessing technique. In this study, (i) the features of the collected vibration signals are extracted using statistical features such as standard deviation, skewness, kurtosis, etc. Then, (ii) the extracted features are used as inputs to the SVM and ANN classifiers. Moreover, the effects of a preprocessing step using DWT prior to the feature-extraction step is also studied. Experimental vibration signals collected from bearings with four health conditions – normal, OR fault, IR fault, and roller fault – are used to validate the proposed method. The results showed that using the simple statistical features extracted from the time domain vibration signals and ANN or SVM, the bearing conditions can be correctly identified. Also, the results demonstrated that preprocessing with DWT improved the performance of both ANN and SVM classifiers in diagnosing rolling bearing faults.

Sreejith et al. (2008), introduced a fault-diagnosis method for rolling bearings using time domain features and ANNs. In this method, normal negative log-likelihood (Nnl) and kurtosis (KURT) are extracted from the time domain vibration signals, and then Nnl and KURT are used as input features for FFNN to diagnose faults in rolling bearings. The time domain vibration signals used in this study are acquired for four different health conditions of the rolling bearings: normal, RE fault, OR fault, and IR fault. The results demonstrated the effectiveness of the proposed method in diagnosing rolling bearing faults.

Li et al. (2009) applied order cepstrum and RBF NN for gear fault detection during a speed-up process. In this method, the time domain vibration signal during a gearbox speed-up process is sampled at constant time increments and then resampled at constant angle increments. Then, cepstrum analysis is used to process the resampled signal. For feature extraction, the order cepstrum with normal condition, wear, and a crack fault is processed. The extracted features are used as inputs to RBF for fault recognition. Experimental vibration data acquired from a gearbox are used to examine the proposed method. The results demonstrated the effectiveness of the proposed method in gear fault detection and identification.

Lei et al. (2009), introduced a method for intelligent fault diagnosis of rotating machinery based on the wavelet packet transform (WPT), empirical mode decomposition (EMD), dimensionless parameters, and RBF NN. In this method, WPT and EMD are used to preprocess the time domain vibration signals. Then, the dimensionless parameters are extracted from the original vibration signal and the preprocessed signals, which are formed into a combined feature set; their sensitivities are evaluated using the distance-evaluation technique. The sensitive features are selected and used as inputs to the RBF NN classifier. Vibration signals acquired from rolling bearings with a normal condition, IR fault, OR fault, and roller fault; and from a heavy oil catalytic cracking unit with a normal condition, a large area of rub fault, and a slight rub fault are used to validate the proposed method. The results showed the effectiveness of the proposed method in fault diagnosis for the rotating machine.

Saravanan et al. (2010) investigated the effectiveness of Morlet wavelet-based features for fault diagnosis of a gearbox using ANNs and proximal support vector machines (PSVMs), and the ability to use a Morlet wavelet in feature extraction. In this study, the Morlet wavelet is used to extract features from the collected vibration signals in the time domain; several statistical features such as kurtosis, standard deviation, peak, etc., are extracted from the Morlet coefficients. Then, the J48 is used to select the best features from the extracted features. Finally, the selected features are used as input to the ANN and PSVM for fault classification. Real vibration signals collected from a gearbox with good condition, gear tooth breakage (GTB), a gear with a crack at the root (GTC), and a gear with face wear are used to examine the proposed methods. The results showed that PSVM has an edge over ANN in the classification of features.

Castejón et al. (2010) developed a method for classifying bearing conditions using multiresolution analysis (MRA) and ANNs. In this study, the authors argued that the WT cannot be used practically by using analytical equations; hence, a discretisation process is needed. MRA is used to perform the discretisation. The mother wavelet, Daubechies-6, is used for feature extraction, and the fifth-level detail coefficients (cD5) are selected as characteristic features, which are normalised in the range [−1 1]. With these features, the MLP NN is used for bearing conditions classification. Four sets of experimental vibrations acquired from a roller bearing experimental system with a normal condition, IR fault, OR fault, and ball fault are used to validate the proposed method. The results showed that the proposed method is sound and detects four bearing conditions in a very incipient stage.

De Moura et al. (2011) evaluated principal component analysis (PCA) and ANN performance for diagnosing rolling bearing faults using vibration signals preprocessed by detrended-fluctuation analysis (DFA) and rescaled-range analysis (RSA). In this study, PCA and ANN are combined with DFA and RSA in a total of four approaches for diagnosing bearing faults. Vibration signals acquired from bearings with four health conditions are used to examine the proposed methods. The results demonstrated that the ANN-based classifier presented performance slightly better that the one based on PCA.

Bin et al. (2012) proposed a method for fault diagnosis of rotating machinery based on wavelet packets, EMD, and ANN. In this method, using vibration signals collected from an experimental rotor-bearing system with ten rotor failures, four stages of investigations are carried out. These are (i) wavelet packet decomposition is used to denoise the collected vibration signal; (ii) EMD is employed to obtain a series of intrinsic mode functions (IMFs) from the denoised signal; (iii) the moment of the energy of IMFs is calculated to express the failure feature; and (iv) a three-layer BPNN with fault feature from the frequency domain is employed as the target input of the NN. The energy in five spectral bandwidths of the vibration spectrum is taken as characteristic parameters, and ten types of representative rotor faults are taken as the output. The results showed the effectiveness of the proposed method in early fault diagnosis of rotating machinery.

Liang et al. (2013) introduced the application of the power spectrum, cepstrum, bispectrum, and ANN for fault pattern extraction from induction motors. This study compared the effectiveness of the power spectrum, cepstrum, and bispectrum for vibration, phase current, and transient speed analyses for detection and diagnostics of induction motor faults, based on experimental results. The authors found that for vibration signals, the power spectrum, cepstrum, and bispectrum presented a better ability to identify induction motors faults if the fault symptoms demonstrated characteristics in rich sidebands and harmonics. In addition, the authors stated that a combination of the power spectrum, cepstrum, and HOS methods along with ANN analysis should undoubtedly provide a better tool for condition monitoring and fault diagnosis of induction motors.

Ertunc et al. (2013) proposed a multistage method for detection and diagnosis of rolling bearing faults based on ANN and adaptive neuro-fuzzy inference system (ANFIS) techniques. In this study, both time and frequency domain parameters extracted from the vibration and current signals are used as inputs to ANN and ANFIS models. Experimental data acquired from a shaft-bearing system is used to examine the performance of the two approaches for diagnosing rolling bearing faults. The experimental results demonstrated that the ANFIS-based approach is superior to the ANN-based approach in diagnosing fault severity.

Zhang et al. (2013) proposed a method for classification of faults and prediction of degradation of components and machines in manufacturing systems using wavelet packet decomposition, FFT, and BPNN. In this method, the collected vibration signals were decomposed into several signals using wavelet.. Then, these signals were transformed to the frequency domain using FFT. Finally, the features extracted in the frequency domain were used as input to the BPNN. The peak values of FFT are selected as features to judge the degradation of the monitored machine. A case study was used to illustrate the proposed method, and the results showed the effectiveness of the proposed method.

Unal et al. (2014), presented a method for diagnostics of rolling bearing faults using envelope analysis, the FFT, and the FFNN. In this study, the authors suggested some methods to extract features using envelope analysis accompanied by the Hilbert transform (HHT) and FFT. Then, the extracted features are used as inputs to GA-based FFNN. Vibration signals, collected from rolling bearings in the experimental setup, are used to validate the proposed method. The experimental results verified the effectiveness of the proposed method in diagnosing faults in the rolling bearing.

Ali et al. (2015) introduced an approach for diagnosing bearing faults based on statistical features, EMD energy entropy, and ANN. In this techniques, (i) 10 statistical features are extracted from the time domain vibration signals. (ii) In addition to these 10-time domain features, EMD is used to extract some other features to form robust and reliable features. Finally, (iii) ANN is adopted to identify bearing health conditions. Also, the authors proposed a health index (HI) for online damage detection at an early stage. Three bearing run-to-failure vibration signals with a roller fault, an IR fault, and an OR fault are used to examine the effectiveness of the proposed method. The results showed that the proposed method achieved high classification accuracy in diagnosing bearing faults.

Bangalore and Tjernberg (2015) introduced a self-evolving maintenance-scheduler framework for maintenance management of wind turbines and proposed an ANN-based condition monitoring technique using data from a supervisory control and data acquisition system (SCADA). The Levenberg–Marquardt back-propagation (LM) training algorithm is used for training the NN. This ANN-based condition monitoring technique is applied to gearbox bearings with real data from onshore wind turbines. The results showed that the proposed technique is capable of identifying damage in the gearbox bearings almost a week before the vibration-based condition monitoring system (CMS) raised an alarm.

Janssens et al. (2016) proposed a feature-learning model for condition monitoring based on convolutional neural networks (CNNs). The CNN model is not applied to extracted features such as kurtosis, skewness, mean, etc. but to the raw amplitudes of the frequency spectrum of the vibration data. By applying CNN on the raw data, the network learns transformations on the data that result in better representation of the data for the fault-classification task in the output layer. Vibration signals collected from rolling bearings with eight health conditions are used to validate the proposed technique. The results showed that the feature-learning system based on CNN significantly outperformed the classical feature-engineering based approach that utilises manually engineered features and a random forest classifier. Another feature-learning technique for condition monitoring is introduced by Lei et al. (2016). In this study, the authors introduced a two-stage learning method for intelligent diagnosis of machines. In the first stage of this method, sparse filtering, which is an unsupervised two-layer NN, is used to directly learn features from the collected mechanical vibration signals. In the second stage, softmax regression, i.e. logistic regression, is employed to classify health conditions based on the learned features. Two vibration datasets acquired from a motor bearing and a locomotive bearing are used to validate the proposed method. The results demonstrated that the proposed method obtained high diagnosis accuracies.

Recently, Han et al. (2018) explored the performances of random forest, SVM, and two advanced ANNs – an extreme learning machine (ELM) and PNN – with different features using two datasets from rotating machinery. The results showed that random forest outperformed the comparative classifiers in terms of recognition accuracy, stability, and robustness of features, in particular with a small training set. Moreover, Ahmed and Nandi (2018) proposed combined compressive sampling (CS) based on a multiple-measurement vector (MMV) and feature-ranking framework to learn optimally fewer features from a large amount of vibration data from which bearing health conditions can be classified. Based on this framework, the authors investigated different combinations of MMV-based CS and feature-ranking techniques to learn features from vibration signals acquired from rolling bearings. Three classification algorithms (multinomial logistic regression (MLR), ANN, and SVM) are tested to evaluate the proposed framework for the classification of bearing faults. The experimental results showed that the proposed framework is able to achieve high classification accuracy in all the faults considered in this study. In particular, results from this proposed framework with MLR and ANN showed the efficiency of this framework with high classification accuracies using different values of the sampling rate (α) and a number of selected features (k) for all the considered CS and feature-selection technique combinations.

12.4 Summary

This chapter has presented essential concepts of ANNs and has described three different types of ANN (i.e. MLP, RBF network, and Kohonen network) that can be used for fault classification. In addition, the applications of these methods and several other types of ANN-based methods in machine fault diagnosis were described. A considerable

Table 12.1 Summary of some of the introduced techniques and their publically accessible software.

Algorithm name	Platform	Package	Function
Perceptron	MATLAB	Deep Learning Toolbox – Define shallow neural network architecture	Perceptron
MLP neural network trained by backpropagation		(Chen 2018)	mlpReg mlpRegPred
Radial basis function with k means clustering		(Shujaat 2014)	RBF
Design probabilistic neural network		Deep Learning Toolbox – Define shallow neural network architecture	newpnn
Train shallow neural network		Deep Learning Toolbox – Function Approximation and Clustering	train
Self-organising map		Deep Learning Toolbox – Function Approximation and Clustering-Self-organising maps	selforgmap
Gradient descent backpropagation			net.trainFcn = 'traingd'

amount of literature has been published on the application of ANNs and variants in machine fault diagnosis. Most of these studies introduced many preprocessing techniques that include normalisation, feature selection, transformation, and feature extraction. The produced data of the preprocessing step represent the final training set that is used as input to ANNs. In order to learn more useful features for machine fault diagnosis, most of the proposed methods combine two or more analysis techniques. For example, the GA algorithm along with various types of time domain statistical features, frequency domain, and time-frequency domain features have been widely used with different types of ANNs. Hence, the challenge will always remain to produce possible approaches to machine condition monitoring capable of improving fault diagnosis accuracy and reducing computations. Most of the introduced techniques and their publicly accessible software are summarised in Table 12.1.

References

Ahmed, H. and Nandi, A.K. (2018). Compressive sampling and feature ranking framework for bearing fault classification with vibration signals. *IEEE Access* 6: 44731–44746.

Ali, J.B., Fnaiech, N., Saidi, L. et al. (2015). Application of empirical mode decomposition and artificial neural network for automatic bearing fault diagnosis based on vibration signals. *Applied Acoustics* 89: 16–27.

Al-Raheem, K.F., Roy, A., Ramachandran, K.P. et al. (2008). Application of the Laplace-wavelet combined with ANN for rolling bearing fault diagnosis. *Journal of Vibration and Acoustics* 130 (5): 051007.

Amato, F., López, A., Peña-Méndez, E.M. et al. (2013). Artificial neural networks in medical diagnosis. *Journal of Applied Biomedicine* 11: 47–58.

Bangalore, P. and Tjernberg, L.B. (2015). An artificial neural network approach for early fault detection of gearbox bearings. *IEEE Transactions on Smart Grid* 6 (2): 980–987.

Bin, G.F., Gao, J.J., Li, X.J., and Dhillon, B.S. (2012). Early fault diagnosis of rotating machinery based on wavelet packets—empirical mode decomposition feature extraction and neural network. *Mechanical Systems and Signal Processing* 27: 696–711.

Bishop, C. (2006). *Pattern Recognition and Machine Learning*, vol. 4. New York: Springer.

Castejón, C., Lara, O., and García-Prada, J.C. (2010). Automated diagnosis of rolling bearings using MRA and neural networks. *Mechanical Systems and Signal Processing* 24 (1): 289–299.

Castro, O.J.L., Sisamón, C.C., and Prada, J.C.G. (2006). Bearing fault diagnosis based on neural network classification and wavelet transform. In: *Proceedings of the 6th WSEAS International Conference on Wavelet Analysis & Multi-Rate Systems*, 16–18. http://www.wseas.us/e-library/conferences/2006bucharest/papers/518-473.pdf.

Chen, M. (2018). MLP neural network trained by backpropagation. Mathworks File Exchange Center. https://uk.mathworks.com/matlabcentral/fileexchange/55946-mlp-neural-network-trained-by-backpropagation.

Chow, T.W. and Fang, Y. (1998). A recurrent neural-network-based real-time learning control strategy applying to nonlinear systems with unknown dynamics. *IEEE Transactions on Industrial Electronics* 45 (1): 151–161.

Chua, L.O. and Yang, L. (1988). Cellular neural networks: theory. *IEEE Transactions on Circuits and Systems* 35 (10): 1257–1272.

Cowan, J.D. (1990). Neural networks: the early days. In: *Advances in Neural Information Processing Systems*, 828–842. http://papers.nips.cc/paper/198-neural-networks-the-early-days.pdf.

D'Antone, I. (1994). A parallel neural network implementation in a distributed fault diagnosis system. *Microprocessing and microprogramming* 40 (5): 305–313.

De Moura, E.P., Souto, C.R., Silva, A.A., and Irmao, M.A.S. (2011). Evaluation of principal component analysis and neural network performance for bearing fault diagnosis from vibration signal processed by RS and DF analyses. *Mechanical Systems and Signal Processing* 25 (5): 1765–1772.

Egmont-Petersen, M., de Ridder, D., and Handels, H. (2002). Image processing with neural networks—a review. *Pattern Recognition* 35 (10): 2279–2301.

Ertunc, H.M., Ocak, H., and Aliustaoglu, C. (2013). ANN-and ANFIS-based multi-staged decision algorithm for the detection and diagnosis of bearing faults. *Neural Computing and Applications* 22 (1): 435–446.

Filippetti, F., Franceschini, G., Tassoni, C., and Vas, P. (2000). Recent developments of induction motor drives fault diagnosis using AI techniques. *IEEE transactions on industrial electronics* 47 (5): 994–1004.

Gish, H. (1992). A minimum classification error, maximum likelihood, neural network. In: *1992 IEEE International Conference on Acoustics, Speech, and Signal Processing, 1992. ICASSP-92*, vol. 2, 289–292. IEEE.

Graupe, D. (2013). *Principles of Artificial Neural Networks*, vol. 7. World Scientific.

Guo, H., Jack, L.B., and Nandi, A.K. (2005). Feature generation using genetic programming with application to fault classification. *IEEE Transactions on Systems, Man, and Cybernetics, Part B (Cybernetics)* 35 (1): 89–99.

Guresen, E., Kayakutlu, G., and Daim, T.U. (2011). Using artificial neural network models in stock market index prediction. *Expert Systems with Applications* 38 (8): 10389–10397.

Hajnayeb, A., Ghasemloonia, A., Khadem, S.E., and Moradi, M.H. (2011). Application and comparison of an ANN-based feature selection method and the genetic algorithm in gearbox fault diagnosis. *Expert Systems with Applications* 38 (8): 10205–10209.

Han, T., Jiang, D., Zhao, Q. et al. (2018). Comparison of random forest, artificial neural networks and support vector machine for intelligent diagnosis of rotating machinery. *Transactions of the Institute of Measurement and Control* 40 (8): 2681–2693.

Hoffman, A.J. and Van Der Merwe, N.T. (2002). The application of neural networks to vibrational diagnostics for multiple fault conditions. *Computer Standards & Interfaces* 24 (2): 139–149.

Hopfield, J.J. (1982). Neural networks and physical systems with emergent collective computational abilities. *Proceedings of the National Academy of Sciences of the United Stated of America* 79 (8): 2554–2558.

Jack, L.B. and Nandi, A.K. (1999). Feature selection for ANNs using genetic algorithms in condition monitoring. In: *ESANN European Symposium on Artificial Neural NetworksBrúges (Belgium)*, 313–318. https://www.researchgate.net/publication/221165522_Feature_selection_for_ANNs_using_genetic_algorithms_in_condition_monitoring.

Jack, L.B. and Nandi, A.K. (2000). Genetic algorithms for feature selection in machine condition monitoring with vibration signals. *IEE Proceedings-Vision, Image and Signal Processing* 147 (3): 205–212.

Jain, A.K., Mao, J., and Mohiuddin, K.M. (1996). Artificial neural networks: a tutorial. *Computer* 29 (3): 31–44.

Janssens, O., Slavkovikj, V., Vervisch, B. et al. (2016). Convolutional neural network based fault detection for rotating machinery. *Journal of Sound and Vibration* 377: 331–345.

Kaewkongka, T., Au, Y.J., Rakowski, R., and Jones, B.E. (2001). Continuous wavelet transform and neural network for condition monitoring of rotodynamic machinery. In: *Instrumentation and Measurement Technology Conference, 2001. IMTC 2001. Proceedings of the 18th IEEE*, vol. 3, 1962–1966. IEEE.

Knapp, G.M. and Wang, H.P. (1992). Machine fault classification: a neural network approach. *International Journal of Production Research* 30 (4): 811–823.

Kohonen, T. (1995). Learning vector quantization. In: *Self-Organizing Maps*, 175–189. Berlin, Heidelberg: Springer.

Kohonen, T., Oja, E., Simula, O. et al. (1996). Engineering applications of the self-organizing map. *Proceedings of the IEEE* 84 (10): 1358–1384.

Lei, Y., He, Z., and Zi, Y. (2009). Application of an intelligent classification method to mechanical fault diagnosis. *Expert Systems with Applications* 36 (6): 9941–9948.

Lei, Y., Jia, F., Lin, J. et al. (2016). An intelligent fault diagnosis method using unsupervised feature learning towards mechanical big data. *IEEE Transactions on Industrial Electronics* 63 (5): 3137–3147.

Li, B., Chow, M.Y., Tipsuwan, Y., and Hung, J.C. (2000). Neural-network-based motor rolling bearing fault diagnosis. *IEEE Transactions on Industrial Electronics* 47 (5): 1060–1069.

Li, B., Goddu, G., and Chow, M.Y. (1998). Detection of common motor bearing faults using frequency-domain vibration signals and a neural network based approach. In: *American Control Conference, 1998. Proceedings of the 1998*, vol. 4, 2032–2036. IEEE.

Li, H., Zhang, Y., and Zheng, H. (2009). Gear fault detection and diagnosis under speed-up condition based on order cepstrum and radial basis function neural network. *Journal of Mechanical Science and Technology* 23 (10): 2780–2789.

Liang, B., Iwnicki, S.D., and Zhao, Y. (2013). Application of power spectrum, cepstrum, higher order spectrum and neural network analyses for induction motor fault diagnosis. *Mechanical Systems and Signal Processing* 39 (1–2): 342–360.

Liano, K. (1996). Robust error measure for supervised neural network learning with outliers. *IEEE Transactions on Neural Networks* 7 (1): 246–250.

Maglogiannis, I., Sarimveis, H., Kiranoudis, C.T. et al. (2008). Radial basis function neural networks classification for the recognition of idiopathic pulmonary fibrosis in microscopic images. *IEEE Transactions on Information Technology in Biomedicine* 12 (1): 42–54.

McCormick, A.C. and Nandi, A.K. (1996a). A comparison of artificial neural networks and other statistical methods for rotating machine condition classification. In: *Colloquium Digest-IEE*, 2-2. IEE Institution of Electrical Engineers.

McCormick, A.C. and Nandi, A.K. (1996b). Rotating machine condition classification using artificial neural networks. In: *Proceedings of COMADEM*, vol. 96, 85–94. Citeseer.

McCormick, A.C. and Nandi, A.K. (1997a). Classification of the rotating machine condition using artificial neural networks. *Proceedings of the Institution of Mechanical Engineers, Part C: Journal of Mechanical Engineering Science* 211 (6): 439–450.

McCormick, A.C. and Nandi, A.K. (1997b). Real-time classification of rotating shaft loading conditions using artificial neural networks. *IEEE Transactions on Neural Networks* 8 (3): 748–757.

McCulloch, W.S. and Pitts, W. (1943). A logical calculus of the ideas immanent in nervous activity. *The Bulletin of Mathematical Biophysics* 5 (4): 115–133.

Meireles, M.R., Almeida, P.E., and Simões, M.G. (2003). A comprehensive review for industrial applicability of artificial neural networks. *IEEE Transactions on Industrial Electronics* 50 (3): 585–601.

Nandi, A.K., Liu, C., and Wong, M.D. (2013). Intelligent vibration signal processing for condition monitoring. In: *Proceedings of the International Conference Surveillance*, vol. 7, 1–15.

Peck, J.P. and Burrows, J. (1994). On-line condition monitoring of rotating equipment using neural networks. *ISA Transactions* 33 (2): 159–164.

Rafiee, J., Arvani, F., Harifi, A., and Sadeghi, M.H. (2007). Intelligent condition monitoring of a gearbox using an artificial neural network. *Mechanical Systems and Signal Processing* 21 (4): 1746–1754.

Rosenblatt, F. (1958). The perceptron: a probabilistic model for information storage and organization in the brain. *Psychological Review* 65 (6): 386.

Samanta, B., Al-Balushi, K.R., and Al-Araimi, S.A. (2001). Use of genetic algorithm and artificial neural network for gear condition diagnostics. In: *Proceedings of COMADEM*, 449–456. Elsevier.

Samanta, B., Al-Balushi, K.R., and Al-Araimi, S.A. (2003). Artificial neural networks and support vector machines with genetic algorithm for bearing fault detection. *Engineering Applications of Artificial Intelligence* 16 (7–8): 657–665.

Sanz, J., Perera, R., and Huerta, C. (2007). Fault diagnosis of rotating machinery based on auto-associative neural networks and wavelet transforms. *Journal of Sound and Vibration* 302 (4–5): 981–999.

Saravanan, N., Siddabattuni, V.K., and Ramachandran, K.I. (2010). Fault diagnosis of spur bevel gear box using artificial neural network (ANN), and proximal support vector machine (PSVM). *Applied Soft Computing* 10 (1): 344–360.

Saxena, A. and Saad, A. (2007). Evolving an artificial neural network classifier for condition monitoring of rotating mechanical systems. *Applied Soft Computing* 7 (1): 441–454.

Shujaat, K. (2014). Radial basis function with k mean clustering. Mathworks File Exchange Center. https://uk.mathworks.com/matlabcentral/fileexchange/46220-radial-basis-function-with-k-mean-clustering?s_tid=FX_rc2_behav.

Shuting, W., Heming, L., and Yonggang, L. (2002). Adaptive radial basis function network and its application in turbine-generator vibration fault diagnosis. In: *International Conference on Power System Technology, 2002. Proceedings. PowerCon 2002*, vol. 3, 1607–1610. IEEE.

Sreejith, B., Verma, A.K., and Srividya, A. (2008). Fault diagnosis of rolling element bearing using time-domain features and neural networks. In: *2008 IEEE Region 10 and the Third International Conference on Industrial and Information Systems*, 1–6. IEEE.

Subrahmanyam, M. and Sujatha, C. (1997). Using neural networks for the diagnosis of localized defects in ball bearings. *Tribology International* 30 (10): 739–752.

Tyagi, C.S. (2008). A comparative study of SVM classifiers and artificial neural networks application for rolling element bearing fault diagnosis using wavelet transform preprocessing. *Neuron* 1: 309–317.

Unal, M., Onat, M., Demetgul, M., and Kucuk, H. (2014). Fault diagnosis of rolling bearings using a genetic algorithm optimized neural network. *Measurement* 58: 187–196.

Vyas, N.S. and Satishkumar, D. (2001). Artificial neural network design for fault identification in a rotor-bearing system. *Mechanism and Machine Theory* 36 (2): 157–175.

Wang, C.C. and Too, G.P.J. (2002). Rotating machine fault detection based on HOS and artificial neural networks. *Journal of Intelligent Manufacturing* 13 (4): 283–293.

Widrow, B. and Hoff, Macian E., 1960. *Adaptive switching circuits*, pp. 96–104.

Wu, S. and Chow, T.W. (2004). Induction machine fault detection using SOM-based RBF neural networks. *IEEE Transactions on Industrial Electronics* 51 (1): 183–194.

Yang, B.S., Han, T., and An, J.L. (2004). ART–KOHONEN neural network for fault diagnosis of rotating machinery. *Mechanical Systems and Signal Processing* 18 (3): 645–657.

Yang, D.M., Stronach, A.F., MacConnell, P., and Penman, J. (2002b). Third-order spectral techniques for the diagnosis of motor bearing condition using artificial neural networks. *Mechanical Systems and Signal Processing* 16 (2–3): 391–411.

Yang, H., Mathew, J. and Ma, L., 2002a. Intelligent diagnosis of rotating machinery faults-a review.

Yang, S., Li, W., and Wang, C. (2008). The intelligent fault diagnosis of wind turbine gearbox based on artificial neural network. In: *International Conference on Condition Monitoring and Diagnosis, 2008. CMD 2008*, 1327–1330. IEEE.

Zhang, Z., Wang, Y., and Wang, K. (2013). Fault diagnosis and prognosis using wavelet packet decomposition, Fourier transform and artificial neural network. *Journal of Intelligent Manufacturing* 24 (6): 1213–1227.

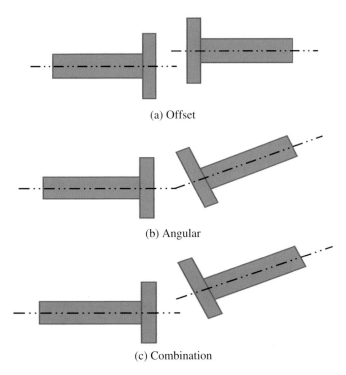

Figure 2.2 Illustrations of offset, angular, and combination shaft misalignment.

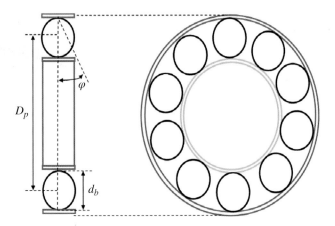

Figure 2.5 Rolling element bearing geometry.

Condition Monitoring with Vibration Signals: Compressive Sampling and Learning Algorithms for Rotating Machines,
First Edition. Hosameldin Ahmed and Asoke K. Nandi.
© 2020 John Wiley & Sons Ltd. Published 2020 by John Wiley & Sons Ltd.

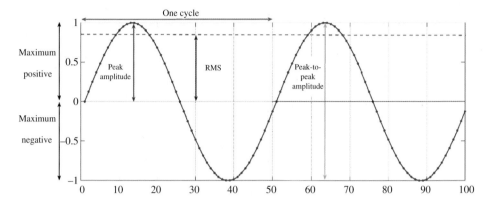

Figure 3.4 A pure sine wave with amplitude of 1 and 100 sample points.

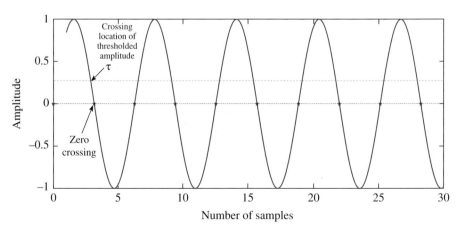

Figure 3.5 Crossing locations of amplitude equal to zero and a τ threshold amplitude.

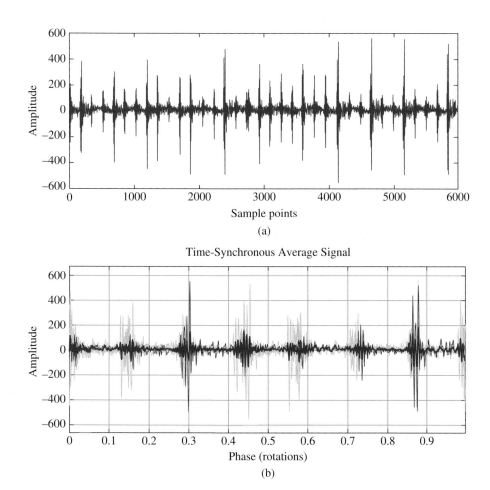

Figure 3.6 (a) Time-domain vibration signal of a roller bearing with an inner race fault condition; (b) its time-synchronous average signal.

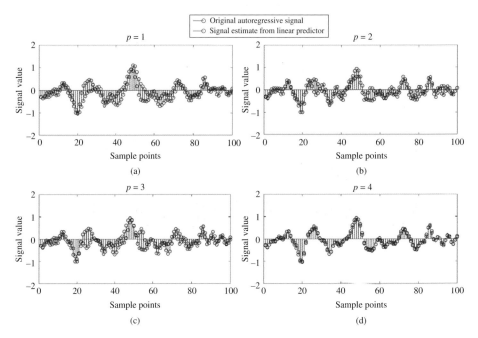

Figure 3.7 Examples of the original autoregressive signal and the linear predictor-based estimated signal of a vibration signal from a brand-new bearing using different values of *p*.

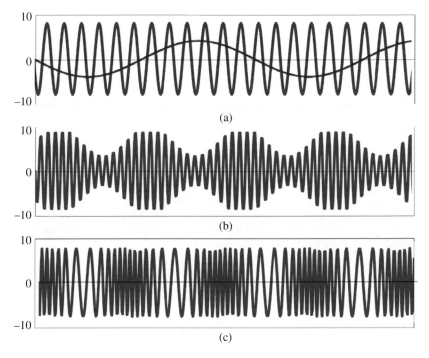

Figure 3.8 Example of sine wave amplitude and frequency modulation.

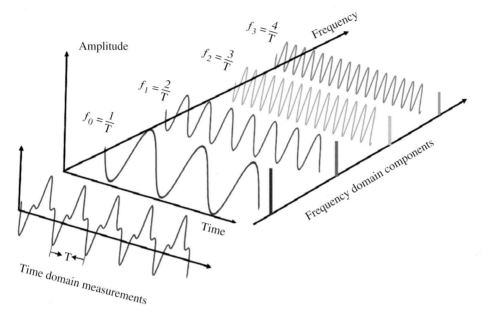

Figure 4.1 Time domain measurements vs. frequency domain measurements. Source: (Brandt 2011).

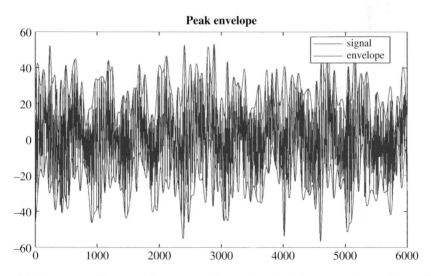

Figure 4.4 The upper and lower peak envelopes of a vibration signal from a worn but undamaged bearing (NW) from Figure 3.9, with length 1200 points.

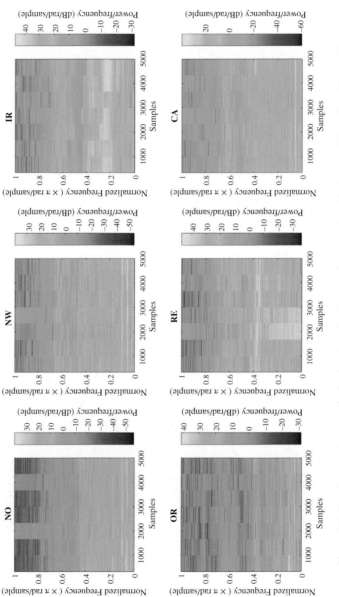

Figure 5.1 Spectrogram images of roller bearing vibrations under normal and faulty conditions, from Figure 3.9.

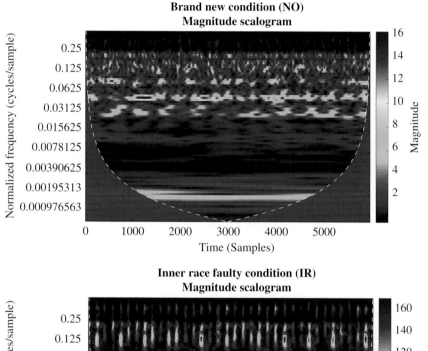

Brand new condition (NO)
Magnitude scalogram

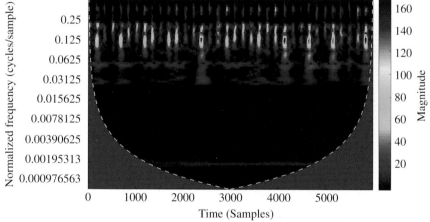

Inner race faulty condition (IR)
Magnitude scalogram

Figure 5.2 Magnitude scalogram images of roller bearing vibrations under normal and faulty conditions, from Figure 3.1.

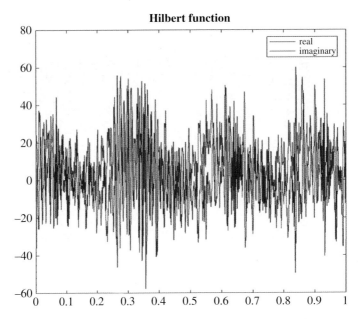

Figure 5.8 The real and imaginary signals of an outer race fault vibration signal produced using the Hilbert transform.

Figure 6.1 The overall framework of vibration-based machine condition monitoring.

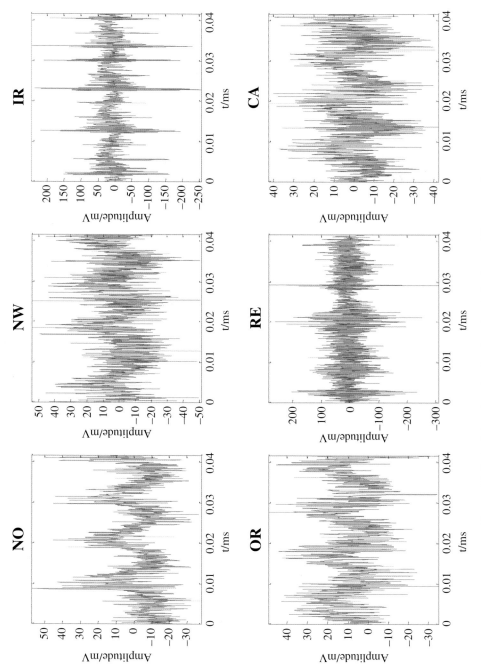

Figure 6.3 Typical time domain vibration signals for the six different conditions.

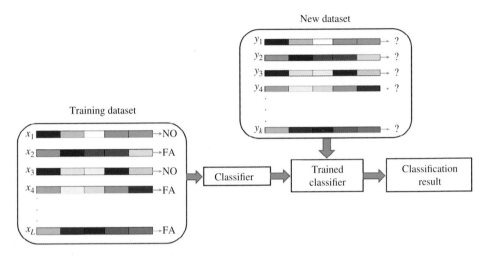

Figure 10.1 A simple example of classification to assign a given vibration signal to the 'normal' or 'fault' category.

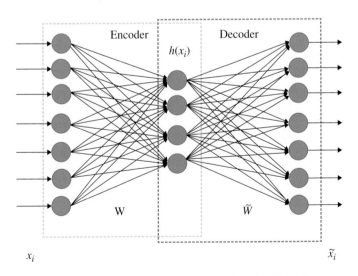

Figure 14.1 Autoencoder architecture (Ahmed et al. 2018).

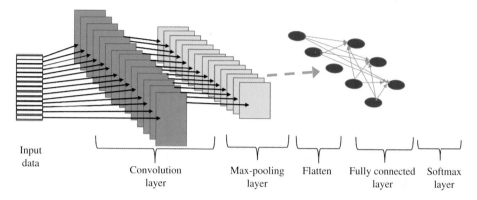

Input
data
Convolution
layer
Max-pooling
layer
Flatten
Fully connected
layer
Softmax
layer

Figure 14.2 Illustration of a one-layer CNN model with one convolutional layer, one max-pooling layer, one fully connected layer, and one softmax layer.

Table 15.1 Sample confusion matrix.

True classes	Predicted classes						Class prediction (%)
	NO	**NW**	**IR**	**OR**	**RE**	**CA**	
NO	200	0	0	0	0	0	100
NW	0	200	0	0	0	0	100
IR	0	0	199		1	0	99.5
OR	0	0	2	198	0	0	99
RE	0	0	5	0	195	0	97.5
CA	0	0	0	0	0	200	100

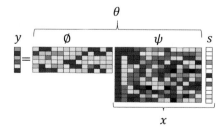

Figure 16.1 Compressive sampling framework (Ahmed and Nandi 2018c).

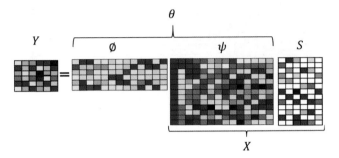

Figure 16.2 The MMV-CS model.

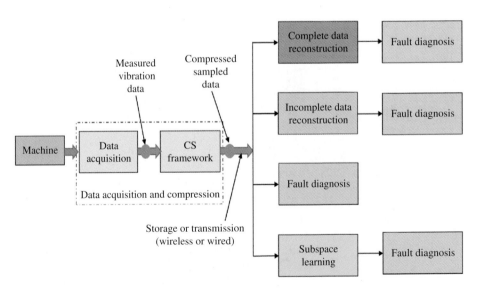

Figure 16.5 Four different compressive signal-processing schemes for machine fault diagnosis.

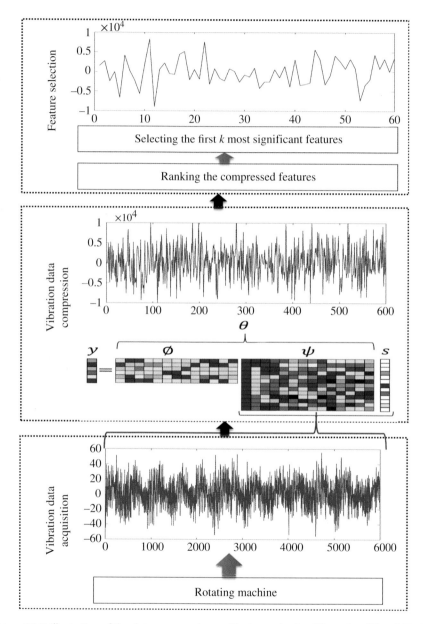

Figure 16.7 Illustration of the data compression and feature selection (Ahmed and Nandi 2018a).

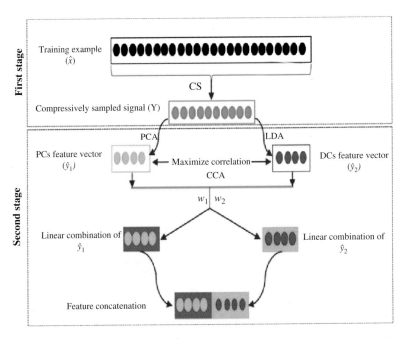

Figure 16.9 Illustration of the training process of the first and second stage of the proposed method (Ahmed and Nandi 2018c).

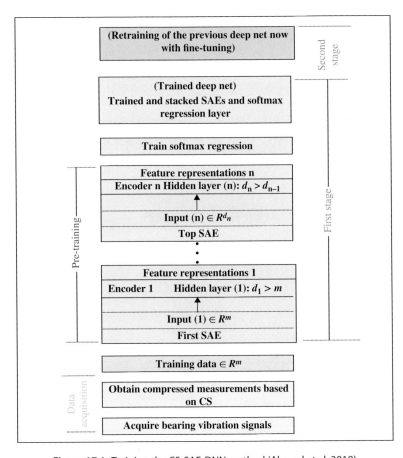

Figure 17.1 Training the CS-SAE-DNN method (Ahmed et al. 2018).

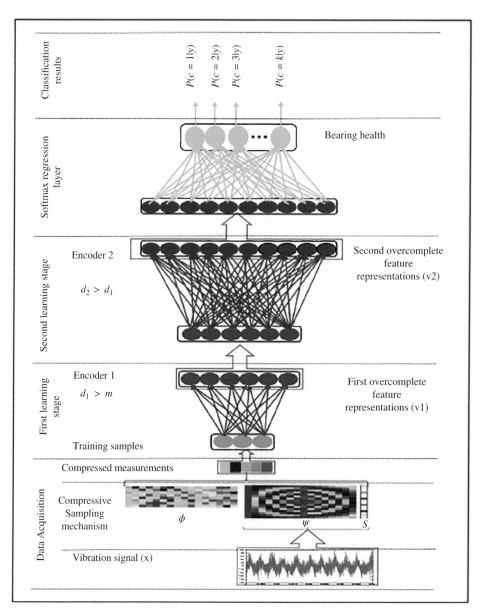

Figure 17.2 Illustration of the proposed method using two hidden layers. Data flow from the bottom to the top (Ahmed et al. 2018).

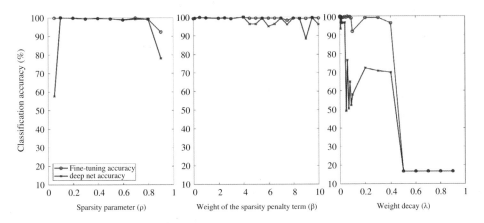

Figure 17.4 Effects of parameterization on classification accuracy (Ahmed et al. 2018).

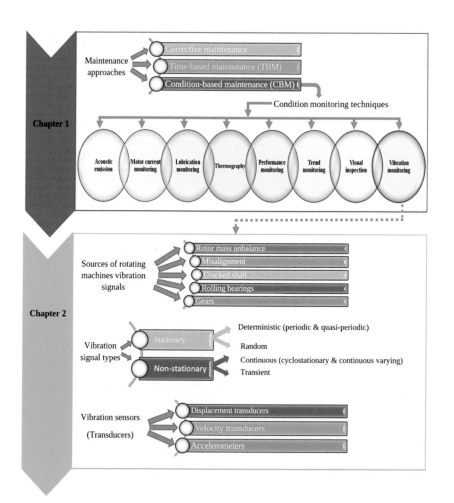

Figure 18.1 Topics covered by Part I of the book.

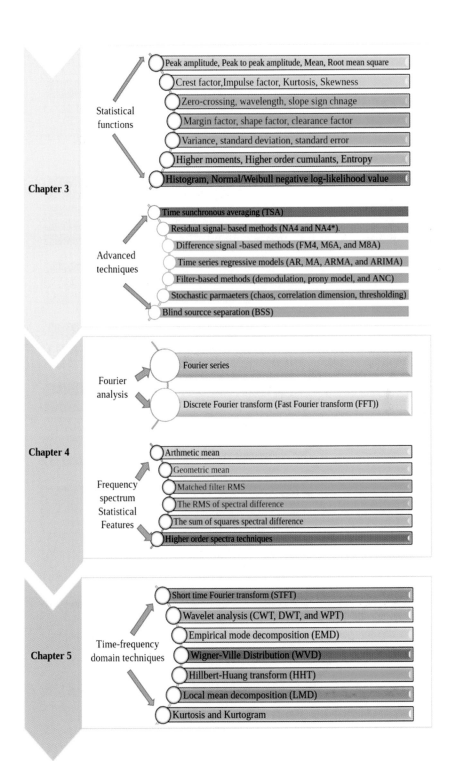

Figure 18.2 Topics covered by Part II of the book.

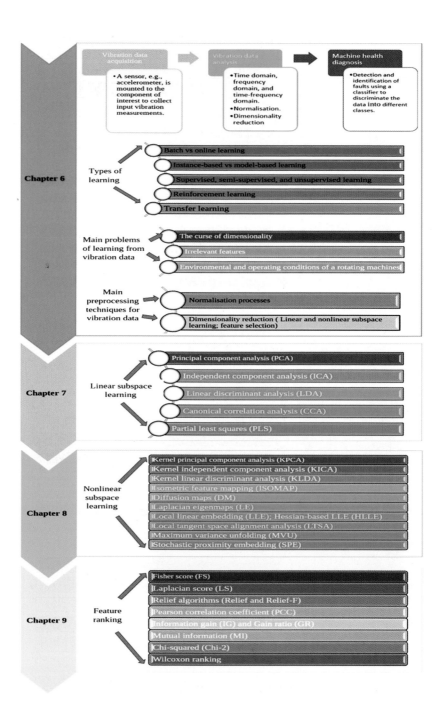

Figure 18.3 Topics covered by Part III of the book.

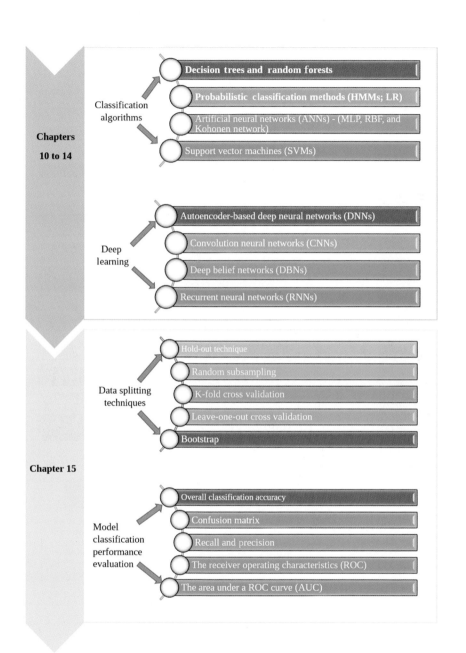

Figure 18.4 Topics covered by Part IV of the book.

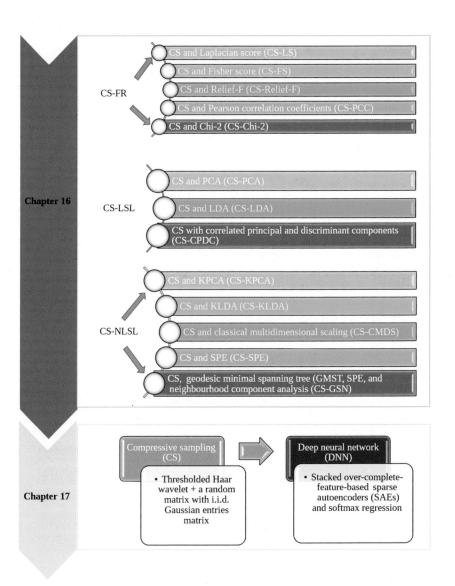

Figure 18.5 Topics covered by Part V of the book.

13

Support Vector Machines (SVMs)

13.1 Introduction

Support vector machines (SVMs) are one of the most popular machine learning methods used to classify machine health conditions using the selected feature space. This is a supervised machine learning method that was first proposed for binary classification problems (Cortes and Vapnik 1995). The SVM is based on a general framework called Vapnik–Chervonenkis (VC) theory, for estimating dependencies from empirical data (Vapnik 1982). Boser et al. (1992) presented a training algorithm that maximises the margin between training patterns and the decision boundary. In this algorithm, the solution is defined as a linear combination of supporting patterns, which is the subset of training patterns that are closest to the decision boundary. Also, Cortes and Vapnik (1995) demonstrated the high generalisation ability of SVMs using polynomial input transformations. The performance of SVMs in these two studies is demonstrated on handwritten digit recognition. In machine fault detection and diagnosis, the SVM is used for learning special patterns from the acquired signal, and then these patterns are classified according to the fault occurrence in the machine.

The basic idea of the SVM is that it can find the best hyperplane(s) to separate data from two different classes such that the distance between the two classes, i.e. the margin, is maximised. Based on the features of the data, the SVM can make linear or nonlinear classifications. Nonlinear classification is often used when the data is not linearly separable. For multiclass classification problems, several SVM classifiers can together deal with the multiclass issues. The basic SVM classifier is constructed from a simple linear maximum margin classifier. Briefly, we present the simplified SVM classifier as follows.

Given a set of examples along with their associated labels in the form of feature-label sets such that,

$$D = (x_1, y_1), (x_2, y_2), \dots, (x_L, y_L) \tag{13.1}$$

$$D = (x_i, y_i), \forall i = 1, 2, \dots, L \tag{13.2}$$

where $x_i \in R^N$ is an N-dimensional input vector and y_i is the associated class label for every $i = 1, 2, \dots, L$, suppose $y_i \in \{+1, -1\}$ for two potential classes *normal condition* (NO) and *fault condition* (FA), which is a binary classification problem. The basic idea of binary classification is to define a function $f(x)$ that can successfully predict the class label y_i for an input vector x_i. If the set of examples can be linearly separable, the function defines a boundary or a hyperplane that can separate the examples in the NO class from

Condition Monitoring with Vibration Signals: Compressive Sampling and Learning Algorithms for Rotating Machines,
First Edition. Hosameldin Ahmed and Asoke K. Nandi.

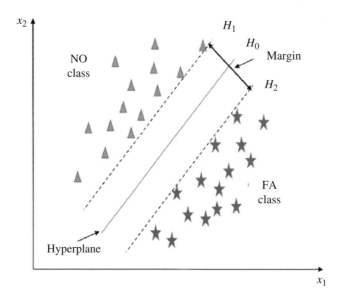

Figure 13.1 An example of a linear classifier for a two-class problem (Ahmed and Nandi 2018).

the examples in the FA class such that,

$$f(x) = sgn(w^T x + b) \tag{13.3}$$

where *sgn* is the sign of $(w^T x + b)$, *w* is an *N*-dimensional vector that defines the boundary, and b is a scalar. To find the best boundary, the following three hypothesises are defined,

$$H_0 = w^T x + b = 0 \tag{13.4}$$

$$H_1 = w^T x + b = -1 \tag{13.5}$$

$$H_2 = w^T x + b = +1 \tag{13.6}$$

As can be seen from Figure 13.1, the distance between H_1 and H_2 is known as a *margin*. The best classifier would be the one with the largest margin, i.e. the larger the margin is, the better is the classifier. Hence, the objective of the training algorithm is to find the parameter vector *w* that maximises the margin (*M*) such that,

$$M^* = \max_{w, \|w\|=1} M \tag{13.7}$$

$$s.t \; y_i(w^T x_i + b) \geq M \; \forall i = 1, \dots, L \tag{13.8}$$

The margin M^* is achieved for those patterns satisfying the following:

$$M^* = \min_i y_i(w^T x_i + b) \tag{13.9}$$

Accordingly, the problem of finding a hyperplane with a maximum margin can be defined as in Eq. (13.10):

$$\max_{w, \|w\|=1} \min_i y_i(w^T x_i + b) \forall i = 1, \dots, L \tag{13.10}$$

Instead of considering the norm of w to pick one of the infinite number of solutions as in Eqs. (13.8) and (13.10), the product of margin M and the norm of w can be defined as follows:

$$M\|w\| = 1 \tag{13.11}$$

Consequently, maximising the margin M is equivalent to minimising the norm of $\|w\|$. As a result, the problem of finding a maximum margin (M) in Eq. (13.7) can be rewritten as Eq. (13.12):

$$M^* = \min_{w} \|w\|^2 \tag{13.12}$$

$$s.t\ y_i(w^T x_i + b) \geq M, \forall i = 1, \ldots, L \tag{13.13}$$

This margin maximisation is meant to be in the direct space. In the dual space, Eq. (13.12) can be transformed by means of the Lagrange multipliers (Burges 1998) such that,

$$\mathcal{L}(w, b, \alpha) = \frac{1}{2}\|w\|^2 - \sum_{i=1}^{L} \alpha_i[y_i(w^T x_i + b) - 1] \tag{13.14}$$

$$s.t\ \alpha_i \geq 0, i = 1, \ldots, L \tag{13.15}$$

where α_i are Lagrange multipliers, also known as Kuhn-Tucker coefficients, that fulfil the following conditions:

$$\alpha_i(y_i(w^T x_i + b) - 1) = 0 \tag{13.16}$$

By taking the partial derivative of the Lagrange in Eq. (13.12) with respect to b and w,

$$\frac{\partial \mathcal{L}(w, b, \alpha)}{\partial b} = \sum_{i=1}^{L} \alpha_i y_i = 0 \tag{13.17}$$

$$\frac{\partial \mathcal{L}(w, b, \alpha)}{\partial w} = w^* - \sum_{i=1}^{L} \alpha_i y_i x_i = 0 \tag{13.18}$$

Hence,

$$w^* = \sum_{i=1}^{L} \alpha_i y_i x_i, \tag{13.19}$$

Substitution of the constraints from Eqs. (13.17) and (13.18) in Eq. (13.14) gives the Lagrangian dual,

$$\mathcal{L}_{Dual} = \sum_{i=1}^{L} \alpha_i - \frac{1}{2} \sum_{ij} \alpha_i \alpha_j y_i y_j x_i x_j \tag{13.20}$$

By solving the dual-optimisation problem, the coefficients α_i can be obtained. Hence, the nonlinear decision function can be defined using the following equation:

$$f(x) = sgn\left(\sum_{i=1}^{L} \alpha_i y_i(x_i x_j) + b\right) \tag{13.21}$$

Based on the features of the data, the SVM can make linear or nonlinear classifications by using different kernel functions that map the data to be classified into a high-dimensional feature space in which linear classification is possible. If the transformation function $(\Phi(x) = \emptyset_1(x), \ldots, \emptyset_p)$ is used to map the N-dimensional input vector x into the p-dimensional feature space, then the linear decision function in dual space can be given as follows:

$$f(x) = sgn\left(\sum_{i=1}^{L} \alpha_i y_i (\Phi(x_i)\Phi(x_j)) + b\right) \tag{13.22}$$

This can be computed by replacing $\Phi(x_i)\Phi(x_j)$ with the kernel functions $K(x_i, x_j)$, i.e. $K(x_i, x_j) = \Phi(x_i)\Phi(x_j)$ such that,

$$f(x) = sgn\left(\sum_{i=1}^{L} \alpha_i y_i K(x_i, x_j) + b\right) \tag{13.23}$$

The training of an SVM is performed by maximising the quadratic form from Eq. (13.20). The literature on SVM theory and application has identified various techniques for maximising the quadratic. For instance, Cortes and Vapnik (1995) in their original technique treated the SVM training as a quadratic programming (QP) problem. However, the QP problem is that it comprises solving $L \times L$ dimensional matrices, where L is the number of examples in the training dataset. Hence, for a large training dataset, i.e. for a large number of examples, this is a significant problem in terms of computation time. Therefore, several techniques have been proposed to overcome the QP problem. For example, Joachims (1998) proposed an algorithm for training the SVM based on a decomposition strategy. Moreover, Platt (1998) proposed a method for training the SVM called sequential minimal optimisation (SMO) that breaks down a large QP problem into a series of smaller QP problems. Furthermore, Campbell and Cristianini (1998) proposed the kernel adatron (KA) algorithm for training the SVM using a validation set (Campbell 2000).

13.2 Multiclass SVMs

The previous section discussed binary classification using SVMs where the class labels can take only two values: 1 and −1. However, in many real applications of machine fault diagnosis, we deal with more than two classes or conditions. For example, in roller bearing condition monitoring, there are several fault conditions, such as outer race (OR) faults, inner race (IR) faults, rolling element (RE) faults, and cage (CA) faults; and in phase induction motors, there are radial and angular shaft misalignment, mechanical looseness, rotor unbalances, etc. Thus, multiclass classification techniques are required to deal with these types of applications. For multiclass classification problems, several binary SVM classifiers can together deal with the multiclass problems. Various techniques based on binary classification used in multiclass problems are as follows:

(1) *One-against-all.* In this technique, several SVMs are used in parallel, e.g. for C classes, we need to build C binary classifiers, where each SVM is used to identify one class against all others.

(2) *One-against-one.* In this technique, for C classes, $C(C-1)/2$ binary classifiers are required to be trained on each subset of distinct classes c_i and c_j (c_i and $c_j \in C$).

(3) *Direct acyclic graph (DAG).* In this technique, rather than using all classifiers to make a prediction, one can perform one classifier at a time to exclude one class, and once $C-1$ runs are completed, there will be only one class left.

A comparison of methods for multiclass SVMs can be found in (Hsu and Lin 2002).

13.3 Selection of Kernel Parameters

The literature on SVMs has highlighted several kernel functions that can be used to perform nonlinear classification, such as linear, polynomial, and Gaussian radial basis functions (RBFs). The formulation of these kernel functions can be represented as follows:

1. Linear kernel:

$$K(x, x_i) = x^T.x_i \tag{13.24}$$

2. Gaussian RBF kernel

$$K(x, x_i) = \exp\left(\frac{-\|x^T - x_i\|^2}{2r^2}\right) \tag{13.25}$$

where $\|x^T - x_i\|^2$ is the squared Euclidean distance between two feature vectors.

3. Polynomial kernel

$$K(x, x_i) = (x^T x_i + 1)^d \tag{13.26}$$

where d is the degree of the classifier.

There are also some other SVM algorithms that have been used in machine fault diagnosis. For example, a wavelet support vector machine (WSVM) has been used in (Chen et al. 2013; Liu et al. 2013), and a least-square support vector machine (LSSVM) has been employed in (Gao et al. 2018).

13.4 Application of SVMs in Machine Fault Diagnosis

Using the greater ability of SVMs in the classification process, researchers have been able to perform detection and diagnosis of rotating machine health conditions. Typically, each fault in a rotating machine yields distinct features. Based on the acquired signals that contain a representation of faults in a rotating machine, to make a sensible deduction using an SVM-based classifier, we often compute certain features of the acquired signal that can describe the signal in essence. Sometimes, multiple features are computed to form a feature set. Depending on the number of features in the set, one may need to perform further filtering of the set using a feature-selection technique. Then, the SVM is used for learning special patterns from the selected signal, and these patterns are classified according to the fault happening in the rotating machine. As we described earlier in this chapter, the basic idea of the SVM is that it can find the best hyperplane(s) to separate data from two different classes such that the distance between the two classes, i.e. the margin, is maximised. Based on the features of the data, the SVM can make linear or nonlinear classifications. Therefore, with a suitable kernel function, one can perform nonlinear classification. Moreover, for multiclass classification problems, several SVM classifiers can together deal with the multiclass issues. For example, the one-against-all and one-against-one methods based on binary classification are commonly used in multiclass classification problems.

A large and growing body of literature has investigated the application of SVMs in diagnosing machine faults. For example, Jack and Nandi (2001) examined the performance of artificial neural network (ANN) and SVM classification algorithms in multiclass fault classification of roller bearings. Vibration signals acquired from roller bearings with six health conditions are used in this comparison. In this study, 18 different statistical features are computed on the basis of the moments and cumulants of the following:

(1) The raw vibration signals
(2) The derivative of the raw vibration signals
(3) The integral of the raw vibration signals
(4) The filtered raw vibration signals using an infinite impulse response (IIR) high-pass filter
(5) The filtered raw vibration signals using an IIR low-pass filter

Furthermore, the raw vibration signals are transformed to the spectral dataset using the fast Fourier transform (FFT). Training is performed using an ANN, and then the same experiments are repeated with SVMs. Three different datasets are used to perform these experiments: the first contains all the statistical features, i.e. the statistical features computed in items 1–5 in the previous list, with 90 features (18 features for each); the second has only spectral data with 66 features; and the third combines the statistical and spectral features with 156 features. The results showed that with all three datasets, the averaged SVM provided results that are worst comparable with the best ANN, and in two of the three cases it outperformed the performance of the ANN.

Jack and Nandi (2002) studied the performance of ANN and SVM classifiers in two-class fault/no-fault classification problems and attempted to improve the overall performance of both techniques through the use of genetic algorithm (GA) based feature selection. In this study, vibration data taken from two machines are used. Raw vibration data with six bearing health conditions are acquired from machine 1, where accelerometers are mounted vertically and horizontally on the bearing housing. A second set of raw vibration data with five health conditions is acquired from an accelerometer on one axis only. The results showed that by applying a GA, both ANNs and SVMs achieved a high classification rate for all three feature sets (statistical feature set, spectral feature set, and combined feature set) from machine 1 using significantly fewer features. Using spectral features extracted from raw vibration data of machine 2, both classifiers achieved a 100% success rate. This observation remained valid when GA feature selection is introduced.

Moreover, a procedure for detection of gear conditions using ANNs and SVMs with GA-based feature selection from time-domain vibration signals is presented in (Samanta 2004). In this study, statistical features are extracted from the raw vibration signals in the time domain, time derivative, and integral of signals, and from high- and low-pass filtering-based signals. Then, the selection of input features and the appropriate classifier parameters are optimised using a GA-based approach. Vibration data collected from a gear under no-fault and fault health condition are used to validate the proposed method. The results showed that the classification accuracy of SVMs was better than ANNs, without the GA algorithm. With GA-based feature election, the performance of both ANNs and SVMs was comparable, at a nearly 100% classification accuracy rate. Rojas and Nandi (2005, 2006) presented a practical SVM-based approach for fault classification of roller bearings. In this study, the training of the SVMs is performed using the SMO algorithm. Two vibration datasets are used to validate the proposed method. The first is collected from a small test rig with a variable-speed DC motor; a number of different roller bearings with six health conditions – OR fault, IR fault, RE fault, CA fault, brand-new, and worn (but undamaged) – are fitted in the rig. The second dataset is acquired from a small electrically driven submersible pump. Five different channels of vibration measurements are acquired from three bearings in the pump casing with a normal condition and OR fault condition. The results showed at least 97% successful classification.

Furthermore, Zhang et al. (2005) used genetic programming (GP) to detect faults in rotating machinery. In this study, three feature sets including statistical features, spectral features, and combined features, described in (Jack and Nandi 2001), are used. The training is performed using GP on three feature sets for machine 1 and a single feature set for machine 2. Each result is achieved in several hours. Compared to GA/ANN and GA/SVM, the computational time was reduced from several hundred hours to several hours. In GA/ANN and GA/SVM, the GA is used to select the optimal combination of input features and the proper size of the hidden layer as well as to select both the size of the kernel radius parameter and the individual inputs to the SVMs. The results showed that for the three feature sets from machine 1, GP with 10 features achieved 100% accuracy on all the training and test datasets. GA/ANNs with 12 features also achieved 100% accuracy, except for 99.9% on the training feature set for the statistical features. On the spectral feature set, GP required 6 features, while GA/ANN needed 9 features to provide a 100% accuracy rate. On the combined feature set, GP and GA/ANN showed the same results. Taken together, these results indicated that GP can play an important role in the field of machine condition monitoring.

Also, Guo et al. (2005) introduced a GP-based feature extractor for the generation/extraction of features from raw vibration data, applied to the problem of bearing fault classification. In this study, the GP-based feature extractor is used to extract useful information from raw vibration data in order to use it as input features for classifiers. Vibration signals collected from a machine over a series of 16 different speeds and 6 roller bearing health conditions are used to validate the proposed method. The results showed a significant improvement in classification results of both ANNs and SVMs using features extracted by GP compared with those using features extracted by classical methods.

Additionally, Yuan and Chu (2006) proposed a one-to-others multiclass SVM algorithm for fault diagnosis of turbopump rotors. In this study, vibration signals in the time domain are transformed into the frequency domain using FFT. Then, the frequencies of the transformed signals are divided into nine bands. The nine-dimensional fault feature vectors are transformed into two-dimensional fault feature vectors using principal component analysis (PCA). Finally, the SVM algorithm is employed to deal with the classification problem. The RBF is used as a kernel function. Vibration data covering 14 typical fault conditions of turbopump rotors are used to validate the proposed method: gear damage, structure resonance, rotor radial touch friction, rotor axial touch friction, shaft crack, bearing damage, body join looseness, bearing looseness, rotor parts looseness, pressure pulse, cavitation, vane rupture, rotor eccentricity, and shaft bend. The classification results demonstrated that the one-to-others multiclass SVM algorithm can effectively diagnosis multiclass faults in the turbopump rotor test bed.

Abbasion et al. (2007) introduced a method for multifault classification of RE bearings based on wavelet denoising and SVMs. In this method, a discrete Meyer wavelet is employed to reduce the effect of noise on the acquired data. Then, the SVM is used to deal with the classification problem. Vibration signals are acquired from a system of an electric motor that has two rolling bearings, one next to the output shaft and the other next to the fan; for each of them, there are four bearing health conditions – one normal and three false forms. The results showed the effectiveness of the proposed method in RE bearing fault classification.

Moreover, Hu et al. (2007) presented a method for diagnosing faults in roller bearings based on an improved wavelet packet transform (IWPT), a distance-evaluation technique, and SVMs ensemble. In this method, (i) IWPT is employed to extract salient frequency-band features from the raw vibration signals. Using these extracted salient frequency-band features, the fault features are detected using envelope spectrum analysis of wavelet package coefficients. (ii) The optimal features are selected from the wavelet package coefficients and statistical features of the raw signals using the distance evaluation technique. Finally, (iii) with these selected features, the SVMs ensemble with AdaBoost technique is utilised to classify the roller bearing health conditions. The vibration-based experimental results showed the efficacy of the SVMs ensemble in diagnosing roller bearing faults.

Furthermore, Widodo et al. (2007) studied the application of independent component analysis (ICA) and SVMs to detect and diagnosis induction motor faults. In this study, (i) statistical features are computed from the time domain and frequency domain of the acquired data. (ii) The ICA is used to reduce the dimensionality. Finally, (iii) the SVM with SMO and the strategy of multiclass SVM-based classification is applied to perform the fault classification. Datasets of vibration and stator current signals are used to validate the proposed method. The results showed the effectiveness of the combination of ICA and SVMs in diagnosing faults in induction motors. Similarly, Widodo and Yang (2007a) studied the application of nonlinear feature extraction and SVMs for fault diagnosis in induction motors. In this study, kernel-based PCA and ICA are employed to extract nonlinear features. Then, with these nonlinear features, a multiclass SVM technique is applied to deal with the classification problem. Vibration signals collected from induction motors under normal condition and with a rotor balance fault, a rotor bar broken fault, a stator fault, a faulty bearing, a bowed rotor, and eccentricity are used to test the proposed method. The results showed that nonlinear feature extraction using kernel PCA or kernel ICA improved the performance of the SVM classifier in induction motor faults diagnosis.

Sugumaran et al. (2007) presented a method for diagnosing roller bearing faults based on decision tree (DT) and proximal support vector machines (PSVMs). In this method, statistical features are first extracted from the acquired vibration signals using the mean, standard error, median, standard deviation, sample variance, kurtosis, skewness, range, minimum, maximum, and sum. Then, J48 is employed to select features from the extracted features. Finally, PSVM is used to classify roller bearing faults. Vibration signals collected from four SKF30206s are used to examine the proposed method. The results verified the effectiveness of the proposed method in classifying roller bearing faults.

Yang et al. (2007) proposed a fault-diagnosis technique for roller bearings based on the intrinsic mode function (IMF) envelope spectrum and SVMs. In this technique, (i) the original signals are decomposed into a number of IMFs using the empirical mode decomposition (EMD) method. (ii) Tthe ratios of amplitudes at the different fault characteristic frequencies in the envelope spectra of some IMFs that contain the main fault information are defined as the characteristic amplitude ratios. Finally, (iii) with these characteristic amplitude ratios, the SVM classifier is used to classify roller bearing health conditions. Vibration signals acquired from roller bearings are used to test the proposed method. The results verified the efficiency of the proposed method in diagnosing roller bearing faults. Additionally, Widodo and Yang (2007b) presented a survey of machine condition monitoring and fault diagnosis using SVMs.

Saravanan et al. (2008) studied the effectiveness of wavelet-based features for fault diagnosis of spur bevel gearboxes using SVMs and PSVMs. In this study, the statistical feature vectors of Morlet wavelet coefficients are classified using the J48 algorithm, and the most important features are used as input for training and testing SVMs and PSVMs. Vibration signals collected from a spur bevel gearbox in good condition and with gear tooth breakage, a gear with a crack at the root, and a gear with face wear are used to validate the proposed method. The results showed that the PSVM has an edge over the SVM in the classification of features.

Also, Xian and Zeng (2009) proposed a method for fault diagnosis of rotating machinery based on wavelet packet analysis (WPA) and hybrid SVMs. In this method, the acquired vibration signals are decomposed by WPA via the Meyer wavelet in the time-frequency domain. Then, a multiclass fault-diagnosis algorithm based on the one-to-others SVM technique is used to deal with the classification problem. Vibration signals from four gear conditions are used to validate the proposed method. The results showed the applicability and effectiveness of the proposed method in fault diagnosis of a rotating machine.

Guo et al. (2009) introduced a method for fault diagnosis of rolling bearings based on the Hilbert envelope spectrum and SVMs. In this method, the Hilbert envelope spectrum is used to extract fault features from the modulation characteristic of rolling bearing vibration signals. Then, a one-to-others SVM algorithm is employed to solve the multiclass classification problem. Real vibration signals measured from rolling bearings with ball faults, IR faults, and OR faults are used to validate the proposed method. The results verified the efficacy of the proposed method in diagnosing rolling bearing faults. Furthermore, Zhang et al. (2010) presented a fault-diagnosis method based on SVMs with parameter optimisation using an ant colony algorithm. In this method, (i) the wavelet transform is adopted to preprocess the raw signals, and then frequency domain statistical features are extracted from the decomposed signals; (ii) the distance-evaluation technique is used to select optimal features amongst all the extracted features; and (iii) the SVM algorithm is applied with parameter optimisation using the ant colony optimisation (ACO) algorithm. Practical vibration signals are measured from locomotive roller bearings under four conditions: normal, OR fault,

IR fault, and roller rub fault. The result demonstrated the efficiency of the proposed method in machine fault diagnosis.

Moreover, Konar and Chattopadhyay (2011) presented a method for detecting bearing faults in induction motors using the continuous wavelet transform (CWT) and SVMs. In this method, first, the CWT is used to extract the local information content of the data. Then, the SVM is adopted to classify the health condition of the roller bearing. Real vibration signals of healthy and faulty motors collected under no load, half load, and full load are used to verify the proposed method. The results showed the efficiency of the CWT-SVM combination in induction motor fault diagnosis.

Baccarini et al. (2011) introduced a method for diagnosing mechanical faults in phase induction motors. In this method, the collected acceleration signals are transformed into the frequency domain using FFT. Then, a filter that rejects vibrations at different frequencies f_r, $2f_r$. $3f_r$, and $4f_r$ is used to select the input data from the rotation frequencies. With this selected input data, the SVM algorithm is utilised to deal with the classification problem. Practical vibration signals from a no-fault condition and three types of mechanical faults including radial and angular shaft misalignment, mechanical looseness, and rotor unbalances, acquired at six positions of the motor, are used to examine the proposed method. The results showed that the best position for signal acquisition and analysis is vertical over the cooling fan.

In addition, Sun and Huang (2011) introduced a method for fault diagnosis for steam turbine-generators (STGS) based on SVMs. In this study, the acquired vibration signals are transformed into discrete spectra using the Fourier transformation. The obtained discrete spectra are partitioned into several frequency bands. Then, the SVM is employed to deal with the diagnosis problem. Vibration signals have eight features representing the vibration amplitudes of different frequency bands, which can be defined as follows: $<0.4f$, $04f - 0.49f$, $0.5f$, $0.51f - 0.99f$, f, $2f$, $3f - 5f$, and $> 5f$. These signals represent six typical mechanical vibration faults – misalignment, unbalance, looseness, rubbing, oil whirl, and steam whirl – acquired from STGS and used to verify the proposed method. The parameters c and σ of the SVM are set to 20 and 80, respectively, as determined by trial-and-error experiments. The results showed the effectiveness of the proposed method in vibration fault diagnosis of STGS.

Li et al. (2012) introduced a method for mechanical fault diagnosis based on the redundant second generation wavelet packet transform (RSGWPT), neighbourhood rough set (NRS), and SVMs. In this method, (i) RSGWPT is utilised to extract features from the sampled vibration data. (ii) For each of the resultant sub-band wavelet packet coefficients, nine statistical characteristics including peak value, mean, standard deviation, root mean square, shape factor, skewness, kurtosis, crest factor, and pulse index are computed. (iii) The NRS is employed to select the key features. Finally, (iv) with these selected features, the SVM is used to accomplish fault classification. Vibration signals acquired from two types of experimental setups including gearbox and valve trains on a gasoline engine are used to validate the proposed method. The results showed the effectiveness of the proposed method in diagnosing machine faults.

Furthermore, Bansal et al. (2013) studied the application of the SVM in multiclass gear-fault diagnosis using the gear vibration data in the frequency domain averaged over a large number of samples. In this study, real vibration signals acquired from a gear under four different health conditions – healthy gear, gear with a chipped tooth, gear with a missing tooth, and worn gear – are used. The authors recognised that the

SVM classifier achieved high classification accuracy when the training data and testing data are at matching angular speeds. Hence, two techniques called *interpolation* and *extrapolation* are proposed by the authors of this study to improve the performance of the SVM classifier for multiclass gear faults, even in the absence of the angular speed of the training and testing datasets. Also, the effects of the size of training, data density, and selection of different kernels and parameters on the overall classification accuracy of the SVM are studied.

Also, Chen et al. (2013) proposed a method for gearbox fault diagnosis based on WSVM and the immune genetic algorithm (IGA). In this study, first, the EMD algorithm is employed to extract features from the measured vibration signals. With these extracted features, WSVM is employed to deal with the fault diagnosis problem. IGA is used to find the optimal parameters for WSVM. Practical vibration signals measured from a gearbox with a no-load condition, a gear surface pitting fault, one tooth broken, a bearing OR fault, and a bearing IR fault are used to examine the proposed method. The results showed that the proposed method outperformed RBF-SVM and RBF-NN.

Additionally, Liu et al. (2013) introduced a multifault classification model for RE bearings based on WSVM and particle swarm optimisation (PSO). In this study, (i) the measured vibration signals are preprocessed using EMD. (ii) A distance-evaluation technique is employed to remove redundant and irrelevant features and select the main features for the classification. Finally, (iii) with these selected features, WSVM is employed to deal with the classification problem. PSO is used to find the optimal parameters for WSVM. The Morlet and Mexican wavelet kernel functions are utilised to construct the WSVM algorithm. Vibration signals acquired from two types of experimental setups including a RE bearings test bench and RE bearings of electric locomotives are used to validate the proposed method. The results demonstrated the efficiency of the proposed method in diagnosing RE bearing faults.

Also, Fernández-Francos et al. (2013) presented a method for diagnosing bearing faults based on the combination of v-SVM, envelope analysis, and a ruled-based expert system. In this study, the energy of different sub-bands of the normal condition spectrum is used as training data. Then, a one-class v-SVM is used to detect a change of behaviour due to the presence of a fault as follows: (i) detection of changes in the conditions of the machine under monitoring, (ii) selection of the frequency band where the defect appears evident in the power spectrum of the signal, and (iii) using envelope analysis to highlight the features of the faulty signal based on the band obtained in step (ii). The experimental results verified the effectiveness of the proposed method in diagnosing RE bearing faults.

Zhang and Zhou (2013) presented a procedure for the multifault diagnosis of rolling bearings based on ensemble empirical mode decomposition (EEMD) and optimised SVMs. In this study, first, the measured vibration signals are decomposed into several IMFs using EEMD. Then, two types of features – the EEMD energy entropy and singular values of the matrix whose rows are IMFs – are extracted. The EEMD entropy is utilised to identify whether the bearing has faults or not. If the bearing has faults, the extracted singular values are used as input to multiclass SVMs optimised using the inter-cluster distance in the feature space (ICDSVM) to recognise the fault type. Vibration signals collected from two roller bearings with 8 normal conditions and 48 fault conditions are used to validate the proposed method. The results demonstrated the effectiveness of the proposed method in diagnosing roller bearing faults.

In addition, Tang et al. (2014) proposed a method for fault diagnosis for wind turbine (WT) transmission systems based on manifold learning and the Shannon wavelet support vector machine (SWSVM). In this study, (i) EMD and autoregressive (AR) model coefficients are used for mixed-domain feature fusion. (ii) Orthogonal preserving embedding (ONPE) is employed to map the high-dimensional feature set into low-dimensional eigenvectors. Finally, (iii) with these low-dimensional eigenvectors, SWSVM is applied to recognise faults. The experimental results of the application of the proposed method in a wind turbine gearbox fault diagnosis problem showed that the proposed method achieved 92% classification accuracy.

Zhu et al. (2014) proposed a method for diagnosing roller bearing faults based on hierarchical entropy (HE) and SVMs with a PSO algorithm. In this study, HE is used to extract features from the original vibration signals. Then, a multiclass SVM is employed to classify bearing faults. The PSO algorithm is utilised to find the optimal parameters of the SVMs. Typical vibration signals measured from roller bearings under 10 different health conditions are used to validate the proposed method. The results showed that the proposed method based on HE and SVMs achieved up to 100% classification accuracy. Likewise, Zhang et al. (2015a) introduced a method for fault diagnosis based on an ant colony algorithm for synchronous feature selection and parameter optimisation in SVMs. In this method, statistical features are extracted from the raw vibration signals and the corresponding FFT spectrums. Then, the ACO is employed for synchronous feature selection and parameter optimisation in the SVM. Practical vibration signals acquired from locomotive roller bearings with nine health conditions – normal; slight fault in the OR; serious fault in the OR; IR fault; roller fault; compound faults in the OR and IR; compound faults in the OR and rollers; compound faults in the IR and rollers; and compound faults in the OR, IR, and rollers are used to test the proposed method in diagnosing roller bearing faults. The results showed that the proposed method can achieve much better classification results in a roller bearing fault diagnosis problem.

Moreover, Zhang et al. (2015a,b) presented a method for fault diagnosis of roller bearings based on a multivariable ensemble-based incremental support vector machine (MEISVM). In this study, statistical features from the raw vibration signals and the corresponding FFT spectrums are extracted. Then, MEISVM is used to perform the fault diagnosis. Two case studies of diagnosing roller bearing faults are used to validate the proposed method. The results illustrated the effectiveness of the proposed method in fault diagnosis of roller bearings from vibration signals.

Soualhi et al. (2015) introduced a method for monitoring ball bearings based on the Hilbert-Huang transform (HHT) and SVMs. In this method, for stationary/nonstationary vibration signals, the HHT is used to extract features representing ball bearing health. Then, with these extracted features, the SVM is used to detect and diagnosis the bearings' health condition. In addition, these features are used for the prognostic process using a one-step time-series prediction based on support vector regression (SVR). Vibration signals measured from three bearings are used to validate the proposed method. The results showed the effectiveness of the combination of the HHT, the SVM, and the SVR in the detection, diagnosis, and prognostics of bearing degradation.

Furthermore, Li et al. (2016) presented an approach for fault diagnosis in a wind turbine (WT) based on a multiclass fuzzy support vector machine (FSVM) classifier. In this method, EMD is first applied to extract feature vectors from the measured vibration signals. Then, the FSVM with RBF kernel is employed as the classifier, where a kernel fuzzy c-means (KFCM) algorithm is applied to determine fuzzy membership values of training samples for FSVM, and PSO is employed to find FSVM parameters. Vibration signals acquired with four health conditions – normal, shaft imbalance, shaft misalignment, and shaft imbalance and misalignment – are used to examine the proposed method. The results showed that the proposed method can achieve up to a 96.7% classification accuracy rate.

Zheng et al. (2017) proposed a method for roller bearing fault detection and diagnosis based on a nonlinear parameter called composite multiscale fuzzy entropy (CMFE) for measuring the complexity of time series, and ensemble support vector machines (ESVMs). In this study, (i) the CMFE of each vibration signal from rolling bearings is calculated. (ii) A number of the first features are selected to represent the fault features of vibration signals. (iii) With these selected features, an ESVM-based multifault classifier is employed to deal with the classification problem. Vibration signals acquired from roller bearings with a normal condition, OR fault, IR fault, and RE fault are used to validate the proposed method. The results demonstrated that the proposed method could efficiently discriminate different fault conditions and severities for rolling bearings.

Additionally, Gangsar and Tiwari (2017) presented a comparative investigation of vibration and current monitoring for prediction of mechanical and electrical faults in induction motors using multiclass support vector machines (MSVMs). In this study, (i) three statistical features, i.e. standard deviation, skewness, and kurtosis, are extracted from the raw vibration and current signals in the time domain. (ii) The MSVM with RBF kernel is employed to deal with the classification problem. Typical vibration signals in three orthogonal directions, and three-phase motor currents acquired from nine faulty and one healthy induction motor, are used to validate the proposed method. The results showed that the MSVM performed effectively based on the vibration-current signal at all the considered operation conditions of the induction motor.

Recently, Gao et al. (2018) proposed a method for fault diagnosis of wind turbines based on integral extension load mean decomposition (IELMD), multiscale entropy (ME), and LSSVM. In this method, (i) the IELMD method is applied to decompose the signal, and the product functions (PFs) are obtained. (ii) The PFs that can reveal fault characteristics are processed by the ME algorithm. Finally, (iii) the ME parameters are used as input to the LSSVM algorithm that deals with the classification problem. The results demonstrated that the combination of IELMD decomposition, ME, and LSSVM can raise the precision of diagnosis and increase identification capability.

Also, Han et al. (2018) explored the performance of random forest (RF), an extreme learning machine (ELM), a probabilistic neural network (PNN), and SVMs with different features using two datasets from rotating machinery. In this study, three types of features are extracted: time domain statistical features, frequency domain statistical features, and multiple-scale features. Then, the classification model can be trained with these extracted features. Two practical vibration datasets are used in this investigation. The first is acquired from roller bearings with a normal condition, IR fault, OR fault, and ball fault. The second dataset is collected from a gearbox under four health conditions: normal, tooth broken, tooth surface spalling, and gear root crack. The overall results showed the superiority of RF in terms of classification accuracy, stability, and robustness to features compared to the other methods.

Table 13.1 summarises the application of SVMs in diagnosing machine faults.

Table 13.1 Summary of the application of SVMs for machine fault detection and diagnosis.

Studies	Items	Feature extraction and/or selection techniques	Solver/Kernel
Jack and Nandi 2001	Roller bearing	Statistical features; spectral features obtained using FFT	QP/RBF
Jack and Nandi 2002	Roller bearing	Statistical features, spectral features obtained using FFT, feature selection using GA	QP/RBF
Samanta 2004	Gear	Statistical features; GA	QP/RBF
Rojas and Nandi 2005	Roller bearing	Statistical features and spectral features obtained using FFT	SMO/RBF
Zhang et al. 2005	Roller bearing	Statistical features, spectral features obtained using FFT, GP, and GA	QP/RBF
Guo et al. 2005	Roller bearing	GP	Polynomial kernel
Yuan and Chu 2006	Turbopump rotor	FFT and PCA	RBF
Abbasion et al. 2007	Roller bearing	Discrete Meyer wavelet	RBF
Hu et al. 2007	Roller bearing	WPT; distance-evaluation technique	RBF
Widodo et al. 2007	Induction motors	Statistical features in the time domain and frequency domain; ICA	SMO/RBF
Widodo and Yang 2007a	Induction motors	KPCA; KICA	SMO/RBF SMO/polynomial
Sugumaran et al. 2007	Roller bearing	Statistical features in the time domain; J48	—
Yang et al. 2007	Roller bearing	EMD	—
Saravanan et al. 2008	spur bevel gearbox	Statistical features in the time domain; J48	—
Xian and Zeng 2009	Gear	WPA	SMO/RBF
Guo et al. 2009	Roller bearing	Hilbert transform–based envelope spectrum analysis	QP/RBF
Zhang et al. 2010	Locomotive roller bearings	Wavelet transform; distance-evaluation technique	QP/RBF
Konar and Chattopadhyay 2011	Induction motor roller bearing	CWT	QP/RBB
Baccarini et al. 2011	Induction motor	FFT; several frequency bands of discrete spectra	RBF
Sun and Huang 2011	Steam turbine-generator (STGS)	FFT; several frequency bands of discrete spectra	QP/RBF

Table 13.1 (Continued)

Studies	Items	Feature extraction and/or selection techniques	Solver/Kernel
Li et al. 2012	Gearbox Valve trains on a gasoline engine	RSGWPT; NSR	SMO/RBF
Bansal et al. 2013	Gear	FFT	C-SVM/Polynomial C-SVM/Linear C-SVM/RBF v-SVM/Polynomial v-SVM/Linear v-SVM/RBF
Chen et al. 2013	Gearbox	EMD	Wavelet kernel
Liu et al. 2013	Roller bearing	E MD; distance-evaluation technique	Wavelet kernel
Fernández-Francos et al. 2013	Roller bearing	Band-pass filter; HHT	v-SVM/RBF
Zhang and Zhou 2013	Roller bearing	EEMD; ICD	Gaussian
Tang et al. 2014	Wind turbine transmission systems	EMD; AR	Shannon wavelet kernel
Zhu et al. 2014	Roller bearing	Hierarchical entropy (HE)	RBF
Zhang et al. 2015a	Locomotive roller bearing	FFT; ACO	RBF
Soualhi et al. 2015	Roller bearing	HHT	Polynomial
Zhang et al. 2015b	Roller bearing	Statistical features in the time domain and frequency domain	QP/Gaussian
Li et al. 2016	Wind turbine	EMD	FSVM/RBF
Zheng et al. 2017	Roller bearing	CMFE	Polynomial RBF Linear
Gangsar and Tiwari 2017	Induction motor	Statistical features	QP/RBF
Gao et al. 2018	Wind turbine	IELMD; ME	LS
Han et al. 2018	Roller bearing Gearbox	Statistical features in the time domain and frequency domain; multiple-scale features	RBF

Solver: a method used to find the separating hyperplane; *QP:* quadratic programming; *SMO:* sequential minimal optimisation; *LS:* least squares.

Kernel function: function used to map the training data into kernel space; *linear:* linear kernel; *polynomial:* polynomial kernel; *RBF*, Gaussian radial basis function.

Table 13.2 Summary of some of the introduced techniques and their publically accessible software

Algorithm name	Platform	Package	Function
Train support vector machine classifier	MATLAB	Statistical and Machine Learning Toolbox	svmtrain
Classify using support vector machine (SVM)			svmclassify
Train binary SVM classifier			fitcsvm
Predict labels using SVM classifier			predict
Fit multiclass models for SVM or other classifiers			fitcecoc
LIBSVM: a library for support vector machines	JAVA, MATLAB, OCTAVE, R, PYTHON, WEKA, SCILAB, C#, etc.	Chang 2011	n/a

13.5 Summary

SVMs are one of the most popular machine learning methods used to classify machine health conditions using the selected feature space. In machine fault detection and diagnosis, SVMs are used for learning special patterns from the acquired signal; then these patterns are classified according to the fault occurrence in the machine. The basic idea of an SVM is that it can find the best hyperplane(s) to separate data from two different classes such that the distance between the two classes, i.e. the margin, is maximised. Based on the features of the data, the SVM can make linear or nonlinear classifications. Nonlinear classification is often used when the data is not linearly separable. For multiclass classification problems, several SVM classifiers can together deal with the multiclass problems. This chapter has presented essential concepts of the SVM classifier by giving a brief description of the SVM model for binary classification. Then, the chapter explained the multiclass SVM approach and different techniques that can be used for multiclass SVMs. A considerable amount of literature has been published on the application of SVMs and variants in diagnosing machine faults. Most of these studies introduced pre-processing techniques that include normalisation, feature extraction, transformation, and feature selection. The data produced during the pre-processing step represent the final training set that is used as input to SVMs. In order to learn more useful features for diagnosing machine faults, most of the proposed methods combine two or more analysis techniques. Most of the introduced techniques and their publicly accessible software are summarised in Table 13.2.

References

Abbasion, S., Rafsanjani, A., Farshidianfar, A., and Irani, N. (2007). Rolling element bearings multi-fault classification based on the wavelet denoising and support vector machine. *Mechanical Systems and Signal Processing* 21 (7): 2933–2945.

Ahmed, H. and Nandi, A.K. (2018). Compressive sampling and feature ranking framework for bearing fault classification with vibration signals. *IEEE Access* 6: 44731–44746.

Baccarini, L.M.R., e Silva, V.V.R., De Menezes, B.R., and Caminhas, W.M. (2011). SVM practical industrial application for mechanical faults diagnostic. *Expert Systems with Applications* 38 (6): 6980–6984.

Bansal, S., Sahoo, S., Tiwari, R., and Bordoloi, D.J. (2013). Multiclass fault diagnosis in gears using support vector machine algorithms based on frequency domain data. *Measurement* 46 (9): 3469–3481.

Boser, B.E., Guyon, I.M., and Vapnik, V.N. (1992). A training algorithm for optimal margin classifiers. In: *Proceedings of the Fifth Annual Workshop on Computational Learning Theory*, 144–152. ACM.

Burges, C.J. (1998). A tutorial on support vector machines for pattern recognition. *Data Mining and Knowledge Discovery* 2 (2): 121–167.

Campbell, C. (2000). Algorithmic approaches to t-raining support vector machines: a survey. In: *Proceedings of ESANN 2000, Bruges*, Paper #ESANN2000-355. https://pdfs .semanticscholar.org/619f/4c6673b7a1943d4f3257c54d063d99c8483e.pdf.

Campbell, C. and Cristianini, N. (1998). *Simple Learning Algorithms for Training Support Vector Machines. University of Bristol*.

Chang, C.C. (2011). LIBSVM: a library for support vector machines. *ACM Transactions on Intelligent Systems and Technology* 2 (3) Software http://www.csie.ntu.edu.tw/~cjlin/ libsvm.

Chen, F., Tang, B., and Chen, R. (2013). A novel fault diagnosis model for gearbox based on wavelet support vector machine with immune genetic algorithm. *Measurement* 46 (1): 220–232.

Cortes, C. and Vapnik, V. (1995). Support-vector networks. *Machine Learning* 20 (3): 273–297.

Fernández-Francos, D., Martínez-Rego, D., Fontenla-Romero, O., and Alonso-Betanzos, A. (2013). Automatic bearing fault diagnosis based on one-class v-SVM. *Computers & Industrial Engineering* 64 (1): 357–365.

Gangsar, P. and Tiwari, R. (2017). Comparative investigation of vibration and current monitoring for prediction of mechanical and electrical faults in induction motor based on multiclass-support vector machine algorithms. *Mechanical Systems and Signal Processing* 94: 464–481.

Gao, Q.W., Liu, W.Y., Tang, B.P., and Li, G.J. (2018). A novel wind turbine fault diagnosis method based on intergral extension load mean decomposition multiscale entropy and least squares support vector machine. *Renewable Energy* 116: 169–175.

Guo, H., Jack, L.B., and Nandi, A.K. (2005). Feature generation using genetic programming with application to fault classification. *IEEE Transactions on Systems, Man, and Cybernetics Part B: Cybernetics* 35 (1): 89–99.

Guo, L., Chen, J., and Li, X. (2009). Rolling bearing fault classification based on envelope spectrum and support vector machine. *Journal of Vibration and Control* 15 (9): 1349–1363.

Han, T., Jiang, D., Zhao, Q. et al. (2018). Comparison of random forest, artificial neural networks and support vector machine for intelligent diagnosis of rotating machinery. *Transactions of the Institute of Measurement and Control* 40 (8): 2681–2693.

Hsu, C.W. and Lin, C.J. (2002). A comparison of methods for multiclass support vector machines. *IEEE Transactions on Neural Networks* 13 (2): 415–425.

Hu, Q., He, Z., Zhang, Z., and Zi, Y. (2007). Fault diagnosis of rotating machinery based on improved wavelet package transform and SVMs ensemble. *Mechanical Systems and Signal Processing* 21 (2): 688–705.

Jack, L.B. and Nandi, A.K. (2001). Support vector machines for detection and characterization of rolling element bearing faults. *Proceedings of the Institution of Mechanical Engineers, Part C: Journal of Mechanical Engineering Science* 215 (9): 1065–1074.

Jack, L.B. and Nandi, A.K. (2002). Fault detection using support vector machines and artificial neural networks, augmented by genetic algorithms. *Mechanical Systems and Signal Processing* 16 (2–3): 373–390.

Joachims, T. (1998). *Making Large-Scale SVM Learning Practical* (No. 1998, 28). Technical Report, SFB 475: Komplexitätsreduktion in Multivariaten Datenstrukturen, . Universität Dortmund.

Konar, P. and Chattopadhyay, P. (2011). Bearing fault detection of induction motor using wavelet and support vector machines (SVMs). *Applied Soft Computing* 11 (6): 4203–4211.

Li, N., Zhou, R., Hu, Q., and Liu, X. (2012). Mechanical fault diagnosis based on redundant second generation wavelet packet transform, neighborhood rough set and support vector machine. *Mechanical Systems and Signal Processing* 28: 608–621.

Li, Y., Xu, M., Wei, Y., and Huang, W. (2016). A new rolling bearing fault diagnosis method based on multiscale permutation entropy and improved support vector machine based binary tree. *Measurement* 77: 80–94.

Liu, Z., Cao, H., Chen, X. et al. (2013). Multi-fault classification based on wavelet SVM with PSO algorithm to analyze vibration signals from rolling element bearings. *Neurocomputing* 99: 399–410.

Platt, J. (1998). Sequential minimal optimization: A fast algorithm for training support vector machines. Res. Tech. Rep. MSR-TR-98-14. Microsoft.

Rojas, A. and Nandi, A.K. (2005). Detection and classification of rolling-element bearing faults using support vector machines. In: *2005 IEEE Workshop on Machine Learning for Signal Processing*, 153–158. IEEE.

Rojas, A. and Nandi, A.K. (2006). Practical scheme for fast detection and classification of rolling-element bearing faults using support vector machines. *Mechanical Systems and Signal Processing* 20 (7): 1523–1536.

Samanta, B. (2004). Gear fault detection using artificial neural networks and support vector machines with genetic algorithms. *Mechanical Systems and Signal Processing* 18 (3): 625–644.

Saravanan, N., Siddabattuni, V.K., and Ramachandran, K.I. (2008). A comparative study on classification of features by SVM and PSVM extracted using Morlet wavelet for fault diagnosis of spur bevel gear box. *Expert Systems with Applications* 35 (3): 1351–1366.

Soualhi, A., Medjaher, K., and Zerhouni, N. (2015). Bearing health monitoring based on Hilbert–Huang transform, support vector machine, and regression. *IEEE Transactions on Instrumentation and Measurement* 64 (1): 52–62.

Sugumaran, V., Muralidharan, V., and Ramachandran, K.I. (2007). Feature selection using decision tree and classification through proximal support vector machine for fault diagnostics of roller bearing. *Mechanical Systems and Signal Processing* 21 (2): 930–942.

Sun, H.C. and Huang, Y.C. (2011). Support vector machine for vibration fault classification of steam turbine-generator sets. *Procedia Engineering* 24: 38–42.

Tang, B., Song, T., Li, F., and Deng, L. (2014). Fault diagnosis for a wind turbine transmission system based on manifold learning and Shannon wavelet support vector machine. *Renewable Energy* 62: 1–9.

Vapnik, V.N. (1982). *Estimation of Dependencies Based on Empirical Data*. New York: Springer.

Widodo, A. and Yang, B.S. (2007a). Application of nonlinear feature extraction and support vector machines for fault diagnosis of induction motors. *Expert Systems with Applications* 33 (1): 241–250.

Widodo, A. and Yang, B.S. (2007b). Support vector machine in machine condition monitoring and fault diagnosis. *Mechanical Systems and Signal Processing* 21 (6): 2560–2574.

Widodo, A., Yang, B.S., and Han, T. (2007). Combination of independent component analysis and support vector machines for intelligent faults diagnosis of induction motors. *Expert Systems with Applications* 32 (2): 299–312.

Xian, G.M. and Zeng, B.Q. (2009). An intelligent fault diagnosis method based on wavelet packer analysis and hybrid support vector machines. *Expert Systems with Applications* 36 (10): 12131–12136.

Yang, Y., Yu, D., and Cheng, J. (2007). A fault diagnosis approach for roller bearing based on IMF envelope spectrum and SVM. *Measurement* 40 (9–10): 943–950.

Yuan, S.F. and Chu, F.L. (2006). Support vector machines-based fault diagnosis for turbo-pump rotor. *Mechanical Systems and Signal Processing* 20 (4): 939–952.

Zhang, X. and Zhou, J. (2013). Multi-fault diagnosis for rolling element bearings based on ensemble empirical mode decomposition and optimized support vector machines. *Mechanical Systems and Signal Processing* 41 (1–2): 127–140.

Zhang, L., Jack, L.B., and Nandi, A.K. (2005). Fault detection using genetic programming. *Mechanical Systems and Signal Processing* 19 (2): 271–289.

Zhang, X.L., Chen, X.F., and He, Z.J. (2010). Fault diagnosis based on support vector machines with parameter optimization by an ant colony algorithm. *Proceedings of the Institution of Mechanical Engineers, Part C: Journal of Mechanical Engineering Science* 224 (1): 217–229.

Zhang, X., Chen, W., Wang, B., and Chen, X. (2015a). Intelligent fault diagnosis of rotating machinery using support vector machine with ant colony algorithm for synchronous feature selection and parameter optimization. *Neurocomputing* 167: 260–279.

Zhang, X., Wang, B., and Chen, X. (2015b). Intelligent fault diagnosis of roller bearings with multivariable ensemble-based incremental support vector machine. *Knowledge-Based Systems* 89: 56–85.

Zheng, J., Pan, H., and Cheng, J. (2017). Rolling bearing fault detection and diagnosis based on composite multiscale fuzzy entropy and ensemble support vector machines. *Mechanical Systems and Signal Processing* 85: 746–759.

Zhu, K., Song, X., and Xue, D. (2014). A roller bearing fault diagnosis method based on hierarchical entropy and support vector machine with particle swarm optimization algorithm. *Measurement* 47: 669–675.

14

Deep Learning

14.1 Introduction

As described in Chapter 6, the core objective of vibration-based machinery condition monitoring is to categorise the acquired vibration signal into the corresponding machine condition correctly, which is generally a multiclass classification problem. In practice, the collected vibration signal usually contains a large collection of responses from several sources in the rotating machine and some background noise. These make it challenging to use the acquired vibration signal directly for diagnosing rotating machine faults, either through manual inspection or automatic monitoring, owing to the curse of dimensionality. As an alternative way of processing the raw vibration signals, the common method is to compute certain features of the raw vibration signal that can describe the signal in essence. In the machine learning community, these features are also called *characteristics*, *signatures*, or *attributes*. Many techniques can be used to classify different vibration type based on the features provided. If the vibration signals' features are carefully devised and the parameters of the classifiers are carefully tuned, it is possible to achieve high accuracy in classification performance. Therefore, if one has extracted useful representations of the vibration data for the purpose of fault classification, it may be easy to achieve high classification accuracy. However, extracting useful features from such a large, noisy vibration dataset is not an easy task. One solution is to use feature-learning techniques, also called representation learning techniques, that have the ability to automatically learn the representations of the data required for feature detection and classification. Of these techniques, deep learning (DL) is often used to attempt to automatically learn representations of data with multiple layers of information-processing modules in hierarchical architectures.

In recent years, there has been an increasing amount of literature on the application of DL techniques in various areas such as bioinformatics (Min et al. 2017), face recognition (Parkhi et al. 2015), speech recognition (Graves et al. 2013), and natural language processing (Collobert and Weston 2008). For the purpose of this book, we will define deep neural network learning techniques that have been used in machine fault diagnosis: autoencoder-based deep neural networks (DNNs), convolutional neural networks (CNNs), deep belief networks (DBNs), and recurrent neural networks (RNNs).

Although supervised learning based artificial neural networks (ANNs) with many hidden layers are found to be difficult in practice, DNNs have been well developed

Condition Monitoring with Vibration Signals: Compressive Sampling and Learning Algorithms for Rotating Machines, First Edition. Hosameldin Ahmed and Asoke K. Nandi.

as a research topic and have been made practically feasible with the assistance of unsupervised learning. Moreover, DNNs have attracted extensive attention by outperforming other machine learning methods. Each layer of a DNN performs a nonlinear transformation of the input samples from the preceding layer to the following one. A good overview of DNNs can be found in (Schmidhuber 2015). Unlike ANNs, DNNs can be trained in a supervised or unsupervised manner (Bengio et al. 2007; Erhan et al. 2010), and they are also appropriate in the general area of reinforcement learning (RL) (Lange and Riedmiller 2010; Salimans and Kingma 2016). The basic idea of training a DNN is that we first train the network layer by layer using an unsupervised learning algorithm, e.g. an autoencoder (AE); this process is called DNN *pretraining*. In this process, the output from each layer will be the input to the succeeding layer. Then the DNN is retrained in a supervised way with a backpropagation algorithm for classification.

We will move on now to consider other types of DL architectures, e.g. CNNs, DBNs, and RNNs, that have been used for machine fault diagnosis. For example, unlike a standard neural network (NN), the architecture of a CNN is usually composed of a convolutional layer and a subsampling layer also called a *pooling layer*. The CNN learns abstract features from alternating and stacking convolutional layers and pooling operation. The convolutional layers convolve multiple local filters with raw input data and generate invariant local features and the pooling layers extract the most-significant features (Ahmed et al. 2018). Furthermore, DBNs are generative NNs that stack multiple restricted Boltzmann machines (RBM) that can be trained in a greedy layer-wise unsupervised way; then they can be further fine-tuned with respect to labels of training data by adding a softmax layer in the top layer. Moreover, RNN build connections between units from a direct cycle. It is able to map target vectors from the entire history of previous inputs in principle. Also, it allows a memory of previous inputs to be kept in the network state. As is the case with DBNs, RNNs can be trained via backpropagation through time for supervised tasks with sequential input data and target outputs. A good overview of these DL architectures can be found in (Zhao et al. 2016; Khan and Yairi 2018).

14.2 Autoencoders

An AE NN provides a means of using an unsupervised learning algorithm that sets the target values, i.e. the outputs, to be equal to the inputs and applies backpropagation (Ng 2011). As shown in Figure 14.1, like many unsupervised feature learning methods, the design of an AE relies on an encoder-decoder architecture, where the encoder produces a feature vector from the input samples and the decoder recovers the input from this feature vector. The encoder part is a feature-extraction function f_θ that computes a feature vector $h(x_i)$ from an input x_i. We define

$$h(x_i) = f_\theta(x_i) \tag{14.1}$$

where $h(x_i)$ is the feature representation. The decoder part is a recovery function g_θ that reconstructs the input space $\widetilde{x}_i$ from the feature space $h(x_i)$ such that

$$\widetilde{x}_i = g_\theta(h(x_i)) \tag{14.2}$$

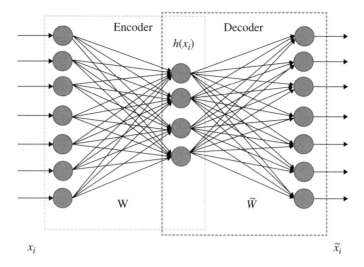

Figure 14.1 Autoencoder architecture (Ahmed et al. 2018). (See insert for a colour representation of this figure.)

The AE is attempting to learn an approximation such that x_i is similar to $\widetilde{x}_i$, i.e. is trying to attain the lowest possible reconstruction error $E(x_i, \widetilde{x}_i)$ that measures the discrepancy between x_i and $\widetilde{x}_i$. Hence the following equation is obtained

$$E(x_i, \widetilde{x}_i) = \|x_i - \widetilde{x}_i\|^2 \tag{14.3}$$

In fact, AEs were mainly developed as multilayer perceptrons (MLPs), and the most commonly used forms for the encoder and decoder are affine transformations that keep collinearity followed by a nonlinearity such that

$$f_\theta(x_i) = s_f(b + W) \tag{14.4}$$

$$g_\theta(h(x_i)) = s_g(c + \widetilde{W}) \tag{14.5}$$

where s_f and s_g are the encoder and decoder activation functions, e.g. sigmoid, identity, hyperbolic tangent etc.; b and c are the encoder and decoder bias vectors; and W and $\widetilde{W}$ are the encoder and decoder weight matrices. The literature on DL has highlighted several activation functions that have been used in various studies (Glorot et al. 2011; He et al. 2015; Schmidhuber 2015; Clevert et al. 2015; Godfrey and Gashler 2015). Table 14.1 presents these activation functions and their corresponding formulas.

The AE is one of the more practical ways to do reinforcement learning. For instance, by forcing some constraints on the AE network such as limiting the number of hidden units and imposing regularisers, the AE may learn about interesting feature structures in the data. Therefore, different constraints give different forms of AEs. A sparse autoencoder (SAE) is an AE that contains a sparsity constraint on the hidden unit's activation that must typically be near 0. This may be accomplished by adding a Kullback–Leibler (KL) divergence penalty term

$$\sum_{j=1}^{d} \mathrm{KL}(\rho \| \hat{\rho}) \tag{14.6}$$

Table 14.1 List of activation functions and their corresponding formulas.

Activation function name	Formula
Sigmoid	$f(x) = \dfrac{1}{1 + e^{-x}}$
Gaussian	$f(x) = e^{-x^2}$
Hyperbolic tangent	$f(x) = \dfrac{e^x - e^{-x}}{e^x + e^{-x}}$
Identity	$f(x) = x$
Bent identity	$f(x) = \dfrac{\sqrt{x^2 + 1} - 1}{2} + x$
Rectified linear unit (ReLU)	$f(x) = \begin{cases} 0 (x < 0) \\ x (x \geq 0) \end{cases}$
Parametric ReLU (PReLU)	$f(x) = \begin{cases} \alpha x \, for \, x < 0 \\ x \, for \, x \geq 0 \end{cases}$
Leaky rectified linear unit (leaky ReLU)	$f(x) = \begin{cases} 0.01x(x < 0) \\ x(x \geq 0) \end{cases}$
Sinusoid	$f(x) = \sin(x)$
Arctangent (Atan)	$f(x) = (\tan(x))^{-1}$
Sinc	$f(x) = \begin{cases} 1 \, for \, x = 0 \\ \dfrac{\sin(x)}{x} \, otherwise \end{cases}$
An exponential linear unit (ELU)	$f(x) = \begin{cases} \alpha(e^x - 1)\,(x < 0) \\ x\,(x \geq 0) \end{cases}$
Soft exponential function	$f(x) = \begin{cases} -\dfrac{\ln(1 - \alpha(x + \alpha))}{\alpha} \, for \, \alpha < 0, \\ x \, for \, \alpha = 0 \\ \dfrac{e^{\alpha x} - 1}{\alpha} + \alpha \, for \, \alpha > 0 \end{cases}$
Soft plus	$f(x) = \ln(1 + e^x)$

where

$$KL(\rho \| \hat{\rho}) = \rho \log \frac{\rho}{\hat{\rho}_j} + (1 - \rho) \log \frac{1 - \rho}{1 - \hat{\rho}_j} \qquad (14.7)$$

and ρ is a sparsity parameter; normally its value can be small and close to zero, e.g. $\rho = 0.2$, while $\hat{\rho}$ is the average threshold activation of hidden units and can be calculated by the following equation:

$$\hat{\rho}_j = \frac{1}{n} \sum_{i=1}^{n} [a_j^2(x_i)] \qquad (14.8)$$

where a_j^2 represents the activation of hidden unit j. By minimising this penalty term, $\hat{\rho}$ is close to ρ, and the overall cost function (CF) can be calculated by the following equation

$$CF_{sparse}(W, b) = \frac{1}{2n} \sum_{i=1}^{n} \|\tilde{x}_i - x_i\|^2 + \lambda \|W\|^2 + \beta \sum_{j=1}^{d} KL(\rho \| \hat{\rho}) \qquad (14.9)$$

where n is the input size, d is the hidden layer size, λ represents the weight decay parameter, and β is the weight of the sparsity penalty term.

14.3 Convolutional Neural Networks (CNNs)

A CNN, also known as a *convNet*, is a multistage NN that was first proposed by LeCun et al. (1990) for image processing. The architecture of a CNN is usually composed of a convolutional layer and a subsampling layer, also called a pooling layer. The CNN learns abstract features from alternating and stacking convolutional layers and pooling operation. The convolutional layers convolve multiple local filters with raw input data and generate invariant local features, and the pooling layers extract the most-significant features (LeCun et al. 1998). CNNs have been proven successful in many applications, e.g. medical imaging (Tajbakhsh et al. 2016), object recognition (Maturana and Scherer 2015), speech recognition (Abdel-Hamid et al. 2014), visual document analysis (Simard et al. 2003) etc. Briefly, we present the simplified CNN model as follows.

Given input data $X = [x_1, x_2, \dots, x_L]$, where L is the number of examples and $x_i \in R^N$, the convolutional layer contains a number of feature maps, where each neuron takes a small subregion of the input feature maps and accomplishes the convolutional operation on the input. The output of the convolutional layer can be calculated using the following equation,

$$C_i^k = \theta(u^k x_{i:i+s-1} + b_k) \tag{14.10}$$

where $C_{i,j}^k$ is the value at i, j points of the convolutional layer's output of the kth feature map, $u^k \in R^s$ is a vector that represents the kth filter, b_k is the bias of the kth feature kernel, and θ is the activation function. The most commonly used activation function is ReLU. Figure 14.2 presents a simple example of a one-layer CNN model with one convolutional layer, one max-pooling layer, one fully connected layer, and one softmax layer, where each convolutional layer is followed by a pooling layer, also called a subsampling layer. The subsampled feature vector in the pooling layer can be computed using the following equation,

$$h_i = \max(c_{(i-1)m}, c_{(i-1)m+1}, \dots, c_{im-1} \tag{14.11}$$

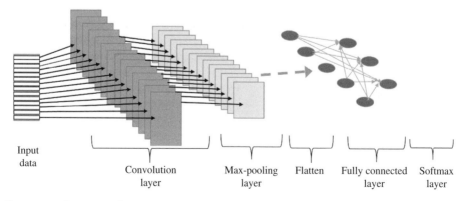

Input
data
Convolution
layer
Max-pooling
layer
Flatten
Fully connected
layer
Softmax
layer

Figure 14.2 Illustration of a one-layer CNN model with one convolutional layer, one max-pooling layer, one fully connected layer, and one softmax layer. (See insert for a colour representation of this figure.)

where m is the pooling length. Then, fully connected layers and softmax layer are often added as top layers to perform predictions.

14.4 Deep Belief Networks (DBNs)

DBNs are generative NNs that stack multiple RBMs that can be trained in a greedy layer-wise unsupervised way; they can be further fine-tuned with respect to labels of training data by adding a softmax layer in the top layer. The RBM is a specific type of Markov random field (MRF), which consists of one hidden layer and one visible layer of binary stochastic units. Thus, the RBM can be described as a two-layer NN that forms a bipartite graph of hidden units H and visible units V. Normally, all visible units are connected to hidden units and there are no connections between units within one group, i.e. the visible units group and hidden units group. In a binary RBM, the weights on the connections and the biases of the individual units describe a probability distribution over the joint states of the visible and hidden units by means of an energy function (Mohamed et al. 2012). The energy of a joint configuration can be computed using the following equation:

$$E(v, h \mid \theta) = \sum_{i=1}^{V} \sum_{j=1}^{H} w_{ij} v_i h_j - \sum_{i=1}^{I} b_i v_i - \sum_{j=1}^{J} a_j h_j \tag{14.12}$$

Here, $\theta = (w, b, a)$ is a model parameter, w_{ij} denotes the symmetric interaction term between visible unit i and hidden unit j, while b_i and a_j are their bias terms.

The probability that an RBM assigns to a visible vector v can be represented as follows,

$$p(v \mid \theta) = \frac{\sum_h e^{-E(v,h)}}{\sum_u \sum_h e^{-E(u,h)}} \tag{14.13}$$

Based on the fact that there are no connections between units in the same group, the conditional properties of hidden and visible units can be computed using the following equations:

$$p(h_j = 1 \mid v, \theta) = \rho \left(\sum_{i=1}^{V} a_j + w_{ij} v_i \right) \tag{14.14}$$

$$p(v_j = 1 \mid h, \theta) = \rho \left(\sum_{j=1}^{H} b_i + w_{ij} h_j \right) \tag{14.15}$$

Here, $\rho(x) = (1 + e^{-x})^{-1}$. For classification with given class labels, we train the RBM using both the dataset and its corresponding class labels. Hence, the energy function can be represented as follows:

$$E(v, I, h, \theta) = - \sum_{i=1}^{V} \sum_{j=1}^{H} w_{ij} v_i h_j - \sum_{y=1}^{L} \sum_{j=1}^{H} w_{yj} h_j l_y - \sum_{j=1}^{H} a_j h_j - \sum_{y=1}^{L} C_y l_y - \sum_{i=1}^{V} b_i v_i$$

$$\tag{14.16}$$

The probability of the label l_y given h and θ in the softmax layer can computed using the following equation:

$$p(l_y = 1 \mid h, \theta) = softmax\left(\sum_{j=1}^{H} w_{yj}h_j + C_y\right) \tag{14.17}$$

The DBN, which was introduced by Hinton in 2006, is a NN built from several layers of RBMs. Here the DBN can be built by stacking multiple RBMs on top of each other, where the hidden layer of one RBM is linked to the visible layer of the next RBM. Like the SAE described earlier, the DBN is often pretrained in an unsupervised way using the training set, which is then followed by a supervised fine-tuning using the same training set. As described in (Chen et al. 2017a,b), the procedure for the DBN algorithm is as follows.

Let $X = \{x_1, x_2, \ldots, x_L\}$ be an input data matrix. The procedure for the DBN algorithm can be described in the following four main steps:

(1) Train an RBM machine on X to get its weight matrix, W, and use this as the weight matrix between the lower two layers of the network.
(2) Transform X with the RBM to produce new data $\acute{X}$.
(3) Repeat this process with $X \leftarrow \acute{X}$ for the next pair of layers, until the top two layers of the network are reached.
(4) Fine-tune all the parameters of this deep architecture under the supervised criterion.

Another type of RBM-based method is the deep Boltzmann machine (DBM), which is regarded as a deep structure of RBMs with hidden units gathered into a hierarchy of layers rather than a single layer (Salakhutdinov and Hinton 2009).

14.5   Recurrent Neural Networks (RNNs)

As described in (Lipton et al. 2015), RNNs are feedforward NNs improved by the insertion of edges that span adjacent time steps, introducing a notion of time to the model. As is the case with feedforward networks, RNNs may not have cycles among conventional edges. However, edges, i.e. connections between units, which connect together time steps and are called recurrent edges or recurrent connections, may form cycles. These include cycles of length one, i.e. of one unit, which are self-connections from a unit to itself across time. Figure 14.3 presents an example of a simple RNN model where at time t, the unit h_t with recurrent connections receive inputs from the current data point x_t and the previous hidden unit values h_{t-1}. This can be represented mathematically using the following equation,

$$h_t = \rho(W_{hx}x_t + W_{hh}h_{t-1} + b_h) \tag{14.18}$$

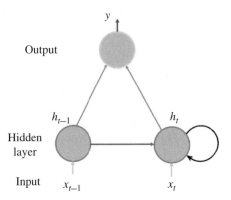

Figure 14.3 Illustration of a basic RNN.

where ρ is nonlinear activation function, W_{hx} is the weight between the hidden layer and the input x, W_{hh} is the recurrent weight between the hidden layer and itself, and b_h is a bias. The output y can be computed as follows,

$$y = softmax(W_{yh}\mathrm{h} + b_y) \tag{14.19}$$

where W_{yh} is the weight between the output and the hidden layer.

RNNs often use their feedback connections, i.e. recurrent connections, to store representations of recent input events in the form of activations. As is the case with DBNs, RNNs can be trained via backpropagation through time for supervised tasks with sequential input data and target outputs. However, it is difficult to train RNNs to capture long-term dependencies as the gradients tend to either vanish or blow up (Hochreiter and Schmidhuber 1997; Chung et al. 2014). Hence, long-short term memory (LSTM) (Hochreiter and Schmidhuber 1997) and gated recurrent units (GRUs) (Chung et al. 2014) were introduced to overcome this problem. In addition, bidirectional RNNs were presented by (Schuster and Paliwal 1997).

14.6 Overview of Deep Learning in MCM

In recent years, there has been an increasing amount of literature on the application of DL in machine condition monitoring (MCM). The following subsections present a brief survey of the application of the previously described DL-based techniques in machine fault diagnosis.

14.6.1 Application of AE-based DNNs in Machine Fault Diagnosis

The use of AE-based techniques in machine fault diagnosis was investigated by several researchers. For instance, Tao et al. (2015) proposed a DL algorithm framework for diagnosing bearing faults based on stacked AE and softmax regression. Vibration signals measured from roller bearings under three types of bearing fault conditions and one normal condition are used to validate the proposed method. The results showed the effectiveness of stacked AE-based DNNs in diagnosing roller bearing faults. Moreover, Lu et al. (2017a) presented a feature-extraction method for rolling bearing fault diagnosis based on a DNN. In this method, the acquired vibration data are divided into segments, and then a fast Fourier transform (FFT) is applied to every segment to obtain its amplitude of frequency. Then, two-layer based DNN structure, i.e. containing two AEs, is used for feature extraction. The proposed method is applied in vibration signals measured from roller bearings under six health conditions. The results showed that the proposed method is a powerful tool to extract features from roller bearing vibration data.

Furthermore, Junbo et al. (2015) presented a method for roller bearings based on a wavelet transform and stacked AE. In this method, (i) the original vibration signal is denoised using the digital wavelet frame (DWF) and nonlinear soft threshold method. (ii) A two-layer stacked AE is used to extract features from the denoised vibration signal. Finally, with these extracted features, the backpropagation (BP) network classifier is used to deal with the classification problem. Typical vibration signals measured from roller bearing with 20 health conditions are used to validate the proposed method. The results showed the effectiveness of the proposed method in diagnosing roller bearing faults. In addition, Huijie et al. (2015) proposed a method for hydraulic pump fault diagnosis based on stacked AE. In this method, first, the spectrum of the hydraulic pump vibration signal is used as the input to the stacked AE-based DNN. Then, the dropout strategy and rectified linear units (ReLU) activation function are both used to improve the performance of traditional stacked AE. Practical vibration signals collected from a hydraulic system where the vibration sensor is installed on the oil outlet of the pump. Four types of faults are considered, namely cylinder fault, valve plate fault, ball fault, and piston fault. The results showed that the proposed method is suitable for hydraulic pump fault diagnosis.

Jia et al. (2016) presented a DNN-based method for faults diagnosis of rotating machinery. In this method, (i) frequency spectra are obtained from the time domain signal. (ii) A DNN is built with multiple hidden layers that are pretrained layer by layer with stacked AEs. Here the number of AEs refers to the number of hidden layers inside the DNN. (iii) The dimension of the output layer is determined based on the number of machinery health conditions. (iv) The BP algorithm is implemented to fine-tune the parameters of the DNN via minimising the error between the output computed from the frequency spectra and health condition labels. Finally, (v) the trained DNN is utilised for machinery fault diagnosis. Two case studies of diagnosing machinery faults are used to validate the efficacy of the proposed method. The first case study is fault diagnosis of rolling element bearings where the bearing vibration dataset is obtained from the experimental system under four health conditions: normal, outer race (OR) fault, inner race (IR) fault, and roller fault. The second case study is fault diagnosis of planetary gearboxes in which the vibration dataset is acquired under seven health conditions: normal, a pitted tooth on the sun gear of the first stage, a cracked tooth on the sun gear of the first stage, a chipped tooth on the planetary gear of the first stage, a chipped tooth on the sun gear of the second stage, a missing tooth on the sun gear of the second stage, and a bearing IR fault of planetary gear of the first stage. The results showed that the proposed method is capable of mining fault characteristics from the frequency spectra adaptively for various diagnosis issues and effectively classifying health conditions of the machinery.

Moreover, Galloway et al. (2016) investigated the use of the DL approach for fault detection in a tidal turbine's generator from vibration data. In this method, first, spectrogram slices are generated from the raw vibration data using a short-time Fourier transform (STFT), allowing the network to potentially learn representations of both stationary and nonstationary signals. A DNN consisting of layers of stacked sparse autoencoders (SAEs) and a softmax classifier is applied. A single AE layer and two AE layer network configurations with various numbers of hidden units are tested. Features are learned from spectrograms of raw vibration data using SAEs. With these learned features, the softmax classifier layer is trained, and classification performance is then improved by retraining the network. Vibration signals measured through a triaxial accelerometer sensor under different operating and weather conditions are used to validate the proposed method. The results demonstrated the effectiveness of the proposed method.

Furthermore, Guo et al. (2016a) introduced a method for roller bearing condition recognition based on multifeature extraction and DNNs. In this method, features are extracted from the raw vibration signals using time domain, frequency domain, and time-frequency domain techniques. The time domain features are extracted using statistical functions including RMS, skewness, kurtosis, peak to peak, crest factor, shape factor, impulse factor, and margin factor. In the frequency domain, a correlation coefficient between two spectral kurtosis of vibration signals is computed. In the time-frequency domain, a wavelet transform (WT) is used to decompose the raw vibration signals into time-frequency space. Then, the extracted features are filtered using a moving-average filter. With these filtered features, an AE-based DNN is used to reduce the feature dimensions. Finally, a softmax regression is used for bearing fault classification. Vibration signals measured from roller bearings under four failure levels – normal, early fault, degradation, and failure – are used to validate the proposed method. The results verified the efficacy of the proposed method in diagnosing roller bearing faults.

Additionally, Sun et al. (2016) presented a DNN approach for induction motor fault diagnosis based on SAE. In this approach, first, the SAE is used to learn concise features from unlabelled induction motor vibration data. Here, the denoising coding is embedded into the SAE to improve the robustness of feature learning and prevent identity transformation. Then, the DNN uses the SAE to extract features that are employed to train a NN classifier for induction motor faults diagnosis. In addition, the authors utilised the 'dropout' technique to overcome the deficiency of overfitting during the process of the DNN. Practical vibration data acquired from a motor under six different operating conditions – normal, broken bar, bowed rotor, defective bearing, stator winding defect, and unbalanced rotor – are used to validate the proposed method. The proposed approach is compared with two classifiers, a support vector machine (SVM) and logistic regression (LR), built on top of the SAE. The results showed that the proposed approach can learn features directly from the raw vibration data using SAE. It also demonstrated the effectiveness of the SAE-based DNN compared to SAE-based SVM and SAE-based LR.

Lu et al. (2017b) proposed a DL method for diagnosing faults in rotary machinery components based on a stacked denoising autoencoder (SDA). In this method, a series of AEs are used to establish a deep network architecture in a layer-by-layer fashion. Then, the deep hierarchical structure with a transmitting rule of greedy training, with sparsity representation and data destruction, is used to map the input into useful, robust, high-order feature representations. With these features, a supervised classifier followed by the global backpropagation optimisation process is used to deal with the classification problem. A softmax regression algorithm is used as the top classifier of the SDA model. Typical vibration signals collected from roller bearings under different operating conditions are used to examine the efficacy of the proposed method. The results showed that the proposed DL method is able to adaptively mine salient fault characteristics and effectively classify the health conditions of the roller bearing with high diagnostic accuracy.

Moreover, Chen and Li (2017) proposed a multisensor feature fusion method for bearing faults using a SAE and DBN. In this method, (i) features are extracted from the acquired vibration data, including 15 time domain features and 3 frequency domain features. (ii) All feature vectors are normalised and rescaled into the range [0, 1]. (iii) Two-layer SAEs are trained through minimising the reconstruction error, and the output of the last hidden layer is considered the fault feature representations. Finally, (iv) with these feature representation, the DBN-based classification model is trained for the purpose of diagnosing bearing faults. Real vibration signals acquired from roller bearings under different health conditions are utilised to validate the proposed method. The results showed that the proposed method achieved high classification accuracy in roller bearing fault classification. Chen et al. (2017a) presented a method for identifying the severity of rolling bearing faults based on a SAE-based DNN with noise-added sample expansion. In this method, the training samples are expanded by adding noise to them to avoid overfitting problem caused by a small number of training samples. Then, a stacked SAE-based DNN with a classifier is trained using the expanded training samples to achieve feature extraction and fault diagnosis. Two bearing vibration datasets – one comprising bearing fault severity data and another comprising bearing life-state data acquired from accelerated bearing life tests – are used to validate the proposed method. The results showed the efficacy of the proposed method in rolling bearing fault severity identification.

Furthermore, Mao et al. (2017) proposed a method for diagnosing bearing faults based on AE and extreme learning machines (ELMs). In this method, the frequency spectrum of the acquired vibration signals is obtained using the FFT algorithm. Then, AE-ELM is used to extract features and identify bearing faults. Vibration signals collected from roller bearings under different health conditions are utilised to examine the proposed method. The results showed the advantage of the proposed method in diagnosing roller bearing faults. Zhou et al. (2017) introduced a multimode fault-classification method based on DNNs. In this method, the time domain original vibration signals are transformed to the frequency domain using the FFT. Then, a hierarchical DNN model with the first hierarchy-based DNN is developed for the purpose of mode partition. In the second hierarchy, a set of DNN classification models is built to extract features separately from different modes and diagnose the fault source. In the third hierarchy, an additional set of DNNs is built to further classify a certain fault in a given mode into several classes with different fault severities. Real vibration signals measured from roller bearings under different operating conditions with a motor load of 0 hp, 1 hp, 2 hp, and 3 hp are used to validate the proposed method. The bearing have four modes, where each mode has four health conditions: IR fault, OR fault, roller fault, and normal. The results showed the effectiveness of the proposed method in diagnosing roller bearing faults compared to the traditional DNN, backpropagation neural network (BPNN), SVM, hierarchical BPNN, and hierarchical SVM.

In addition, Sohaib et al. (2017) proposed a two-layered bearing fault diagnosis approach based on a stacked SAE-based DNN. In this approach, first, time domain features, envelope power spectrum features, and wavelet energy features are extracted from the raw vibration signals. Then, these features are used in combination with the stacked SAE-based DNN to extract discriminating information that can be used for diagnosing bearing faults. Typical vibration signals acquired from roller bearings under different health conditions – normal bearings and bearings with IR, OR, and rolling element (RE) faults – are used to examine the efficacy of the proposed method in diagnosing roller bearing faults. The results demonstrated the efficacy of the proposed method.

Shao et al. (2017a) proposed a deep AE feature-learning method for rotating machinery fault diagnosis. This method includes three parts: (i) the maximum correntropy is used as a loss function in place of multiscale entropy (MSE) in the standard AE; (ii) the artificial fish swarm algorithm (AFSA) is used to optimise the deep AE parameters and finally, (iii) the learned features of the deep AE are used as the input of the softmax classifier for fault classification. Two vibration datasets acquired from a gearbox under five different operating conditions and roller bearings from an electric locomotive under nine different operating conditions are used to validate the proposed method. The results confirmed the effectiveness of the proposed method for rotating machinery fault diagnosis.

Shao et al. (2017b) introduced an enhanced deep feature fusion method for rotating machinery fault diagnosis. In this method, (i) a deep AE is constructed with a denoising autoencoder (DAE) and a contractive autoencoder (CAE) to enhance feature learning. (ii) A locality-preserving projection (LPP) is employed to fuse the learned features. (iii) For the purpose of machinery fault diagnosis, the fused features are used as input to the softmax classifier. The proposed method is applied to the fault diagnosis of rotors and bearings. The rotor vibration dataset is measured under seven health conditions. The bearing vibration dataset is acquired from an electrical locomotive bearing in the experimental setup under nine health conditions. The results demonstrated the efficacy of the proposed method in diagnosing machinery faults.

Moreover, Qi et al. (2017) introduced a machine fault diagnosis method based on a stacked autoencoder. In this method, the acquired signals are first pre-processed using ensemble empirical mode decomposition (EEMD) and autoregressive (AR) models. Here, the proposed method extracts AR parameters of the first IMFs and converts them into the input vectors. Then, the constructed SAE is trained using the training dataset of the input vectors to learn high-level features that are representative enough for fault classification. With these learned features, the softmax classifier is used to deal with the classification problem. Two case studies of machine fault diagnosis are used to validate the proposed method. The vibration data of the first case study is measured from roller bearings under four different operating conditions, and the vibration data for the second case study is measured from a gearbox with four health conditions. The results showed better diagnostic performance for the proposed method compared with EEMD + AR + SVM, SAE, SVM, and ANN.

Recently, Sun et al. (2018) introduced a method for diagnosing bearing faults based on compressive sampling (CS) and deep learning (DL). In this method, CS is first used to extract features from the original raw vibration signals. Then, a DNN is constructed using stacked SAEs, where each SAE is pretrained to initialise the weights of the DNN and then retrained using the BP algorithm to advance the performance of the DNN. Practical vibration signals acquired from roller bearings under different health conditions are used to test the proposed method. Two vibration datasets measured from the planetary gearbox and roller bearings are used to validate the proposed method. The results verified that the efficacy of the proposed method in diagnosing roller bearing faults.

Moreover, Jia et al. (2018) presented a local connection network (LCN) method constructed with a normalised sparse autoencoder (NSAE), called NSAE-LCN, and its application to machine fault diagnosis. In this method, a LCN is constructed from four layers – i.e. input layer, local layer, feature layer, and output layer – and the NSAE is used to train the local layer to learn meaningful features from the vibration signals. Then, in the feature layer, shift-invariant features are obtained from the learned features in the local layer. Finally, with these shift-invariant features, the output layer classifier, i.e. the softmax regression, is used to identify mechanical health conditions. Two case studies are used to validate the proposed method. The dataset used in the first case study is collected from a two-stage planetary gearbox under 7 health conditions, and the dataset used in the second case study is acquired from roller bearings under 10 health conditions. The results of the gearbox vibration data obtained using features + softmax, features + SVM, principal component analysis (PCA) + SVM, stacked SAE + softmax, SAE – LCN, and NSAE – LCN showed that NSAE – LCN achieved the highest classification accuracy. In addition, with the roller bearing dataset, the superiority of the proposed NSAE – LCN is demonstrated compared to other methods.

In addition, Shao et al. (2018a) proposed a method for diagnosing roller bearing faults, called ensemble deep autoencoders (EDAEs). In this method, (i) different activation functions are employed to design a series of AEs with different characteristics. (ii) With various AEs, the EDAEs are constructed to learn features in an unsupervised manner from the acquired vibration signals. Finally, (iii) the learned features of each AE are fed into softmax classifiers for accurate and stable fault classification based on a combination strategy. A vibration dataset acquired from roller bearings under 12 health conditions is used to test the proposed method. Also, Li et al. (2018) presented a method for diagnosing roller bearing faults based on a fully connected, winner-take-all autoencoder (FC-WTA). This method explicitly places lifetime sparsity constraints on a hidden layer by keeping only the $k\%$ largest activations of each neuron across all samples in a mini-batch. Then, a soft-voting technique is implemented to increase accuracy and stability. A vibration dataset collected from roller bearings under four health conditions and a simulated dataset generated by adding white Gaussian noise to the original collected data is used to validate the proposed method. The results demonstrated the effectiveness of the proposed method in diagnosing bearing faults.

Ahmed et al. (2018) investigated the effects of sparse AE-based overcomplete sparse representations on the classification performance of highly compressed measurements of roller bearing vibration signals. In this investigation, the CS method is used to produce highly compressed measurements of the original bearing dataset. Then, an effective stacked SAE-based DNN is utilised for learning overcomplete sparse representations of the compressed measurements. Finally, the fault classification is performed using two stages: pretraining classification-based stacked AEs and the softmax regression layer, and fine-tuning by retraining the classification based on the backpropagation algorithm. Practical vibration signals acquired from roller bearings with different operating conditions are used to validate the proposed method. The results demonstrated that the proposed approach is able to achieve high levels of classification accuracy with extremely compressed measurements.

Table 14.2 presents a summary of the application of stacked AE-based DNN methods for machine fault detection and diagnosis.

14.6.2 Application of CNNs in Machine Fault Diagnosis

Many researchers have used CNN algorithms to process vibration signals for the purpose of machine fault diagnosis. For instance, Chen et al. (2015) introduced DL techniques for gearbox fault diagnosis based on CNN algorithms. In this method, the original time domain vibration signals are first pre-processed to obtain RMS values, standard deviation, skewness, kurtosis, rotation frequency, and applied load measurements, which are used to form vectors of features that represent the original vibration signals. These feature vectors are used as input for the CNN model. Two types of CNN architecture are investigated in this study: (i) two convolutional layers and two subsampling layers, and (ii) one convolutional layer and one subsampling layer. Then, the softmax layer is used to classify gearbox health conditions. Vibration signals measured from a gearbox fault experimental platform under 12 different health conditions are used to validate the proposed method. The results demonstrated the effectiveness of the proposed method in diagnosing gearbox faults.

Dong and colleagues proposed a front-end speed-controlled wind generator (FSCWG) small-fault diagnosis method based on deep CNN (DCNN). This method built a learning model with multiple hidden layers, through the layer-wise initialisation and fine-tuning training mechanism. In this way, the authors were able to obtain useful characteristics from the original vibration signals, which were acquired from a wind farm, and which have the ability to improve the diagnosis accuracy compared to the NN and SVM methods (Dong et al. 2016). Moreover, Lee et al. (2016) investigated the usage of CNNs on raw vibration signals for diagnosing bearing faults. Two vibration datasets collected from roller bearings with different health conditions are used in these investigations. The authors tested the accuracy of CNNs as classifiers on these bearing fault datasets by varying the configurations of CNN from one layer to a deep three-layer model. Also, the impact of deep architectures when noise is present in the vibration signals is examined. Guo et al. (2016b) presented a hierarchical learning rate adaptive deep CNN (ADCNN) model for diagnosing bearing faults. This model has two hierarchical arranged components: a fault-pattern determination layer and a fault-size evaluation layer. Vibration signals measured from bearings under different health conditions with different fault sizes are used to validate the proposed method. The results demonstrated the effectiveness of the proposed method in diagnosing bearing faults.

Furthermore, Janssens et al. (2016) proposed a feature learning model for condition monitoring based on a 2D CNN model. The main goal of this model is to autonomously learn useful features for bearing fault detection from the raw frequency spectrum of vibration data. In this method, the discrete Fourier transform (DFT) is utilised to obtain the frequency spectrum from the time domain vibration data. Then, features are learned using a CNN model of one convolutional layer and one fully connected layer. As a final point, the softmax layer is employed to deal with the classification problem. Practical vibration signals acquired from bearings in the test setup under different health conditions are used to validate the proposed method. The results showed that the CNN-based method achieved an overall increase in classification accuracy of approximately 6% compared with a classical manual feature-extraction method. In addition,

Table 14.2 Summary of the application of stacked AE-based DNN methods for machine fault detection and diagnosis.

Studies	Items	Pre-processing techniques	No. of AEs used	Activation function
Tao et al. 2015	Roller bearing	n/a	2	Sigmoid
Lu et al. 2015	Roller bearing	FFT	2	Sigmoid; identity
Junbo et al. 2015	Roller bearing	Wavelet transform; nonlinear soft threshold	2	Sigmoid
Huijie et al. 2015	Hydraulic pump	FFT	4	ReLU
Sun et al. 2016	Induction motor	n/a	1	Sigmoid
Galloway et al. 2016	Tidal turbine	STFT	1 and 2	Sigmoid
Jia et al. 2016	Roller bearing Planetary gearbox	Frequency spectra of time domain series signals	3	Hyperbolic tangent
Guo et al. 2016b	Roller bearing	Statistical features in the time domain; spectral kurtosis; and WT	2	Sigmoid
Lu et al. 2017b	Roller bearing	n/a	3	Sigmoid
Zhou et al. 2017	Roller bearing	FFT	5, 4, and 3	Sigmoid
Chen et al. 2017	Roller bearing	n/a	4	Sigmoid
Sohaib et al. 2017	Roller bearing	Time domain statistical features; envelope power spectrum features; and wavelet energy features	1	—
Mao et al. 2017	Roller bearing	FFT	1	—
Chen & Li 2017	Roller bearing	Statistical features in time domain and frequency domain	2 for SAE 3 for DBN	Sigmoid
Shao et al. 2017a	Gearbox Roller bearing	n/a	3	Sigmoid
Shao et al. 2017b	Gearbox Roller bearing	n/a	3	—
Qi et al. 2017	Roller bearing Gearbox	EEMD; AR	3	Sigmoid
Sun et al. 2018	Roller bearing	Compressive sampling	2	sigmoid
Jia et al. 2018	Roller bearing Planetary gearbox	n/a	1	ReLU
Shao et al. 2018a	Roller bearing	n/a	3	Identity, ArcTan, TanH, Sinusoid, Softsign, ReLU, LReLU, PReLU, ELU, SoftExponential, Sinc Sigmoid, Gaussian
Li et al. 2018	Roller bearing	Min-max normalisation	1	ReLU
Ahmed et al. 2018	Roller bearing	Haar wavelet transform; compressive sampling	2, 3, 4	Sigmoid

Li et al. (2017) introduced a method called an improved Dempster–Shafer-CNN (IDSCNN) for diagnosing bearing faults based on an ensemble of deep CNNs and an improved Dempster-Shafer theory-based evidence fusion algorithm. In this method, first, the selected time domain vibration signal is multiplied by a Hanning window, and then the FFT is applied to obtain the frequency spectrum. At that point, the RMS is calculated over a sub-band of the frequency spectrum. Then, the IDSCNN prediction model composed of an ensemble of classifiers is trained with two sensor signals. The outputs of the CNN classifiers are then fused using the IDS fusion algorithm. Typical vibration signals measured from roller bearings with four health conditions are used to validate the proposed method. The results showed that the proposed method achieved higher diagnosis accuracy by fusing signals from two sensors compared with other methods, e.g. SVM, MLP, and DNN.

Also, Wang et al. (2017) proposed a hybrid method for gearbox health monitoring based on the integration of a wavelet transform and deep CNN. In this method, one-dimensional time series vibration signals are transformed into time-scale images using wavelet analysis. Then, these images are processed by a deep CNN for fault diagnosis. Vibration signals measured from four gearbox health conditions are used to examine the proposed method for gearbox fault diagnosis. Moreover, Ding and He proposed a technique for diagnosis of spindle bearings based on wavelet packet energy (WPE) and a deep CNN (Ding and He 2017). In this method, the WPT algorithm is first combined with a phase-space reconstruction to rebuild a 2D WPE image of the frequency subspace. Then, the deep ConvNet is used to further learn identifiable characteristics of the 2D WPE image. Six vibration datasets collected from bearings under different health conditions are used to validate the proposed method. The results demonstrated the effectiveness of the proposed method in diagnosing bearing faults.

Additionally, Verstraete et al. (2017) proposed a method for RE bearing fault diagnosis based on time-frequency image analysis and a deep CNN model. In this method, time-frequency representations of the raw data are used to generate image representations of the raw signals. Then, these image representations are used as input to a deep CNN model for the purpose of fault diagnosis. Three time-frequency analysis methods, including STFT, WT, and Hilbert-Huang transform (HHT), are investigated in this study for their representation effectiveness in fault-diagnosis accuracy. Two vibration datasets acquired from roller bearings under different health conditions are used to validate the efficacy of the proposed method in diagnosing bearing faults. Xie and Zhang presented an approach for rotating machinery fault diagnosis with a feature-extraction algorithm based on EMD and CNN (Xie and Zhang 2017). In this method, features are extracted from the raw vibration signals in the time domain using (i) time domain statistical functions, including mean value, standard deviation, kurtosis, skewness, and root mean square; (ii) the combination of FT and CNN with four convolutional layers and two subsampling layers along with the ReLU activation function; and (iii) the energy entropy of the first five IMFs obtained using EMD. These features are combined together to form feature vectors that can be used for fault diagnosis. Two classifiers, SVM and softmax, are employed to deal with the classification problem. Practical vibration signals collected from roller bearings under different health conditions are used to test the efficacy of the proposed method in fault diagnosis.

Furthermore, Jing et al. (2017) presented a method for gearbox fault detection based on a CNN model. In this method, FFT is used to obtain frequency data segments from the data segments of the raw time domain data. Then, a CNN is used to learn features from the frequency data directly and detect gearbox faults. In this study, several key parameters of the CNN including the size of data segments, size filters, number of filters of the convolutional layer, and number of nodes in the fully connected layer, are examined. Two vibration datasets acquired from gearboxes with different health conditions are used to test the proposed method. The results demonstrated the effectiveness of the proposed method in learning features adaptively from frequency data and achieving high diagnostic accuracy.

You et al. proposed a hybrid technique based on CNNs and support vector regression (SVR) for rotating machinery fault diagnosis (You et al. 2017). In this method, a CNN is used to learn features from the raw signals, and SVR is employed to deal with the multiclass classification problem. The proposed method is validated using real acoustic signals collected from locomotive bearings and vibration signals acquired from an automobile transmission gearbox. The results showed that the proposed method is able to detect faults in both bearing and gears.

Liu et al. (2017) developed a method called dislocated time series CNN (DTS-CNN) for fault diagnosis of electric machines. The architecture of DTS-CNN is composed of a dislocating layer, a convolutional layer, a sub-sampling layer, and a fully connected layer. The simple idea of DTS-CNN is that the periodic fault information between nonadjacent signals can be extracted by continuously dislocating the input raw signals. Two experiments on an electric machine fault simulator with different operating conditions are used to test the proposed method: (i) operating under nine health conditions with constant speed, and (ii) operating under six motor conditions with variable speed. Also, Lu et al. (2017) presented a DL method for diagnosing bearing faults based on a CNN algorithm. In this method, the CNN model is used to learn high-level feature representations from the input samples directly utilising supervised DL where the convolutional and pooling layers are set consecutively for greedy learning. ReLU, local contrast normalisation, and weight replication are utilised to represent elementary features during convolutional computations. Typical vibration signals acquired from bearings under different fault conditions are used to validate the proposed method.

In their case study of gear fault diagnosis, Sun et al. (2017) introduced a method based on a dual-tree complex wavelet transform (DTCWT) and CNN. The DTCWT is used to obtain the multiscale signal's features. Then, a CNN is employed to automatically recognise a fault feature from the multiscale signal features from which the gear fault can be identified using a softmax classifier. Vibration signals acquired from the gearbox test rig under four health conditions are used to validate the proposed method. The results showed that the proposed method is able to distinguish the four types of gear faults. In another study that involves the use of a wavelet transform as a pre-processing step, Guo et al. (2018) presented a method for fault diagnosis of rotating machinery based on the CWT technique and a CNN model. In this method, first, the CWT is used to form a continuous wavelet transform scalogram (CWTS) from the original raw vibration signals. Then, CWTS cropping is introduced to use part of the CWTS as the CNN input. A CNN with two convolutional layers, two subsampling layers, and one fully connected layer along with the sigmoid function as the activation function is utilised to process the obtained part of the CWTS for fault diagnosis. Typical vibration signals acquired from the rotor test bed under four different fault modes – rotor imbalance, rotor misalignment, bearing block looseness, and contact rubbing – are utilised to examine the efficacy of the proposed method in diagnosing rotating machinery faults.

Recently, Cao et al. (2018) introduced a deep CNN-based transfer learning approach for gear fault diagnosis using small datasets. The architecture of the proposed transfer learning contains two parts: the first part is built with a pretrained DNN to extract features automatically from the input, and the second part is a fully connected stage to deal with the classification problem. Vibration signals measured from a gear under nine different conditions – healthy, missing tooth, root crack, spalling, and chipping tip with five different levels of severity – are used to validate the proposed method. Moreover, Wen et al. (2018) presented a LeNet-5-based CNN method for fault diagnosis. In this method, the time domain raw vibration signals are converted into images. Then, a CNN is trained to classify these images. Three datasets of fault diagnosis – the motor bearing fault dataset, the self-priming centrifugal pump dataset, and an axial piston hydraulic pump dataset – are used to validate the proposed method.

Table 14.3 presents a summary of the application of CNN methods for machine fault detection and diagnosis.

14.6.3   Application of DBNs in Machine Fault Diagnosis

A considerable amount of literature has been published on the application of DBNs in machine fault diagnosis. For example, Chen et al. (2016) presented a method based on DBNs. In this method, a DBN composed of three layers of RBMs is used to process vibration signals acquired from a healthy bearing and two sets of bearings with different defects. To further validate the effectiveness of the proposed DBN diagnostic model to process the raw vibration data, six different input features are used with the same softmax layer classifier used in the proposed DBN method. These include features obtained using univariate feature techniques, multivariate features techniques, and image features. The results showed that the DBN can be used directly in the analysis and processing of the raw vibration data to diagnosis bearing faults. Moreover, Yin et al. (2016) presented a combined assessment model (CAM) for assessing machine health. In this model, 38 features are extracted from the acquired vibration signal using time domain analysis, frequency domain analysis, and a wavelet packet transform (WPT). Then, the ISOMAP algorithm is utilised for dimensionality reduction and extraction of more representative features. Next, the low-dimensional features are used to train a DBN model for the purpose of evaluating the machine's performance status. Typical vibration signals measured from bearings are used to examine the proposed method.

Furthermore, Tao et al. (2016) presented a method for diagnosing bearing faults using multivibration signals and DBN. In this method, (i) multiple vibration signals are acquired from various fault bearings. (ii) Several time domain statistical features are extracted from the original vibration signals. Finally, (iii) with these extracted features, the DBN with two hidden layers is trained for the purpose of fault diagnosis. The results showed that the proposed DBN with multisensor vibration signals achieved 97.5% classification accuracy, which is 10% higher than a single sensor. In addition, compared to other methods, e.g. SVM, k-nearest neighbour (KNN), and BPNN, the results showed that the proposed DBN is more effective and stable for the identification of rolling bearing faults.

Table 14.3 Summary of the application of CNN methods for machine fault detection and diagnosis.

Studies	Items	Pre-processing techniques	No. of Convolutional layers used	No. of subsampling layers used
Chen et al. 2015	Gearbox	RMS values, standard deviation, skewness, kurtosis, rotation frequency, and applied load measurements	1 and 2	1 and 2, respectively
Lee et al. 2016	Bearings	n/a	1, 2, and 3	1, 2, and 3, respectively
Guo et al. 2016	Bearings	n/a	3	3
Janssens et al. 2016	Bearings	DFT	1	1
Li et al. 2017	Bearings	Hanning window, FFT, and RMS	3	n/a
Wang et al. 2017	Gearbox	WT	2	2
Ding and He 2017	Bearings	WPE	3	2
Verstraete et al. 2017	Bearings	STFT, WT, and HHT	6	3
Xie and Zhang 2017	Bearings	FFT	4	2
Jing et al. 2017	Gearboxes	FFT	1	1
You et al. 2017	Bearing Gear	n/a	3	3
Liu et al. 2017	Induction motor	Dislocating time series	4	2
Lu et al. 2017b	Bearings	n/a	2	2
Sun et al. 2017	Gear	DTCWT	2	2
Guo et al. 2018	Rotor	CWT	2	2
Cao et al. 2018	Gear	TSA	5	4
Wen et al. 2018	Bearing; self-priming centrifugal pump; and an axial piston hydraulic pump.	Time domain raw vibration signals are converted into images	4	4

Additionally, Han et al. (2017) introduced a method for diagnosing roller bearing faults based on the Teager-Kaiser energy operator, particle swarm optimisation (PSO), and SVM with DBN. In this method, (i) the Teager-Kaiser energy operator is used to demodulate the acquired vibration signals. (ii) Statistic features from the time domain and frequency domain are obtained from the demodulated signals, and the DBN is used to learn features in the time domain and the frequency domain. Finally, (iii) with these learned features, a PSO-based optimised SVM is used to deal with the classification problem. Typical vibration signals acquired from roller bearings with different health conditions are used to validate the proposed method. The results demonstrated the effectiveness of the proposed method in fault diagnosis of roller bearings.

Furthermore, a DL approach for fault diagnosis of induction motors in manufacturing is introduced in (Shao et al. 2017c). In this, approach, a DBN model is used to learn features from the frequency distribution of the collected vibration signals for the purpose of characterising the work status of induction motors. Here, the collected vibration signals are transformed from the time domain to the frequency domain using the FFT. Practical vibration signals with six different motor operating conditions are used to validate the proposed method. Also, He et al. (2017) proposed a method for fault diagnosis of a gear transmission chain based on a DBN where the genetic algorithm (GA) is used to optimise the structural parameters of the network. In this method, the acquired signals are directly segmented to form sample sets. Then, a DBN model with several RBMs based on GA is constructed and trained using the sample sets. Finally, the fully trained DBN is used for fault classification of a gear transmission chain. Two typical vibration signals datasets measured from bearings and a gearbox are used to test the efficacy of the proposed method. The results showed that the proposed method is able to achieve fault-classification accuracy up to 99.26% for roller bearings and 100% for the gearbox.

In a study investigating the combination of a DBN and dual-tree complex wavelet packet (DTCWPT), Shao et al. (2017d) proposed a technique called adaptive DBN with DTCWPT for rolling bearing fault diagnosis. In this method, the DTCWPT is used to decompose vibration signals into eight components with different frequency bands. Then, nine features are extracted from the eight decomposed frequency-band signals. Next, an adaptive DBN model is constructed based on a series of trained adaptive RBMs. A vibration dataset measured from roller bearings under different operating conditions is used to examine the efficacy of the proposed method.

Recently, a method for fault detection using continuous DBN (CDBN) with locally linear embedding is introduced in (Shao et al. 2018b). In this method, (i) six time-domain features are extracted from the acquired vibration data. (ii) The feature fusion method local linear embedding (LLE) is used to define a comprehensive feature index from the extracted features. (iii) The comprehensive feature index values during normal operation are selected as a training dataset used to train the CDBN predictor. Here, the main parameters of the trained predictor are optimised using GA. Experimental bearing signals are used to test the proposed method. The results demonstrated that the proposed method is able to detect the system's dynamic behaviour stably and accurately.

14.6.4 Application of RNNs in Machine Fault Diagnosis

Turning now to the experimental evidence on the application of the RNN technique in machine fault diagnosis, Tse and Atherton (1999) studied the prediction of machine deterioration using vibration-based faults trends and RNNs. In this study, the use of an RNN demonstrated that it is able to learn the trend of nonlinear temporal data and then predict the future trend. Şeker et al. (2003) studied the capability of RNNs for condition monitoring and diagnosis in nuclear power plant systems and rotating machinery. Two applications are presented in this study: (i) the use of RNNs for detecting anomalies introduced from simulated power operation of a high-temperature gas-cooled nuclear reactor, and (ii) the use of RNNs for detecting motor bearing faults using a coherence function technique that is defined between the motor current and vibration signals for induction motors. In the second application, the proposed method presented an auto-associative structure based on RNNs to follow changes in the coherence signals.

Moreover, Zhao et al. (2017) developed a DNN structure called convolutional bidirectional LSTM networks (CBLSTM) for monitoring machine health. CBLSTM first uses a CNN to extract local features from the input signal. Then, bidirectional LSTM is used to encode temporal information. Next, two fully connected dense layers are stacked together to process the output of LSTMs. Finally, a linear regression layer is employed to predict the actual tool wear. The results showed that the proposed method is able to capture and discover meaningful features under the sensory signal for machine health monitoring.

Lee et al. (2018) introduced a multi-layered approach for detecting bearing anomalies in rotating machinery based on CNNs and bidirectional and unidirectional LSTM RNNs. In this approach, the stacked CNNs are used to extract features for the acquired vibration signals. Then, with these extracted features, the stacked bidirectional LSTMs (SB-LSTMs) are used for feature learning. Next, the stacked unidirectional LSTMs (SU-LSTMs) are used to enhance the feature learning. Finally, with these enhanced features, a regression layer is employed to deal with the detection problem. Practical vibration signals acquired from bearings are utilised to validate the proposed method. Furthermore, Liu et al. proposed a method for diagnosing bearing faults based on RNNs in the form of an AE (Liu et al. 2018). In this method, multiple vibration values for the rolling bearings during the next period are predicted from the previous period using a GRU-based denoising AE. The bearing fault diagnosis is performed by comparing reconstruction errors generated by different trained GRU-based nonlinear predictive denoising AEs (GRU-NP-DAEs) from multiple-dimension time-series data. Vibration datasets acquired from bearings with different health operating conditions are used to test the effectiveness of the proposed method. The results verified that the robustness of the trained model is satisfactory in diagnosing roller bearing faults, even under the condition of a low SNR and fluctuant.

Recently, Zhang et al. (2019) introduced a data-driven method for assessing bearing performance degradation based on LSTM RNN. In this study, a degradation simulation model based on the vibration response mechanism is built for feature verification. Many time domain features are adopted to explore the relationship between the bearing's running conditions and vibration signals. In addition, a waveform entropy (WFE) indicator is introduced. Then, the WFE and other conventional indicators are used as input to the LSTM RNN for the purpose of recognizing the bearing's running state. Here, the PSO algorithm is used to optimise the parameters of the network structure. Experimental results demonstrated the effectiveness of the proposed method in bearing degradation state recognition and predicting remaining useful life.

14.7 Summary

A great deal of previous research into vibration-based fault diagnosis has focused on designing and applying methods that have the ability to extract and select good representations of raw vibration data that are expected to build a better classification model. If the vibration signals' features are carefully devised and the parameters of the classifiers are carefully tuned, it is possible to achieve high accuracy in classification performance. However, extracting useful features from such a large and noisy vibration data is not an easy task. One solution is to use feature-learning techniques, also called representation learning techniques, which have the ability to automatically learn representations of the data required for feature detection and classification. Of these techniques, deep learning (DL) is often used to automatically learn representations of data with multiple layers of information-processing modules in hierarchical architectures.

This chapter has presented several deep neural techniques and their applications in machine fault diagnosis. These include AE-based DNNs, CNNs, DBNs, and RNNs. Unlike ANNs, DNNs can be trained in a supervised or unsupervised manner, and they are also appropriate in the general area of RL. The basic idea of training DNNs is that we first train the network layer by layer using an unsupervised learning algorithm, e.g. an AE; this process is called DNN pretraining. In this process, the output from each layer will be the input to the succeeding layer. Then the DNN is retrained in a supervised way with a backpropagation algorithm for classification. The convolutional deep neural network (CNN) architecture is usually composed of a convolutional layer and a sub-sampling layer, also called a pooling layer. The CNN learns abstract features from alternating and stacking convolutional layers and pooling operation. The convolutional layers convolve multiple local filters with raw input data and generate invariant local features, and the pooling layers extract most significant features. Furthermore, DBNs are generative NNs that stack multiple RBMs that can be trained in a greedy layer-wise unsupervised way; then they can be further fine-tuned with respect to the labels of training data by adding a softmax layer in the top layer. Moreover, RNNs build connections between units from a direct cycle, map from the entire history of previous inputs to target vectors in principle, and allow the memory of previous inputs to be kept in the network state. As is the case with DBNs, RNNs can be trained via backpropagation through time for supervised tasks with sequential input data and target outputs.

Most of the introduced techniques and their publicly accessible software are summarised in Table 14.4.

Table 14.4 Summary of some of the introduced techniques and their publically accessible software.

Algorithm name	Platform	Package	Function
Train an autoencoder	MATLAB	Deep Learning Toolbox	trainAutoencoder
Train a softmax layer for classification			trainsoftmaxlayer
Encode input data			encode
Decode encoded data			decode
Stack encoders from several autoencoder			stack
Rectified linear unit (ReLU) layer			reluLayer
Leaky rectified linear unit (ReLU) layer			leakyReluLayer
2D convolutional layer		Neural Network Toolbox	convolution2dLayer
Fully connected layer			fullyConnectedLayer
Max pooling layer			maxPooling2dLayer
Layer recurrent neural network			layrecnet
Long short-term memory (LSTM) layer			lstmLayer
Dropout layer			dropoutLayer

References

Abdel-Hamid, O., Mohamed, A.R., Jiang, H. et al. (2014). Convolutional neural networks for speech recognition. *IEEE/ACM Transactions on Audio, Speech, and Language Processing* 22 (10): 1533–1545.

Ahmed, H.O.A., Wong, M.L.D., and Nandi, A.K. (2018). Intelligent condition monitoring method for bearing faults from highly compressed measurements using sparse over-complete features. *Mechanical Systems and Signal Processing* 99: 459–477.

Bengio, Y., Lamblin, P., Popovici, D., and Larochelle, H. (2007). Greedy layer-wise training of deep networks. In: *Advances in Neural Information Processing Systems*, 153–160.

Cao, P., Zhang, S., and Tang, J. (2018). Preprocessing-free gear fault diagnosis using small datasets with deep convolutional neural network-based transfer learning. *IEEE Access* 6: 26241–26253.

Chen, Z. and Li, W. (2017). Multisensor feature fusion for bearing fault diagnosis using sparse autoencoder and deep belief network. *IEEE Transactions on Instrumentation and Measurement* 66 (7): 1693–1702.

Chen, Z., Li, C., and Sanchez, R.V. (2015). Gearbox fault identification and classification with convolutional neural networks. *Shock and Vibration* 2015.

Chen, Z., Zeng, X., Li, W., and Liao, G. (2016). Machine fault classification using deep belief network. In: *Instrumentation and Measurement Technology Conference Proceedings (I2MTC), 2016 IEEE International*, 1–6. IEEE.

Chen, R., Chen, S., He, M. et al. (2017a). Rolling bearing fault severity identification using deep sparse auto-encoder network with noise added sample expansion. *Proceedings of the Institution of Mechanical Engineers, Part O: Journal of Risk and Reliability* 231 (6): 666–679.

Chen, Z., Deng, S., Chen, X. et al. (2017b). Deep neural networks-based rolling bearing fault diagnosis. *Microelectronics Reliability* 75: 327–333.

Chung, J., Gulcehre, C., Cho, K., and Bengio, Y. (2014). Empirical evaluation of gated recurrent neural networks on sequence modeling. arXiv preprint arXiv:1412.3555.

Clevert, D.A., Unterthiner, T., and Hochreiter, S. (2015). Fast and accurate deep network learning by exponential linear units (elus). arXiv preprint arXiv:1511.07289.

Collobert, R. and Weston, J. (2008). A unified architecture for natural language processing: Deep neural networks with multitask learning. In: *Proceedings of the 25th international conference on Machine learning*, 160–167. ACM.

Ding, X. and He, Q. (2017). Energy-fluctuated multiscale feature learning with deep convnet for intelligent spindle bearing fault diagnosis. *IEEE Transactions on Instrumentation and Measurement* 66 (8): 1926–1935.

Dong, H.Y., Yang, L.X., and Li, H.W. (2016). Small fault diagnosis of front-end speed controlled wind generator based on deep learning. *WSEAS Transactions on Circuits and Systems* (9): 15.

Erhan, D., Bengio, Y., Courville, A. et al. (2010). Why does unsupervised pre-training help deep learning? *Journal of Machine Learning Research* 11: 625–660.

Galloway, G.S., Catterson, V.M., Fay, T. et al. (2016). Diagnosis of tidal turbine vibration data through deep neural networks. In: *Proceedings of the 3rd European Conference of the Prognostic and Health Management Society*, 172–180. PHM Society.

Glorot, X., Bordes, A., and Bengio, Y. (2011). Deep sparse rectifier neural networks. In: *Proceedings of the fourteenth international conference on artificial intelligence and statistics*, 315–323.

Godfrey, L.B. and Gashler, M.S. (2015). A continuum among logarithmic, linear, and exponential functions, and its potential to improve generalization in neural networks. In: *Knowledge Discovery, Knowledge Engineering and Knowledge Management (IC3K), 2015 7th International Joint Conference on*, vol. 1, 481–486. IEEE.

Graves, A., Mohamed, A.R., and Hinton, G. (2013). Speech recognition with deep recurrent neural networks. In: *Acoustics, Speech and Signal Processing (ICASSP), 2013 IEEE International Conference on*, 6645–6649. IEEE.

Guo, L., Gao, H., Huang, H. et al. (2016a). Multifeatures fusion and nonlinear dimension reduction for intelligent bearing condition monitoring. *Shock and Vibration* 2016.

Guo, X., Chen, L., and Shen, C. (2016b). Hierarchical adaptive deep convolution neural network and its application to bearing fault diagnosis. *Measurement* 93: 490–502.

Guo, S., Yang, T., Gao, W., and Zhang, C. (2018). A novel fault diagnosis method for rotating machinery based on a convolutional neural network. *Sensors (Basel, Switzerland)* 18 (5).

Han, D., Zhao, N., and Shi, P. (2017). A new fault diagnosis method based on deep belief network and support vector machine with Teager–Kaiser energy operator for bearings. *Advances in Mechanical Engineering* 9 (12) p. 1687814017743113.

He, K., Zhang, X., Ren, S., and Sun, J. (2015). Delving deep into rectifiers: Surpassing human-level performance on imagenet classification. In: *Proceedings of the IEEE International Conference on Computer Vision*, 1026–1034.

He, J., Yang, S., and Gan, C. (2017). Unsupervised fault diagnosis of a gear transmission chain using a deep belief network. *Sensors* 17 (7): 1564.

Hinton, G.E., Osindero, S., and Teh, Y.W. (2006). A fast learning algorithm for deep belief nets. *Neural Computation* 18 (7): 1527–1554.

Hochreiter, S. and Schmidhuber, J. (1997). Long short-term memory. *Neural Computation* 9 (8): 1735–1780.

Huijie, Z., Ting, R., Xinqing, W. et al. (2015). Fault diagnosis of hydraulic pump based on stacked autoencoders. In: *Electronic Measurement & Instruments (ICEMI), 2015 12th IEEE International Conference on*, vol. 1, 58–62. IEEE.

Janssens, O., Slavkovikj, V., Vervisch, B. et al. (2016). Convolutional neural network based fault detection for rotating machinery. *Journal of Sound and Vibration* 377: 331–345.

Jia, F., Lei, Y., Lin, J. et al. (2016). Deep neural networks: A promising tool for fault characteristic mining and intelligent diagnosis of rotating machinery with massive data. *Mechanical Systems and Signal Processing* 72: 303–315.

Jia, F., Lei, Y., Guo, L. et al. (2018). A neural network constructed by deep learning technique and its application to intelligent fault diagnosis of machines. *Neurocomputing* 272: 619–628.

Jing, L., Zhao, M., Li, P., and Xu, X. (2017). A convolutional neural network based feature learning and fault diagnosis method for the condition monitoring of gearbox. *Measurement* 111: 1–10.

Junbo, T., Weining, L., Juneng, A., and Xueqian, W. (2015). Fault diagnosis method study in roller bearing based on wavelet transform and stacked auto-encoder. In: *Control and Decision Conference (CCDC), 2015 27th Chinese*, 4608–4613. IEEE.

Khan, S. and Yairi, T. (2018). A review on the application of deep learning in system health management. *Mechanical Systems and Signal Processing* 107: 241–265.

Lange, S. and Riedmiller, M.A. (2010). Deep auto-encoder neural networks in reinforcement learning. In: *IJCNN*, 1–8.

LeCun, Y., Boser, B.E., Denker, J.S. et al. (1990). Handwritten digit recognition with a back-propagation network. In: *Advances in Neural Information Processing Systems*, 396–404.

LeCun, Y., Bottou, L., Bengio, Y., and Haffner, P. (1998). Gradient-based learning applied to document recognition. *Proceedings of the IEEE* 86 (11): 2278–2324.

Lee, D., Siu, V., Cruz, R., and Yetman, C. (2016). Convolutional neural net and bearing fault analysis. In: *Proceedings of the International Conference on Data Mining series (ICDM)*, 194–200. Barcelona.

Lee, K., Kim, J.K., Kim, J. et al. (2018). Stacked convolutional bidirectional LSTM recurrent neural network for bearing anomaly detection in rotating machinery diagnostics. In: *2018 1st IEEE International Conference on Knowledge Innovation and Invention (ICKII)*, 98–101. IEEE.

Li, S., Liu, G., Tang, X. et al. (2017). An ensemble deep convolutional neural network model with improved DS evidence fusion for bearing fault diagnosis. *Sensors* 17 (8): 1729.

Li, C., Zhang, W., Peng, G., and Liu, S. (2018). Bearing fault diagnosis using fully-connected winner-take-all autoencoder. *IEEE Access* 6: 6103–6115.

Lipton, Z.C., Berkowitz, J., and Elkan, C. (2015). A critical review of recurrent neural networks for sequence learning. arXiv preprint arXiv:1506.00019.

Liu, R., Meng, G., Yang, B. et al. (2017). Dislocated time series convolutional neural architecture: An intelligent fault diagnosis approach for electric machine. *IEEE Transactions on Industrial Informatics* 13 (3): 1310–1320.

Liu, H., Zhou, J., Zheng, Y. et al. (2018). Fault diagnosis of rolling bearings with recurrent neural network-based autoencoders. *ISA Transactions* 77: 167–178.

Lu, W., Wang, X., Yang, C., and Zhang, T. (2015). A novel feature extraction method using deep neural network for rolling bearing fault diagnosis. In: *Control and Decision Conference (CCDC), 2015 27th Chinese*, 2427–2431. IEEE.

Lu, C., Wang, Z., and Zhou, B. (2017a). Intelligent fault diagnosis of rolling bearing using hierarchical convolutional network based health state classification. *Advanced Engineering Informatics* 32: 139–151.

Lu, C., Wang, Z.Y., Qin, W.L., and Ma, J. (2017b). Fault diagnosis of rotary machinery components using a stacked denoising autoencoder-based health state identification. *Signal Processing* 130: 377–388.

Mao, W., He, J., Li, Y., and Yan, Y. (2017). Bearing fault diagnosis with auto-encoder extreme learning machine: a comparative study. *Proceedings of the Institution of Mechanical Engineers, Part C: Journal of Mechanical Engineering Science* 231 (8): 1560–1578.

Maturana, D. and Scherer, S. (2015). Voxnet: a 3d convolutional neural network for real-time object recognition. In: *Intelligent Robots and Systems (IROS), 2015 IEEE/RSJ International Conference on*, 922–928. IEEE.

Min, S., Lee, B., and Yoon, S. (2017). Deep learning in bioinformatics. *Briefings in Bioinformatics* 18 (5): 851–869.

Mohamed, A.R., Dahl, G.E., and Hinton, G. (2012). Acoustic modeling using deep belief networks. *IEEE Trans. Audio, Speech & Language Processing* 20 (1): 14–22.

Ng, A. (2011). Sparse autoencoder. CS294A lecture notes. Stanford University.

Parkhi, O.M., Vedaldi, A., and Zisserman, A. (2015). Deep face recognition. In: *Proceedings of the British Machine Vision Conference (BMVC) 2015*, vol. 1, No. 3, 6.

Qi, Y., Shen, C., Wang, D. et al. (2017). Stacked sparse autoencoder-based deep network for fault diagnosis of rotating machinery. *IEEE Access* 5: 15066–15079.

Salakhutdinov, R.R. and Hinton, G.E. (2009). Deep Boltzmann machines. International Conference on Artificial Intelligence and Statistics.

Salimans, T. and Kingma, D.P. (2016). Weight normalization: a simple reparameterization to accelerate training of deep neural networks. In: *Advances in Neural Information Processing Systems*, 901–909.

Schmidhuber, J. (2015). Deep learning in neural networks: An overview. *Neural networks* 61: 85–117.

Schuster, M. and Paliwal, K.K. (1997). Bidirectional recurrent neural networks. *IEEE Transactions on Signal Processing* 45 (11): 2673–2681.

Şeker, S., Ayaz, E., and Türkcan, E. (2003). Elman's recurrent neural network applications to condition monitoring in nuclear power plant and rotating machinery. *Engineering Applications of Artificial Intelligence* 16 (7–8): 647–656.

Shao, H., Jiang, H., Wang, F., and Wang, Y. (2017a). Rolling bearing fault diagnosis using adaptive deep belief network with dual-tree complex wavelet packet. *ISA Transactions* 69: 187–201.

Shao, H., Jiang, H., Wang, F., and Zhao, H. (2017b). An enhancement deep feature fusion method for rotating machinery fault diagnosis. *Knowledge-Based Systems* 119: 200–220.

Shao, H., Jiang, H., Zhao, H., and Wang, F. (2017c). A novel deep autoencoder feature learning method for rotating machinery fault diagnosis. *Mechanical Systems and Signal Processing* 95: 187–204.

Shao, S.Y., Sun, W.J., Yan, R.Q. et al. (2017d). A deep learning approach for fault diagnosis of induction motors in manufacturing. *Chinese Journal of Mechanical Engineering* 30 (6): 1347–1356.

Shao, H., Jiang, H., Li, X., and Liang, T. (2018a). Rolling bearing fault detection using continuous deep belief network with locally linear embedding. *Computers in Industry* 96: 27–39.

Shao, H., Jiang, H., Lin, Y., and Li, X. (2018b). A novel method for intelligent fault diagnosis of rolling bearings using ensemble deep auto-encoders. *Mechanical Systems and Signal Processing* 102: 278–297.

Simard, P.Y., Steinkraus, D., and Platt, J.C. (2003). Best practices for convolutional neural networks applied to visual document analysis. In: *Proc. Seventh Int'l Conf. Document Analysis and Recognition*, vol. 2, 958–962.

Sohaib, M., Kim, C.H., and Kim, J.M. (2017). A hybrid feature model and deep-learning-based bearing fault diagnosis. *Sensors* 17 (12): 2876.

Sun, W., Shao, S., Zhao, R. et al. (2016). A sparse auto-encoder-based deep neural network approach for induction motor faults classification. *Measurement* 89: 171–178.

Sun, W., Yao, B., Zeng, N. et al. (2017). An intelligent gear fault diagnosis methodology using a complex wavelet enhanced convolutional neural network. *Materials* 10 (7): 790.

Sun, J., Yan, C., and Wen, J. (2018). Intelligent bearing fault diagnosis method combining compressed data acquisition and deep learning. *IEEE Transactions on Instrumentation and Measurement* 67 (1): 185–195.

Tajbakhsh, N., Shin, J.Y., Gurudu, S.R. et al. (2016). Convolutional neural networks for medical image analysis: full training or fine tuning? *IEEE Transactions on Medical Imaging* 35 (5): 1299–1312.

Tao, S., Zhang, T., Yang, J. et al. (2015). Bearing fault diagnosis method based on stacked autoencoder and softmax regression. In: *Control Conference (CCC), 2015 34th Chinese*, 6331–6335. IEEE.

Tao, J., Liu, Y., and Yang, D. (2016). Bearing fault diagnosis based on deep belief network and multisensor information fusion. *Shock and Vibration* 2016.

Tse, P.W. and Atherton, D.P. (1999). Prediction of machine deterioration using vibration based fault trends and recurrent neural networks. *Journal of Vibration and Acoustics* 121 (3): 355–362.

Verstraete, D., Ferrada, A., Droguett, E.L. et al. (2017). Deep learning enabled fault diagnosis using time-frequency image analysis of rolling element bearings. *Shock and Vibration* 2017: 17.

Wang, P., Yan, R., and Gao, R.X. (2017). Virtualization and deep recognition for system fault classification. *Journal of Manufacturing Systems* 44: 310–316.

Wen, L., Li, X., Gao, L., and Zhang, Y. (2018). A new convolutional neural network-based data-driven fault diagnosis method. *IEEE Transactions on Industrial Electronics* 65 (7): 5990–5998.

Xie, Y. and Zhang, T. (2017). Fault diagnosis for rotating machinery based on convolutional neural network and empirical mode decomposition. *Shock and Vibration* 2017: 12.

Yin, A., Lu, J., Dai, Z. et al. (2016). Isomap and deep belief network-based machine health combined assessment model. *Strojniski Vestnik/Journal of Mechanical Engineering* 62 (12): 740–750.

You, W., Shen, C., Guo, X. et al. (2017). A hybrid technique based on convolutional neural network and support vector regression for intelligent diagnosis of rotating machinery. *Advances in Mechanical Engineering* 9 (6) p. 1687814017704146.

Zhang, B., Zhang, S., and Li, W. (2019). Bearing performance degradation assessment using long short-term memory recurrent network. *Computers in Industry* 106: 14–29.

Zhao, R., Yan, R., Chen, Z. et al. (2016). Deep learning and its applications to machine health monitoring: A survey. arXiv preprint arXiv:1612.07640.

Zhao, R., Yan, R., Wang, J., and Mao, K. (2017). Learning to monitor machine health with convolutional bi-directional lstm networks. *Sensors* 17 (2): 273.

Zhou, F., Gao, Y., and Wen, C. (2017). A novel multimode fault classification method based on deep learning. *Journal of Control Science and Engineering* 2017: 14.

15

Classification Algorithm Validation

15.1 Introduction

As described in Chapter 6, classification is an essential task in the machine fault diagnosis framework for assigning class labels, i.e. machine health conditions, to new vibration signal examples. There is a large volume of published studies describing the role of classification algorithms in machine fault diagnosis. Previous chapters introduced various types of commonly used classification algorithms; accordingly, various techniques can be used to classify different vibration types based on the features provided. If the vibration signals' features are carefully devised and the parameters of the classifiers are carefully tuned, it is possible to achieve high classification accuracies. Therefore, model selection and performance estimation are the two main tasks we need to consider when dealing with classification problems. In fact, these tasks can be achieved by using model-validation techniques. The cross-validation technique is usually recommended as a superior test of the classification model compared to testing the predictive validity of a model on the same data that were utilised to estimate its parameters, which often leads to bias (Cooil et al. 1987). To assess the efficiency of the classification algorithms introduced in the previous chapters, this chapter describes different validation techniques that can be used to evaluate and verify the classification model's performance before proceeding to its application and implementation in machine fault diagnosis problems.

The most common approach to cross-validation is data-splitting, which can be done using one of the following techniques:

- The hold-out technique
- Random subsampling
- K-fold cross-validation
- Leave-one-out cross-validation
- Bootstrapping

Furthermore, various measures can be used to evaluate the performance of a classification model, including overall classification accuracy, a confusion matrix, the receiver operating characteristic (ROC), precision, and recall. The first part of this

Condition Monitoring with Vibration Signals: Compressive Sampling and Learning Algorithms for Rotating Machines, First Edition. Hosameldin Ahmed and Asoke K. Nandi.

chapter describes the data-splitting techniques mentioned. The second part is devoted to a description of the commonly used classifiers' performance measures.

15.2 The Hold-Out Technique

The hold-out is a simple technique for model validation. In this technique, a sub-sample, which is called the *training* or *estimation set*, of the available total set of observations is selected (usually randomly). The remainder of the total observations after the training set is selected are often called the *testing set*. The training set is used to train the classification algorithm, and the testing set is used to test the trained classification model. The procedure of this technique can be described in three steps as follows:

(1) The total set of observations, i.e. the full amount of available of data, is split into training and testing sets. In the literature on classification studies, many classification algorithms have been proposed, and their performance has been examined using different sizes of training and testing sets including the following percentages: 90% and 10%; 80% and 20%; 70% and 30%; 60% and 40%; and 50% and 50% for the training set and testing set, respectively.
(2) The selected classification algorithm is then trained using the training set to obtain a trained classification model.
(3) The obtained model is used to predict the class labels of the testing set.

Suppose that we have sufficient data to apply the hold-out technique to validate a classification algorithm: Figure 15.1 shows an illustration of a data-splitting example using the hold-out technique with a training set of size 70% and a testing set size of 30% of the total observations.

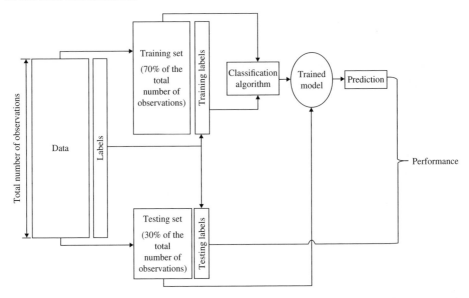

Figure 15.1 Data-splitting example using the hold-out technique with a training set of size 70% and a testing set size of 30% of the total observations.

15.2.1 Three-Way Data Split

The procedure just described, which splits the data into two subsamples, can be used to validate the performance of a trained classification model with new data. Before proceeding to test the trained model using the testing set, one may evaluate the model using different parameters to select the best model using a validation test. This technique is called the *three-way data split* validation technique. The procedure of this technique can be described as follows,

1. The available total number of observations are divided into training, validation, and test sets.
2. The training parameters are selected.
3. The training set is used to train the classification model.
4. The trained classification model is evaluated using the validation set.
5. Using different training parameters, the classification model training and evaluation in previous two steps are repeated.
6. The best model is selected and trained using data combining the training and validation set.
7. The final trained model is then tested using the testing set.

Numerous studies have assessed the effects of the training set size on classification model performance. For example, Raudys and Jain (1991) showed that small sample sizes for observations make the problem of designing a classification model very difficult; and Guyon (1997) derived a formula for splitting the training database into a training set and a validation set valid for large training databases. Kavzoglu and Colkesen (2012) investigated the effect of training set size in the performance of support vector machines (SVMs) and decision trees (DTs) classification methods. Moreover, Beleites et al. (2013) compared the sample sizes needed to train good classifiers. In fact, if a small number of observations are available, which is not helpful in the hold-out validation technique, which requires splitting the total number of observations into a training set and a testing set, the result is very few observations in the training set and testing set to be used for training a classification algorithm and testing the created classification model, respectively. Also, a small training set may result in a large variance in parameter estimations, and a small testing set may lead to a large variance in performance. To overcome the limitations of the hold-out technique, the following subsections introduce four resampling techniques that are widely used for model validation.

15.3 Random Subsampling

Random subsampling, which also called *Monte Carlo cross-validation* (MCCV) or *repeated hold-out*, is a technique that repeats the hold-out technique k times randomly, i.e. performs k data splits (Xu and Liang 2001; Lendasse et al. 2003). In this technique, each split randomly selects a number of examples without replacement; for each data split, the training set is used to train the classification algorithm and the testing set is used to test the performance of the created classification model. The performance of the model is calculated by taking the average performance over the k repetitions of the data splits. Figure 15.2 shows an illustration of this technique using $k = 5$.

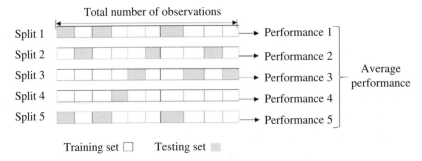

Figure 15.2 An illustration of a randomly repeated hold-out technique with $k = 5$.

15.4 K-Fold Cross-Validation

K-fold cross validation is a commonly used technique to estimate the performance of a classification algorithm or compare performance between two classification algorithms. The procedure for this technique can be described as follows:

- The total available observations, i.e. data, are divided into k disjoint folds.
- For each k training experiments, $k - 1$ folds are used for training and one fold for testing.
- The performance of the classification algorithm is evaluated by the average of the k accuracies obtained from the k-fold cross validation.

The following four factors can affect the accuracy estimated using k-fold cross validation (Wong 2015):

- Number of folds
- Number of instances in a fold
- Level of averaging
- Repetition of k-fold cross validation

Figure 15.3 shows an illustration of the k-fold technique using $k = 5$, i.e. five fold. It can be seen in the figure that all the observations in the data are utilised for both training and testing through the five folds used to split the data.

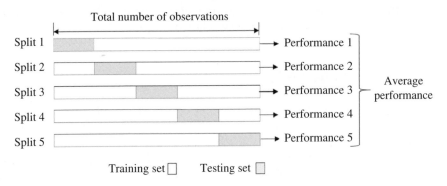

Figure 15.3 An illustration of the k-fold technique using $k = 5$, i.e. fivefold.

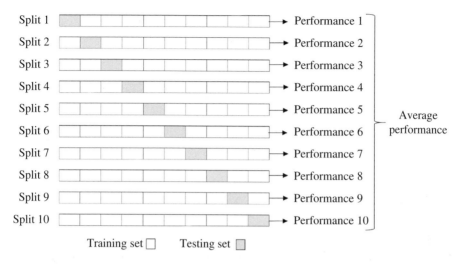

Figure 15.4 An illustration of leave-one-out cross-validation used for data with 10 observations.

15.5 Leave-One-Out Cross-Validation

Leave-one-out cross-validation (LOOCV) is a special case of the k-fold technique where k is selected to be the total number of observations. In this technique, for a dataset $X \in R^N$ with N observations, we select $K = N$, i.e. we need to run N experiments for training and testing, where $N - 1$ observations are used for training and one observation is left out to be used for testing. Figure 15.4 shows an example of LOOCV used for data with 10 observations.

15.6 Bootstrapping

Bootstrapping is a resampling technique with replacement (Efron and Tibshirani 1994). The basic idea is to generate new data using repeated sampling from the original data with replacement. For given data $X \in R^n$ with n observations, a bootstrap sample is generated by sampling n observations from X with replacement. This step is repeated for m chosen bootstrap series. The generated m bootstrap samples are used to train a classification model and compute the resubstitution accuracy, i.e. training accuracy. The model accuracy is then computed by averaging the m bootstrap accuracy estimates. Figure 15.5 illustrates four random bootstrap samples, i.e. $m = 4$, generated from data with six observations $X = \{x_1, x_2, x_3, x_4, x_5, x_6\}$. As the data is sampled with replacement, the probability of any given observation not being selected after n samples is given by $\left(1 - \frac{1}{n}\right)^n \approx e^{-1} \approx 0.368$. The predictable number of different observations from the original data appearing in the test set is thus $0.632n$ (Kohavi 1995).

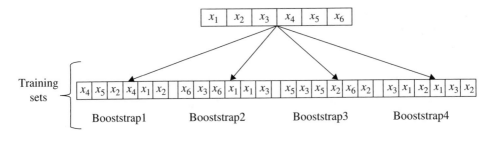

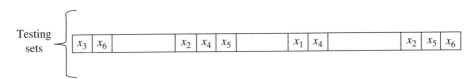

Figure 15.5 An illustration of four random bootstrap samples, i.e. $m = 4$, generated from data with six observations $X = \{x_1, x_2, x_3, x_4, x_5, x_6\}$.

Assume acc_i to be the accuracy estimate for bootstrap sample i. Then the 0.632 bootstrap acc_{boot} can be defined using the following equation,

$$acc_{boot} = \frac{1}{m} \sum_{i=1}^{m} (0.632.acc_i + 0.368.acc_s) \tag{15.1}$$

where acc_s is the accuracy on the training set.

15.7 Overall Classification Accuracy

Overall classification accuracy, also called the *classification rate*, is the most widely used measure of classifier performance. It can be defined as the ratio of the number of correct predictions (N_p) over the total number of predictions (N). This can be represented mathematically using the following equation:

$$classification\ accuracy = \frac{N_p}{N} \tag{15.2}$$

Alternatively, one may use the error rate as a measure of the classification model. The error rate can be defined using the following equation:

$$error\ rate = 1 - classification\ accuracy = 1 - \left(\frac{N_p}{N}\right) \tag{15.3}$$

In classification problems, we usually run the classification algorithm several times, and therefore averages and standard deviations are commonly used to describe the overall classification accuracy. Hence, the overall classification accuracy can be represented as follows,

$$overall\ classification\ accuracy = \frac{\sum_{i}^{q} \frac{N_{pi}}{N}}{q} \tag{15.4}$$

where q represents the number of trials and N_{pi} is the number of correct predictions obtained in trial i. The overall classification accuracy can be also represented as a percentage rather than a ratio such that,

$$overall\ classification\ accuracy(\%) = \left(\frac{\sum_i^q \frac{N_{pi}}{N}}{q} \right) \times 100 \tag{15.5}$$

The standard deviation (SD) can be computed using the following equation,

$$SD = \sqrt{\frac{\sum_i^q (N_{pi} - \overline{N}_p)}{q - 1}} \tag{15.6}$$

where $\overline{N}_p$ is the average number of correct predictions obtained in all trials.

15.8 Confusion Matrix

The confusion matrix, also called the *error matrix* or *contingency table*, is a matrix that contains information about actual classes and predicted classes obtained by a classification algorithm. The performance of the classification model can be evaluated using information from this matrix (Story and Congalton 1986). To compute the confusion matrix, we need to obtain a set of predictions that can be compared to the target values. The basic information that can be derived from this matrix is to count the number of times examples of class c_1 are classified as class c_2. A typical example of a confusion matrix can be seen in Table 15.1, where the classification is performed using roller bearing vibration data with six health conditions: two normal conditions (brand new [NO] and worn but undamaged [NW]) and four fault conditions (inner race [IR] fault, outer race [OR] fault, rolling element [RE] fault, and cage [CA] fault). From the table, it can be clearly seen that only one IR (0.5 of the IR testing examples) is likely to be confused with RE, two OR (1% of the OR testing examples) may be classified as IR, and five RE (2.5% of the testing examples) are expected to be classified as IR.

Table 15.1 Sample confusion matrix. (See insert for a colour representation of this table.)

True classes	Predicted classes						Class prediction (%)
	NO	NW	IR	OR	RE	CA	
NO	200	0	0	0	0	0	100
NW	0	200	0	0	0	0	100
IR	0	0	199		1	0	99.5
OR	0	0	2	198	0	0	99
RE	0	0	5	0	195	0	97.5
CA	0	0	0	0	0	200	100

15.9 Recall and Precision

Recall and precision are commonly used as evaluation measures for classification models (Fawcett 2006; Powers 2011). Briefly, we describe recall and precision measures as follows.

Consider a classification problem for two classes, normal condition (NO) and fault condition (FA). Given a classifier and a test set, the classifier predicts the class of each instance of the test set. For each instance, there are four possible outcomes:

(1) If the instance is NO and it is classified as NO, it is counted as a true NO.
(2) If the instance is FA and it is classified as FA, it is counted as a true FA.
(3) If the instance is NO and it is classified as FA, it is counted as a false FA.
(4) If the instance is FA and it is classified as NO, it is counted as a false NO.

Table 15.2 shows a confusion matrix that represents these four possible outcomes. From the table, we can calculate the following classifier measures:

(1) The true No rate ($Tr_{NO}rate$), also called *recall*, can be calculated using the following equation:

$$recall = Tr_{NO}\text{rate} = \frac{Tr_{NO}}{A} \tag{15.7}$$

(2) The false NO rate (Fa_{NO} rate), also called the *false alarm rate*, can be computed using the following equation:

$$Fa_{NO}\text{rate} = \frac{Fa_{NO}}{B} \tag{15.8}$$

(3) The precision, also called *confidence*, can be computed using the following equation:

$$Precision = confidence = \frac{Tr_{NO}}{C} \tag{15.9}$$

(4) The F-measure, also called the *F1 score* or *F-score*, which is a measure of test accuracy that combines recall and precision, can be calculated using the following equation:

$$F-measure = \frac{2}{\frac{1}{precision} + \frac{1}{recall}} = 2\left(\frac{precision.recall}{precision + recall}\right) \tag{15.10}$$

Table 15.2 Sample confusion matrix for a classification problem with two classes, normal condition (NO) and fault condition (FA).

True classes	Predicted classes		Total
	NO	**FA**	
NO	True NO (Tr_{NO})	False FA (Fa_{FA})	A
FA	False NO (Fa_{NO})	True FA (Tr_{FA})	B
Total	C	D	

15.10   ROC Graphs

The ROC graph is a two-dimensional graph that plots the false positive rate, i.e. the false prediction rate of a class, on the x-axis and true positive rate, i.e. the true prediction rate of a class, on the y-axis (Fawcett 2006). To plot the ROC curve, we need first to compute

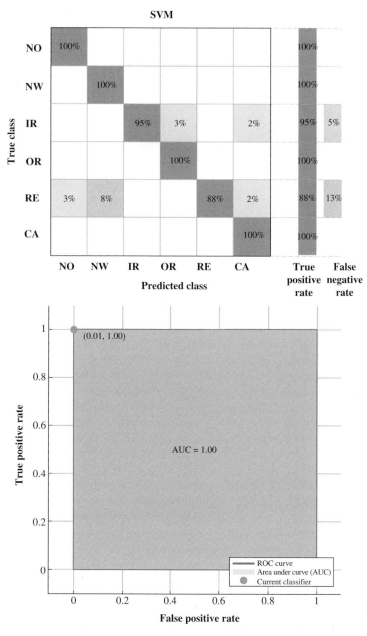

Figure 15.6 A typical example of a confusion matrix and its corresponding ROC graphs for roller bearing fault classification results obtained using an SVM classifier.

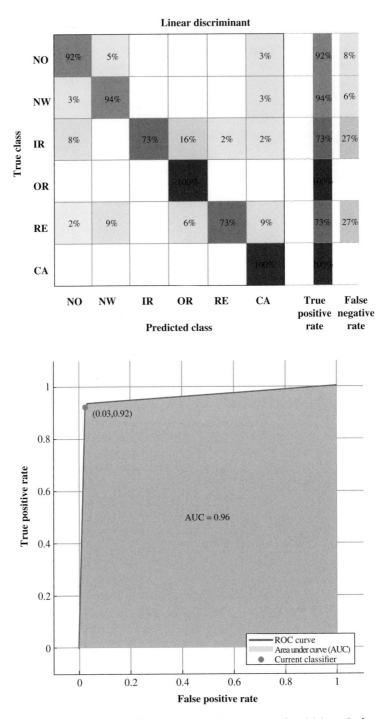

Figure 15.7 A typical example of a confusion matrix and its corresponding ROC graphs for roller bearing fault classification results obtained using a linear discriminant classifier.

the true positive rate and the false positive rate. There are several important points on the ROC graph that can be identified as follows:

- The point (0,0) is the lower-left point on the graph that shows an approach that never delivers a positive classification, e.g. a classifier that does not deliver false positive errors and, on the other hand, also provides no true positives.
- The point (1,1) is the opposite of the point (0,0), i.e. an approach that provides only positive classifications.
- The point (0,1) represents the best classification.

In general, the points' positions on this graph are used to describe the performance of classifiers. For example, when the point that describes the current classifier is positioned at upper left, this means that the classifier provides high true positive predictions and low false positive predictions. Figures 15.6 and 15.7 present two typical examples of confusion matrices and their corresponding ROC graphs for two different classifier, SVM and a linear discriminant classifier, applied on a feature set extracted from roller bearing vibration data with six health conditions: two normal conditions (NO and NW) and four fault conditions (IR fault, OR fault, RE fault, and CA fault).

As Figure 15.6 shows, the SVM classifier achieved a 100% true positive rate for NO, NW, OR, and CA. However, it achieved 95% and 88% true positive rates for IR and RE, respectively. Hence, the corresponding ROC chart shows the current classifier position at point (0.01, 1). In Figure 15.7, the linear discriminant classifier provided 92%, 95%, 73%, 100%, 73%, and 100% true positive rates for NO, NW, IR, OR, RE, and CA, respectively. The corresponding ROC graph of these results shows that ROC shows the current classifier position at point (0.03, 0.92). Taken together, the current position of the SVM classifier, which is at point (0.01,1), is closer to the perfect classification point at (0,1), as explained earlier, than the linear discriminant classifier at point (0.03, 0.92).

Another measure that can be used to compare classifiers is the area under the ROC curve (AUC). As can be seen in Figures 15.6 and 15.7, the AUC is all or part of the unit square: hence it is always between 0 and 1. Figure 15.6 shows that the AUC has a value of 1.0, which is the greatest value of the AUC and therefore indicates better average performance compared to the AUC in Figure 15.7, which has a value of 0.96.

Useful discussions about different validation techniques can be found in (Arlot and Celisse 2010; Esbensen and Geladi 2010; Raschka 2018.).

15.11 Summary

Model selection and performance estimation are the two main tasks we need to consider when dealing with classification problems. These tasks can be achieved by using model-validation techniques. The cross-validation technique is usually recommended as a superior test of the classification model compared to testing the predictive validity of a model on the same data that were utilised to estimate its parameters, which often leads to bias. To assess the efficiency of the classification algorithms introduced in the previous chapters, this chapter described different validation techniques that can be used to evaluate and verify a classification model's performance before proceeding to its application and implementation in machine fault diagnosis problems. These include the most commonly used data-splitting techniques, the hold-out technique,

Table 15.3 Summary of some of the introduced techniques and their publically accessible software.

Algorithm name	Platform	Package	Function
Classification confusion matrix	MATLAB	Neural Network Toolbox– Functions	confusion(targets, outputs)
Receiver operating characteristic		Neural Network Toolbox– Functions	Roc(targets, outputs)
Compare predictive accuracies of two classification models		Statistics and Machine Learning Toolbox –Functions	testcholdout
Compare accuracies of two classification models by repeated cross-validation		Statistics and Machine Learning Toolbox –Functions	testckfold
Create cross validation partition for data		Statistics and Machine Learning Toolbox –Functions	cvpartition
Random sample		Statistics and Machine Learning Toolbox-Resampling Techniques	randsample
Bootstrap sampling		Statistics and Machine Learning Toolbox-Resampling Techniques	bootstrp
Evaluate classifier performance		Bioinformatics Toolbox	classperf
Receiver operating characteristic (ROC) curve or another performance curve for classifier output		Statistics and Machine Learning Toolbox-Classification- Model building and assessment	perfcurve

random subsampling, k-fold cross-validation, leave-one-out cross-validation, and boot-strapping. Furthermore, various measures that can be used to evaluate the performance of a classification model were described in this chapter, including overall classification accuracy, the confusion matrix, the ROC curve, the area under a ROC (AUC), precision, and recall. Most of the introduced techniques and their publicly accessible software are summarised in Table 15.3.

References

Arlot, S. and Celisse, A. (2010). A survey of cross-validation procedures for model selection. *Statistics Surveys* 4: 40–79.

Beleites, C., Neugebauer, U., Bocklitz, T. et al. (2013). Sample size planning for classification models. *Analytica Chimica Acta* 760: 25–33.

Cooil, B., Winer, R.S., and Rados, D.L. (1987). Cross-validation for prediction. *Journal of Marketing Research*: 271–279.

Efron, B. and Tibshirani, R.J. (1994). *An Introduction to the Bootstrap*. CRC press.

Esbensen, K.H. and Geladi, P. (2010). Principles of proper validation: use and abuse of re-sampling for validation. *Journal of Chemometrics* 24 (3–4): 168–187.

Fawcett, T. (2006). An introduction to ROC analysis. *Pattern Recognition Letters* 27 (8): 861–874.

Guyon, I. (1997). *A Scaling Law for the Validation-set Training-Set Size Ratio*, 1–11. AT&T Bell Laboratories.

Kavzoglu, T. and Colkesen, I. (2012, July). The effects of training set size for performance of support vector machines and decision trees. In: *Proceeding of the 10th International Symposium on Spatial Accuracy Assessment in Natural Resources and Environmental Sciences*, 1013.

Kohavi, R. (1995). A study of cross-validation and bootstrap for accuracy estimation and model selection. *International Joint Conference on Artificial Intelligence* 14 (2): 1137–1145.

Lendasse, A., Wertz, V., and Verleysen, M. (2003). Model selection with cross-validations and bootstraps—application to time series prediction with RBFN models. In: *Artificial Neural Networks and Neural Information Processing—ICANN/ICONIP 2003*, 573–580. Berlin, Heidelberg: Springer.

Powers, D.M. (2011). Evaluation: From precision, recall and F-measure to ROC, informedness, markedness and correlation. *Journal of Machine Learning Research* 2 (1): 37–63.

Raschka, S. (2018). Model evaluation, model selection, and algorithm selection in machine learning. arXiv:1811.12808.

Raudys, S.J. and Jain, A.K. (1991). Small sample size effects in statistical pattern recognition: Recommendations for practitioners. *IEEE Transactions on Pattern Analysis & Machine Intelligence* (3): 252–264.

Story, M. and Congalton, R.G. (1986). Accuracy assessment: a user's perspective. *Photogrammetric Engineering and Remote Sensing* 52 (3): 397–399.

Wong, T.T. (2015). Performance evaluation of classification algorithms by k-fold and leave-one-out cross-validation. *Pattern Recognition* 48 (9): 2839–2846.

Xu, Q.S. and Liang, Y.Z. (2001). Monte Carlo cross-validation. *Chemometrics and Intelligent Laboratory Systems* 56 (1): 1–11.

Part V

New Fault Diagnosis Frameworks Designed for MCM

16

Compressive Sampling and Subspace Learning (CS-SL)

16.1 Introduction

Fault diagnosis is an important part of a machine condition monitoring (MCM) system and plays a key role in condition-based maintenance (CBM). The key motive for applying MCM is to produce useful and accurate information on the current health condition of the machine. In consequence, with this type of dependable information, the correct decision regarding whether maintenance activities (if any) are required can be made to avoid machines breakdowns. The literature on vibration-based MCM identifies important developments in the field of fault detection and classification as a result of the numerous computational methods from the disciplines of feature learning and pattern recognition that have been proposed for fault diagnosis. Nevertheless, the performance of these methods is limited by the large amounts of sampled vibration data that need to be acquired from rotating machines to achieve the anticipated accuracy of machine fault detection and classification. In fact, sampling theorems including the Shannon-Nyquist theorem are at the core of the current sensing systems. However, a Nyquist sampling rate that is at least twice the highest frequency contained in the signal is high for some modern developing applications, e.g. industrial rotating machines (Eldar 2015). One aspect of much of the literature on using the Nyquist sampling rate is that it may result in measuring a large amount of data that needs to be transmitted, stored, and processed. Moreover, for some applications that include wideband, it is very costly to collect samples at the necessary rate. It is clear that acquiring a large amount of data requires large storage and considerable time for signal processing, and this also may limit the number of machines that can be monitored remotely across a wireless sensor network (WSNs) due to bandwidth and power constraints.

For this reason, it is currently becoming of essential importance to develop new MCM methods that not only have the ability to achieve accurate detection and identification of machine health but can also address two main challenges: (i) the cost of learning from a large amount of vibration data, i.e. transmission costs, computation costs, and power needed for computations; and (ii) the demand for more accurate and sensitive MCM systems that have the ability to provide information on the current health condition of the machine more accurately and more quickly than existing ones. The more accurate and quicker MCM system, the more correct maintenance decisions are made and the more time is available to plan and perform maintenance before machine breakdowns; thus machines will always run in a healthy condition and provide satisfactory work.

Condition Monitoring with Vibration Signals: Compressive Sampling and Learning Algorithms for Rotating Machines,
First Edition. Hosameldin Ahmed and Asoke K. Nandi.
© 2020 John Wiley & Sons Ltd. Published 2020 by John Wiley & Sons Ltd.

A reasonable approach to tackle the challenges involved in dealing with too many samples could be to compress the data. One of the best-known recent advanced techniques for signal compression is transform coding, which depends on mapping the signal samples into bases that provide sparse or compressible representations of the signal of interest (Rao and Yip 2000). Recent advances in techniques beyond bandlimited sampling offer lower sampling rates and a reduced amount of data (Eldar 2015). New advances in transform coding techniques have facilitated the investigation of the compressive sampling (CS) framework (Donoho 2006; Candès and Wakin 2008), which relies on linear dimensionality reduction. CS supports sampling below the Nyquist rate for signals that have a sparse or compressible description. Accordingly, if the signal has a sparse representation in a known basis, then one is able to reduce the number of measurements that need to be stored, transmitted, and processed. CS is being considered in a large diversity of applications including medical imaging, seismic imaging, radio detection and ranging, and communications and networks (Holland et al. 2010; Qaisar et al. 2013; Merlet and Deriche 2013; Rossi et al. 2014). The basic idea of CS is that a finite-dimensional signal having a sparse or compressible representation can be reconstructed from fewer linear measurements far below than the Nyquist sampling rate.

In this chapter, a fault-diagnosis framework called *compressive sampling and subspace learning* (CS-SL) is introduced. CS-SL based techniques combine CS and subspace learning techniques to learn optimally fewer features from a large amount of vibration data. With these learned features, a machine's health can be classified using a machine learning classifier. CS-SL receives a large amount of vibration data as input and produces fewer features as output, which can be used for fault diagnosis. Based on the CS-SL framework, we introduce the following four techniques.

(1) A recent fault-diagnosis framework called *compressive sampling and feature ranking* (CS-FR) (Ahmed et al. 2017b; Ahmed and Nandi 2017, 2018a). CS-FR is introduced to learn optimally fewer features from a large amount of vibration data from which machine health can be classified. The CS model is the first step in this framework and provides compressively sampled signals based on compressed sampling rates. In the second step, the search for the most important features of these compressively sampled signals is performed using feature ranking and selection techniques that assess feature importance based on certain characteristics of the data, e.g. feature correlation. Several feature ranking and selection techniques – three similarity-based techniques (Fisher score [FS], Laplacian score [LS, and Relief-F), one correlation-based technique (Pearson correlation coefficients [PCCs]), and one independence test technique (Chi-square [Chi-2]) are used to select fewer features that can sufficiently represent the original vibration signals.

(2) A fault-diagnosis framework called *compressive sampling and linear subspace learning* (CS-LSL) (Ahmed et al. 2017a; Ahmed and Nandi 2018b,c). CS-LSL based techniques receive a large amount of vibration data and produce fewer features that can be used for classification of machine faults. It uses the CS model to produce compressively sampled signals, i.e. compressed data $Y = \{y_1, y_2, ..., y_L\} \in R^m$, which have enough information from the original vibration raw data $X = \{x_1, x_2, ..., x_L\} \in R^n$. To extract feature representations of these compressively sampled signals, CS-LSL based techniques perform a linear transformation to map the m-dimensional space of the compressively sampled vibration to a lower-dimensional feature space. Two

linear subspace learning techniques – unsupervised principal component analysis (PCA) and supervised linear discriminant analysis (LDA) – are used to select fewer features from the compressively sampled signals. Moreover, an advanced technique called *compressive sampling with correlated principal and discriminant components* (CS-CPDC) (Ahmed and Nandi 2018c) is introduced. CS-CPDC is a three-stage hybrid method for classification of machine faults. In the first stage, it uses the CS model to obtain compressively sampled vibration signals. In the second stage, it employs a multi-step approach of PCA, LDA, and canonical correlation analysis (CCA) to extract features from the obtained compressively sampled signals. In the third stage, it applies a support vector machine (SVM) to classify a machine's health using the learned features from the previous stage.

(3) A fault-diagnosis framework called *compressive sampling and nonlinear subspace learning* (CS-NLSL). Unlike CS-LSL based techniques that perform a linear transformation to map the m-dimensional space of the compressively sampled vibration to a lower-dimensional feature space, CS-NLSL based techniques perform a nonlinear transformation to do so.

16.2   Compressive Sampling for Vibration-Based MCM

The basic idea of CS is that a finite-dimensional signal having a sparse or compressible representation can be reconstructed from fewer linear measurements far below the Nyquist sampling rate. In the last few years, there has been a growing interest in the application of CS in machine fault diagnosis since machine vibration signals have a compressible representation in several domains, e.g. the frequency domain. The advantages of CS in vibration-based MCM can be summarised as follows:

- *Reduced computations.* CS is able to reduce a large amount of the acquired vibration data. A larger reduction in the amount of vibration data results in more reduction in the computations.
- *Reduced transmission costs.* In the cases of having to send vibration data from remote places by wireless (e.g. in the case of offshore wind turbines) or wired transmission, the cost of transmission will be less as CS reduces the amount of vibration data.
- *Benefits to the environment.* As the application of CS results in reduced computations, it helps to reduce the amount of power needed for both computations and transmissions. As a consequence, CS offers considerable benefits to the environment.
- *Increased number of machines that can be monitored remotely across WSNs.* It is clear that acquiring a large amount of data limits the number of machines that can be monitored remotely across WSNs due to bandwidth and power constraints. Therefore, a large reduction in the amount of data will increase the number of machines that can be monitored remotely.

16.2.1   Compressive Sampling Basics

In place of processing the high-dimensional collected vibration data, the common methodology is to identify a lower-dimensional features space that can represent the large amount of acquired vibration signals while retaining the important information from the machine conditions. Quite recently, considerable attention has been paid to CS for its ability to allow one to sample far below the Nyquist sampling rate and yet be able to reconstruct the original signal when needed. CS (Donoho 2006; Candès and Wakin 2008) is an extension of a sparse representation and a special case of it. The simple idea of CS is that many real-world signals have sparse or compressible representations in some domain, e.g. the Fourier transform (FT), and can be recovered from fewer measurements under certain conditions. In fact, CS is based on two principles: (i) sparsity of the signal of interest, and (ii) the measurements matrix that satisfies the minimal data information loss, i.e. fulfills the restricted isometry property (RIP) (Candes and Tao 2006). Concisely, we describe sparsity as follows.

Assume that $x \in R^{n \times 1}$ is an original time-indexed signal. Given a sparsifying transform matrix $\psi \in R^{n \times n}$ whose columns are the basis elements $\{\psi_i\}_{i=1}^n$, x can be represented using the following equation,

$$x = \sum_{i=1}^{n} \psi_i s_i \tag{16.1}$$

or, more efficiently,

$$x = \psi s \tag{16.2}$$

Here, s is a $n * 1$ column vector of coefficients. If the basis ψ produces q-sparse representations of x, i.e. x of length n can be represented with $q \ll n$ nonzero coefficients, then Eq. (16.1) can be rewritten as Eq. (16.3),

$$x = \sum_{i=1}^{q} \psi_{ni} s_{ni} \tag{16.3}$$

where ni is the index of the basis elements and the coefficients corresponding to the q nonzero elements. So, $s \in R^{n \times 1}$ is a vector column with only q nonzero elements and represents the sparse representation vector of x.

Based on the CS framework, $m \ll n$ projections of the vector x with a group of measurement vectors $\{\varnothing_j\}_{j=1}^m$ and the sparse representations s of x can be can be computed using Eq. (16.4),

$$y = \varnothing \psi s = \theta s \tag{16.4}$$

where y is a $m * 1$ column vector of the compressed measurements and $\theta = \varnothing \psi$ is the measurement matrix. Based on CS theory, the original signal x can be recovered from the compressed measurements y using recovery algorithm. This can be done by first recovering the sparse representation vector s and then using the inverse of the sparsifying transform ψ to recover the original signal x. Figure 16.1 shows an illustration of the CS framework.

To recover the sparse representations $s \in R^n$ from its compressed measurement vector $y \in R^m$, one may think of using the simple technique l_0 minimization that search for a sparse vector consistent with the measured data $y = \theta s$ such that,

$$\hat{s} = \arg \min_{z} \|z\|_0 \text{ s.t } \theta z = y \tag{16.5}$$

However, l_0 minimization is a NP hard problem, i.e. computationally intractable. One way to make this problem tractable is to use the convex optimisation $\|.\|_1$ in place of $\|.\|_0$

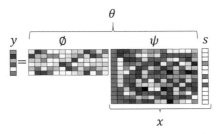

Figure 16.1 Compressive sampling framework (Ahmed and Nandi 2018c). (See insert for a colour representation of this figure.)

such that,

$$\hat{s} = \arg\min_{z}\|z\|_1 \text{ s.t } \theta z = y \tag{16.6}$$

The sparse representation s can be recovered by solving the convex program in Eq. (16.6), provided that the measurement matrix θ satisfies the minimal data information loss, i.e. satisfies the RIP.

Definition 16.1 The measurement matrix θ satisfies the rth restricted isometry property (RIP) if there exists a $\delta_r \ll 1$ such that

$$(1 - \delta_r)\|s\|_{l_2}^2 \le \|\theta s\|_{l_2}^2 \le (1 + \delta_r)\|s\|_{l_2}^2 \tag{16.7}$$

Founded along the idea of compressive sensing, when Φ and Ψ are incoherent, the original signal can be recovered from $m = O\ (q\log\ (n))$ Gaussian measurements or $q \le C.\ m/\log\ (n/m)$ Bernoulli measurements (Baraniuk and Wakin 2009), here, C is constant and q is the sparsity level. Both a random matrix with independent and identically distributed (*i.i.d.*) Gaussian entries and a Bernoulli (± 1) matrix satisfy the RIP. The size of the measurement matrix $(m*n)$ depends on the compressive sampling rate (α) (i.e. $m = \alpha*n$).

Another type of technique commonly used for sparse representation recovery are greedy/iterative techniques, such as orthogonal matching pursuit (OMP) (Tropp and Gilbert 2007), stagewise OMP (StOMP) (Donoho et al. 2006), subspace pursuit (SP) (Dai and Milenkovic 2009), compressive sampling matching pursuit (CoSaMP) (Needell and Tropp 2009), etc.

The original vector x can be reconstructed using the inverse of the sparsifying transform ψ and the estimated sparse representation $\hat{s}$ such that

$$\hat{x} = \psi^{-1}\hat{s} \tag{16.8}$$

The model described here is meant to be single measurement vector compressive sampling (SMV-CS) that recovers one vector from its corresponding compressed measurement vector. But multiple measurement vector compressive sampling (MMV-CS) is considered for signals that are represented as a matrix with a set of jointly sparse vectors. Figure 16.2 shows an illustration of the MMV-CS framework. Based on MMV-CS, the compressed data matrix can be computed using Eq. (16.9),

$$Y = \theta S \tag{16.9}$$

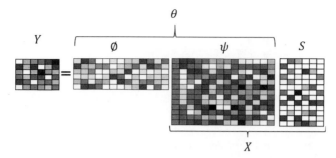

Figure 16.2 The MMV-CS model. (See insert for a colour representation of this figure.)

where $Y \in R^{m \times L}$, m is the number of compressed measurements and L is the number of observations, $\theta \in R^{m \times n}$ is a dictionary, and $S \in R^{n \times L}$ is a sparse representation matrix. Several studies have been conducted to reconstruct jointly sparse signals (S) given multiple compressed measurement vectors (Chen and Huo 2006; Sun et al. 2009). Then, the original signal matrix X can be recovered using the inverse of the sparsifying transform ψ and the estimated sparse representations $\hat{S}$ such that

$$\hat{X} = \psi^{-1}\hat{S} \tag{16.10}$$

Here, $\hat{X}$ and $\hat{S}$ are the estimation of X and S, respectively. The better signal reconstruction indicates that the compressed samples possess the quality of the original signal. In this book, MMV-CS has been used to obtain compressively sampled signals since the dataset consists of a matrix of multiple measurements. Also, since it is possible to recover the original signal (X) from the compressed data (Y), this indicates that Y possesses the quality of the original signal X. Hence, we may use the compressed measurements directly without recovering the original data.

16.2.2 CS for Sparse Frequency Representation

The CS framework requires that the signal of interest have a sparse or compressible representation in a known transform domain. A commonly utilised sparse basis for vibration signal is the fast Fourier transform (FFT) matrix (Rudelson and Vershynin 2008; Zhang et al. 2015a; Wong et al. 2015; Yuan and Lu 2017). We assume the time domain vibration signal of the rotating machines is compressible in the frequency domain and that the FFT-based frequency representation of vibration signals preserves a compressible structure. Given an acquired vibration dataset $X = \{x_1, x_2, \ldots, x_L\} \in R^n$, the process to obtain a compressively sampled dataset using the MMV-CS framework in Eq. (16.9) is as follows. First, the compressible representations ($S \in R^{n \times L}$), which consist of only a small number $q \ll n$ of nonzero coefficients, are obtained from the raw vibration signals ($X \in R^{n \times L}$) using the FFT algorithm, which computes the n-point complex discrete Fourier transform (DFT) of signal X. In our case, we used the magnitude of the DFT, i.e. the absolute value of DFT of signal X, to obtain S. Then, the obtained (S) is projected into a suitable measurement matrix ($\Theta \in R^{m \times n}$) that satisfies the RIP. In this book, a random matrix with i.i.d. Gaussian entries matrix and a compressed sampling rate (α) is used to generate the compressively sampled signals ($Y \in R^{m \times L}$), where m is the number of compressed signal elements (i.e. $m = \alpha * n$). These procedures are summarized in Algorithm 16.1.

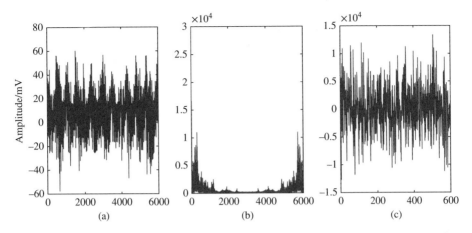

Figure 16.3 An example of (a) an outer race fault time domain signal x_{OR}; (b) the corresponding absolute values of Fourier coefficients for x_{OR}; and (c) the obtained compressively sampled signal of x_{OR}.

Algorithm 16.1   Compressive Sampling with FFT

Input: $X \in R^{n \times L}; \Theta \in R^{m \times n}$; and α
Output: $Y \in R^{m \times L}$
$abs\,(FFT(X)) \to S \in R^{n \times L}$
Project S into Θ with compressed sampling rate α to obtain compressively sampled signal $Y \in R^{m \times L}$

Figure 16.3 shows an example of an obtained compressively sampled signal of a bearing outer race (OR) fault signal x_{OR} using Algorithm 16.1 with $\alpha = 0.1$.

16.2.3   CS for Sparse Time-Frequency Representation

The wavelet transform that decomposes the signal into low- and high-frequency levels is used to obtain the sparse components of the vibration signal that are demanded by the compressive sensing framework. One of the choices is the Haar wavelet basis used as sparse representations for vibration signals in (Bao et al. 2011). Given an acquired vibration dataset $X = \{x_1, x_2, \ldots, x_L\} \in R^n$, the process to obtain a compressively sampled dataset using the MMV-CS framework in Eq. (16.9) is as follows: First the compressible representations ($S \in R^{n \times L}$), which consist of only a small number $q \ll n$ of nonzero coefficients, are obtained from raw vibration signals ($X \in R^{n \times L}$) using the thresholded Haar wavelet basis with five decomposition levels as a sparsifying transform. For example, the wavelet coefficients of the bearing OR vibration signal x_{OR} are displayed in Figure 16.4b. After applying the hard threshold, the input is preserved if it is bigger than the threshold τ; else, it is set to zero (Chang et al. 2000); the wavelet coefficient is sparse in the Haar wavelet domain, as shown in Figure 16.4c. Then, the obtained (S) is projected into a random matrix, and an i.i.d. Gaussian entries matrix with the compressed sampling rate (α) is used to generate the compressively sampled signals ($Y \in R^{m \times L}$), where m is the number of compressed signal elements. These procedures are summarized in Algorithm 16.2.

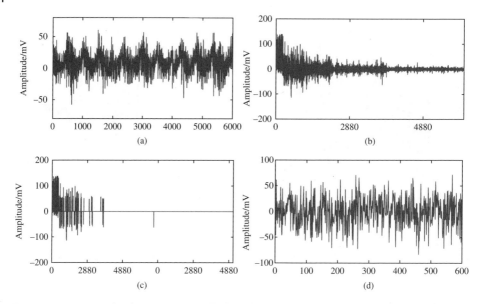

Figure 16.4 An example of (a) an outer race fault time domain signal x_{OR}; (b) the corresponding Haar WT coefficients of x_{OR}; (c) the corresponding thresholded Haar WT coefficients of x_{OR}; and (d) the obtained compressively sampled signal of x_{OR}.

Algorithm 16.2   Compressive Sampling with Thresholded Haar WT

Input: $X \in R^{n \times L}$; $\Theta \in R^{m \times n}$; and α
Output: $Y \in R^{m \times L}$
Thresholded Haar WT(X)) $\rightarrow S \in R^{n \times L}$
Project S into Θ with compressed sampling rate α to obtain compressively sampled signal $Y \in R^{m \times L}$

Figure 16.4d shows an example of an obtained compressively sampled signal for a bearing OR fault signal x_{OR} using Algorithm 16.2 with $\alpha = 0.1$.

16.3   Overview of CS in Machine Condition Monitoring

A typical CS-based framework for machine fault diagnosis often uses the CS model to produce compressed sensed data from a large amount of acquired vibration data. This compressed sensed data then becomes available for further processing. Compressive signal processing for machine fault diagnosis can be categorised into four schemes (see Figure 16.5).

16.3.1   Compressed Sensed Data Followed by Complete Data Construction

The possibility to diagnosis machine faults from complete reconstructed signals based on CS has been validated in (Li et al. 2012; Wong et al. 2015), where the CS-based compressively sampled signals are used and then followed by signal reconstruction. In order

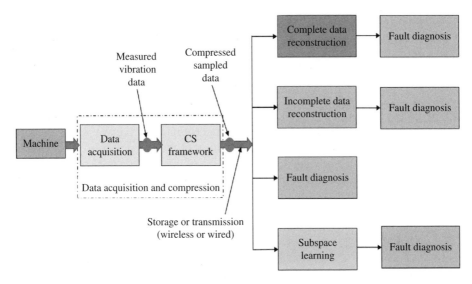

Figure 16.5 Four different compressive signal-processing schemes for machine fault diagnosis. (See insert for a colour representation of this figure.)

to overcome problematic issues (i.e. unnecessary cost of energy and network bandwidth) related to network monitoring technology used in fault detection of a train's rolling bearings, Li et al. (2012) proposed a CS-based method for fault diagnosis. In this method, first, a CS model with a wavelet domain-based sparse representation and a Gaussian matrix as the measurement matrix is used to compress the data. Then, the OMP algorithm is used to reconstruct the original signal from the compressed signal. Finally, fault diagnosis is performed using EMD and envelope analysis. Similarly, Wong et al. (2015) classify roller element bearing fault types under a CS framework. In this framework, first, vibration signals of roller element bearings are acquired in the time domain and resampled with a random Bernoulli matrix to match the CS mechanism. Then, the reconstruction from the compressed sensed vector is performed using a convex optimisation technique based on l_1-norm. Finally, for classifying the reconstructed signals, entropic features and SVM are used. They have demonstrated that it is possible to sample the vibration data of roller bearings at less than the Nyquist rate and recover the signal for fault classification. However, signal reconstruction techniques may not be practical in all applications and make no attempt to address the question of whether or not it is possible to learn in the compressed domain rather than having to recover the original signals. For instance, a rotating machine vibration signal is always acquired for fault detection and estimation, and as long as it is possible to detect faulty signals in the measurement domain, then it is not necessary to recover the original signal to identify faults. Moreover, signal reconstruction is a complex computational problem that depends on the sparsity of the measured vibration signal. Therefore, CS-based signal-recovery methods may not be useful in reducing computational complexity for MCM.

16.3.2 Compressed Sensed Data Followed by Incomplete Data Construction

Most research in compressive sensing-based methods has emphasized the use of sparse representations, compressed measurements, and incomplete signal reconstruction for machine fault diagnosis. For example, Tang et al. (2015a) developed a sparse classification strategy based on a compressive sensing strategy. In this strategy, first, a target sample training set E is constructed, established by c categories of bearing fault signals. Then, the test signal and a Gaussian random matrix are selected as the measurement matrix to perform the dimensionality reduction. Using the CS model, the compressed signal can be calculated. To acquire the sparse feature vector, sparse promotion via a SP greedy algorithm is employed. Finally, the fault category is determined by the minimum redundant error between the calculated compressed signal and the estimated compressed signal obtained using a decomposition of the sparse feature vector. Zhang et al. (2015a) suggested a bearing fault diagnosis method based on the low-dimensional compressed vibration signal by training several overcomplete dictionaries that can be effective in signal sparse decomposition for each vibration signal state. This method starts by training the overcomplete dictionaries, $D_0, D_1, ..., D_C$, corresponding to each bearing health condition using high-dimensional vibration signals. Then, thresholds τ_0, $\tau_1, ..., \tau_C$ are set for signal representation errors corresponding to each bearing health condition. Low-dimensional signals obtained from the high-dimensional signals using the CS model are represented on all the trained overcomplete dictionaries, and for each representation the error (δ_i) is calculated $(i = 0, 1, ..., c)$. The smallest error of the calculated errors is used to represent the signal condition. Finally, the smallest error (δ) is used to estimate the state of the bearing based on the threshold values of each representation error, e.g. $\delta \leq \tau_i$ can be used to determine the bearing health condition is in fault state i.

Another learning dictionary basis, sparse extraction of impulses by adaptive dictionary (SpaEIAD), for extracting impulse components embedded in noisy signals, is described by Chen et al. (2014). This method relies on the sparse model of CS, comprising the sparse dictionary learning and redundant representation over the learned dictionary, which produces compressed noisy observations. In the first stage of the dictionary learning, the sparse coefficient of the noisy observations is obtained using the greedy reconstruction algorithm OMP. Then, the dictionary learning of the obtained sparse coefficient is created using the K-SVD algorithm, which searches for the best possible dictionary of the sparse representation. SpaEIAD is validated on vibration signals of motor bearings and a gearbox. In a study investigating the characteristic harmonics of sparse measurements reconstructed using the CS model, Tang et al. (2015b) reported that a sampling rate varying from 5–80% achieved 72–80% detection accuracy using this technique, where fault features may be directly detected from only a few samples without complete reconstruction. In this technique, the CS model is used to generate a compressed vibration signal. Then a reconstruction process of the compressed vibration signals is pursued as attempts are conducted to detect the characteristic harmonics from the sparse representation using the iterative algorithm CoSaMP. So, reconstruction and detection may proceed simultaneously without complete recovery.

16.3.3 Compressed Sensed Data as the Input of a Classifier

The possibility to diagnosis machine faults from the compressed sensed data directly as the input of a classifier has been investigated by Ahmed et al. (2017a). In this study, Ahmed et al. demonstrated that roller bearing fault diagnosis can be conducted directly from the compressed sensed data obtained from the CS model. For classification purposes, a logistic regression classifier (LRC) is applied on the compressed sensed data as shown in Eq. (16.11),

$$y \xrightarrow{LRC} h_\theta^i(y) = p(class = i \mid y; \theta) \tag{16.11}$$

where $h_\theta^i(y)$ is a LRC that can be trained for each class $i = 1, 2, 3, \ldots, c$ to predict the probability (p) that the class is equal to i and c is the number of classes. A significant analysis and discussion on the subject of how to solve a range of signal detection and estimation problems given compressed measurements without reconstructing the original signal can be found in (Davenport et al. 2006).

Even though the efficiency of CS in machine fault diagnosis has been validated in these studies, there are two main problems with these studies: (i) CS-based sparse signal reconstruction is a complex computational problem that depends on the sparsity of the measured vibration signal. Therefore, CS-based signal recovery methods may not be useful in reducing computational complexity for condition monitoring of rotating machines. And (ii) most of the methods that are based on learning directly from the compressed measurements achieved good classification accuracy by only increasing the sampling rate, thereby requiring higher computational complexity.

16.3.4 Compressed Sensed Data Followed by Feature Learning

While projections obtained from the CS model help to recover the original signal from low-dimension features, they may not be the best from a discriminant point of view. Furthermore, the size of the CS projections may still represent a large amount of data collected in real operating conditions. Consequently, techniques to learn fewer features of the CS projections are required.

Aiming to address the challenges of learning from high-dimensional data and embrace the efficiency of CS in improving MCM, the next subsections will introduce three frameworks for feature learning including CS and feature ranking (CS-FR), CS and linear subspace learning (CS-LSL), and CS and nonlinear subspace learning (CS-NLSL), which can be used for vibration-based fault diagnosis of rotating machines.

16.4 Compressive Sampling and Feature Ranking (CS-FR)

The compressive sampling and feature ranking framework (CS-FR) has been proposed recently to avoid the burden of storage requirements and processing time of a tremendously large amount of vibration data acquired for the purpose of MCM (Ahmed and Nandi 2018a). This framework combines CS based on multiple measurement vectors (MMV) and feature ranking and selection techniques to learn optimally fewer features from a large amount of vibration data. With these learned features, a machine's health can be classified using a machine learning classifier. CS-FR receives a large amount of vibration data as input and produces fewer features as output, which can be used for fault diagnosis. As shown in Figure 16.6, CS-FR first compresses the vibration data and then ranks the features of the compressed data from which the most significant features can be selected to be used for classification. The details are as follows:

(1) *Vibration data compression,* with the aim to reduce computations, transmission cost, and demands on the environment compared to other techniques. CS-FR

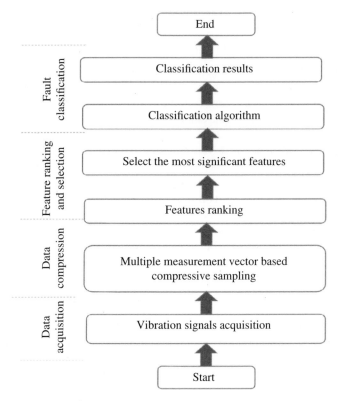

Figure 16.6 Flowchart summarizing the steps of the CS-FR framework (Ahmed and Nandi 2018a).

employs the MMV-CS model to produce compressively sampled signals, i.e. compressed data $Y = \{y_1, y_2, ..., y_L\} \in R^m$ that have enough information from the original bearing raw data $X = \{x_1, x_2, ..., x_L\} \in R^n$ where $m << n$.

(2) *Feature ranking and selection.* As long as the compressively sampled signals produced by the CS model have enough information about the original vibration signals, we may further filter the compressively sampled signals using feature ranking and selection techniques to rank and select fewer features from the compressively sampled signals that can sufficiently represent characteristics of machine health conditions.

(3) *Fault classification.* With these fewer selected features, a classifier is used to classify rotating machine health conditions.

Figure 16.7 shows an illustration of the data compression and feature-selection process in CS-FR.

16.4.1 Implementations

Based on CS-FR, two techniques of feature selection are considered to select fewer features of the compressively sampled signals. These are:

(1) *Similarity-based methods* that assign similar values to the compressively sampled signals that are close to each other. Three algorithms – FS, LS, and Relief-F – were

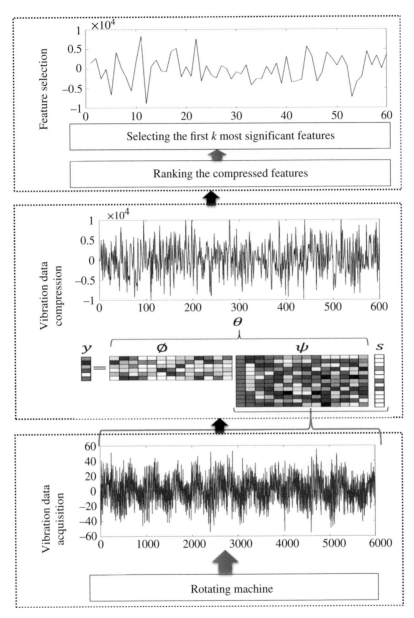

Figure 16.7 Illustration of the data compression and feature selection (Ahmed and Nandi 2018a). (See insert for a colour representation of this figure.)

investigated to select fewer features based on similarity. These methods measure the importance of features by their capability to preserve data similarity. Given a CS-based compressively sampled data matrix $Y \in R^{m \times L}$ with L examples and m compressed data points, assume the pairwise similarity matrix among examples is $S \in R^{L \times L}$. To select the k most relevant features s, the utility function $U(s)$, which measures the ability of the selected feature set to preserve the data similarity structure, needs to be maximised such that,

$$\max_{S} U(s) = \max_{S} \sum_{f \in s} U(f) = \max_{S} \sum_{f \in s} \hat{f}' \hat{S} \hat{f} \tag{16.12}$$

where $U(s)$ is the utility function of feature set s, $U(f)$ is the utility function of feature $f \in S$, $\hat{f}$ is the transformed feature of feature f, and $\hat{S}$ is the transformed matrix of the similarity matrix S. The transformation of f and S can be performed using different techniques based on the method used to select the features (Li et al. 2017).

(2) *Statistical-based methods* that measure the importance of features of the compressively sampled signals using different statistical measures. Two algorithms, PCC and Chi-2, were investigated to select fewer features based on correlation and independence tests, respectively.

16.4.1.1 CS-LS

The compressive sampling and Laplacian score feature-selection (CS-LS) method receives a large amount of vibration data as input and produces fewer features with smaller Laplacian scores as output, which can be used for fault classification of rotating machines (Ahmed et al. 2017b). The CS-LS first obtains the compressively sampled data (Y) and then uses LS to rank the features of the compressively sampled data. The LS ranks the features depending on their locality-preserving power. Given a dataset $Y = [y_1, y_2, ..., y_n]$, where $Y \in R^{m \times L}$, suppose the LS of the rth feature is L_r and f_{ri} represents the ith sample of the rth feature, where $i = 1, ..., m$ and $r = 1, ..., L$. First the LS algorithm constructs the nearest-neighbour graph G with m nodes, where the ith node corresponds to y_i. Next, an edge between nodes i and j is placed: if y_i is among the k nearest neighbours of y_j or vice versa, then i and j are connected. The elements of the weight matrix of graph G is S_{ij} and can be defined as follows:

$$S_{ij} = \begin{cases} e^{-\frac{\|y_i - y_j\|^2}{t}}, & i \text{ and } j \text{ are connected} \\ 0, & otherwise \end{cases} \tag{16.13}$$

Here, t is a suitable constant. The LS L_r for each sample can be computed as follows:

$$L_r = \frac{\tilde{f}_r^T L \tilde{f}_r}{\tilde{f}_r^T D \tilde{f}_r} \tag{16.14}$$

where $D = diag(S1)$ is the identity matrix, $\mathbf{1} = [1, ..., 1]^T$, $L = D - S$ is the graph Laplacian matrix, and $\tilde{f}_r$ can be calculated using the following equation:

$$\tilde{f}_r = f_r - \frac{f_r^T D\mathbf{1}}{\mathbf{1}^T D\mathbf{1}} \tag{16.15}$$

where $f_r = [f_{r1}, f_{r2}, ..., f_{rm}]^T$.

CS-FR selects the k features with the smaller LSs (L_r) that can be used for fault classification of rotating machines ($k < m$).

16.4.1.2 CS-FS

The compressive sampling and Fisher score feature-selection (CS-FS) method receives a large amount of vibration data as input and produces fewer features with the larger Fisher scores as output, which can be used for fault classification of rotating machines

(Ahmed and Nandi 2017). Given a dataset of rotating machine vibrations $X \in R^{n \times L}$, CS-FS first uses CS based on the MMV model to produce compressively sampled signals, i.e. compressed data $Y = \{y_1, y_2, \ldots, y_L\} \in R^m$, that have enough information from the original rotating machine raw data $X = \{x_1, x_2, \ldots, x_L\} \in R^n$, where $m << n$. Then, it employs FS to rank the features of the compressively sampled data (Y). The main idea of FS is to compute a subset of features with a large distance between the compressively sampled data points in different classes and a small distance between data points in the same class. The Fisher score of the ith feature can be computed using the following equation:

$$FS(Y^i) = \frac{\sum_{c=1}^{C} L_c(\mu_c^i - \mu^i)^2}{(\sigma^i)^2} \tag{16.16}$$

Here, $Y^i \in R^{1 \times L}$, L_c is the size of the cth class, $(\sigma^i)^2 = \sum_{c=1}^{C} L_c(\sigma_c^i)^2$, μ_c^i and σ_c^i are the mean and standard deviation of the cth class corresponding to the ith feature; and μ^i and σ^i are the mean and standard deviation of the entire dataset corresponding to the ith feature. Finally, CS-FR selects the k features with larger Fisher scores that can be used for fault classification of rotating machines ($k < m$).

Moreover, FS can be defined as a special case of LS, providing that the within-class data similarity matrix (S_w) satisfies the following equation:

$$S_w(i, j) = \begin{cases} \frac{1}{L_c} & if \, y_i = y_j = c \\ 0 & otherwise \end{cases} \tag{16.17}$$

Accordingly, FS can be computed using the following equation:

$$FS(Y^i) = 1 - \frac{1}{L_r(Y^i)} \tag{16.18}$$

16.4.1.3 CS-Relief-F

The compressive sampling and Relief-F feature-selection (CS-Relief-F) method receives a large amount of vibration data as input and produces fewer features as output, which can be used for fault classification of rotating machines. Similar to the CS-FS and CS-LS methods, CS-Relief-F first obtains the compressively sampled data ($Y \in R^{m \times L}$) and then uses Relief-F to rank the features of the compressively sampled data. The Relief-F technique uses a statistical approach to select important features from the compressively sampled data based on their weight W. The main idea of Relief-F is to estimate attributes according to how well their values distinguish amongst compressively sampled examples (Y) that are near to each other by randomly computing examples from Y and then calculating their nearest neighbours from the same class (also called the *nearest hit*) and the other nearest neighbours from different classes (also called the *nearest miss*). Relief-F can be applied using the *relieff* function in the Statistics and Machine Learning Toolbox in MATLAB. This function returns the ranks and weights of features of the input matrix data and class labels vector c with d nearest neighbours. CS-Relief-F selects the k most important features, which can be used for fault classification of rotating machines where $k < m$. The procedure for the CS-Relief-F algorithm for feature ranking of the compressively sampled signals is summarized in Algorithm 16.3.

16.4.1.4 CS-PCC

The compressive sampling and Pearson correlation coefficients feature-selection (CS-PCC) method receives a large amount of vibration data as input and produces fewer features as output, which can be used for fault classification of rotating machines. Similar to the CS-FS, CS-LS, CS-Relief-F methods, CS-PCC first obtains the compressively sampled data (Y) and then uses PCC to rank the features of the compressively sampled data. PCC examines the relationship between two variables according to their correlation coefficient (r), $-1 \leq r \leq 1$. Here, the negative values indicate inverse relations, the positive values indicate a correlated relation, and the value 0 indicates no relation. The PCC can be computed using the following equation:

$$r(i) = \frac{cov(y_i, c)}{\sqrt{var(y_i) * var(c)}} \tag{16.19}$$

Here, y_i is the ith variable and c is the class labels. CS-PCC selects the k features that are correlated with the class labels, which can be used for fault classification of rotating machines where $k < < m$.

Algorithm 16.3   Relief-F

Input: l learning instances, m features, and c classes; probabilities of classes py; Sampling parameter a; number of nearest instances from each class d.
Output: For each feature fi, a feature weight $-1 \leq W[i] \leq 1$;
for i = 1 to m do W[i] = 0.0; end for;
for h = 1 to a do
randomly compute an instance yk with class yk;
for y = 1 to c do
find d nearest instances y[j, c] from class c, j = 1 . . . d;
for i = 1 to m do
for j = 1 to d do
if y = yk {nearest hit}
then W[i] = W[i] − diff (i, yk, y[j, c])/ (a*d);
else W[i] = W[i]+ py / (1 - pyk) * diff(i, yk, y[j, y])/ (a*d);
end if;
end for; {j} end for; {i}
end for; {y} end for ; {h}
return (W);

16.4.1.5 CS-Chi-2

The compressive sampling and Chi-square feature-selection (CS-Chi-2) method receives a large amount of vibration data as input and produces fewer features as output, which can be used for fault classification of rotating machines. Similar to the CS-FS and CS-LS methods, CS-Chi-2 first obtains the compressively sampled data (Y) and then uses Chi-2 to rank the features of the compressively sampled data. The $\chi2$ value for each feature f in a class label group c of Y can be computed using Eq. (16.20),

$$\chi^2(f, c) = \frac{L(E_{cf}E - E_cE_f)^2}{(E_{cf} + E_c)(E_f + E)(E_{cf} + E_f)(E_c + E)} \tag{16.20}$$

where L is the total number of examples in Y, $E_{c,f}$ is the number of times f and c co-occur, E_f is the number of time feature f occurs without c, E_c is the number of times c occurs without f, and E is the number of times neither f nor c occurs. A bigger value of χ^2 indicates that the features are highly related. Chi-2 can be applied using the cross-tabulation function in the Statistics and Machine Learning Toolbox in MATLAB. The *cross-tabulation* function returns the Chi-square statistic, and the obtained values of Chi-2 are sorted in descending order to create a new feature vector with ranked features. CS-Chi-2 selects the k features with bigger values of χ^2 that can be used for fault classification of rotating machines ($k < m$).

16.5 CS and Linear Subspace Learning-Based Framework for Fault Diagnosis

The CS-LSL based techniques receive a large amount of bearing vibration data and produce fewer features that can be used for fault classification of rotating machines. CS-LSL based techniques first use the MMV-CS model to produce compressively sampled signals, i.e. compressed data $Y = \{y_1, y_2, \ldots, y_L\} \in R^m$, which has enough information from the original machine raw data $X = \{x_1, x_2, \ldots, x_L\} \in R^n$. To extract feature representations of these signals, CS-LSL based techniques perform a linear transformation to map the m-dimensional space of the compressively sampled vibration to a lower-dimensional feature space, using the following equation:

$$\hat{y}_r = W^T y_r \tag{16.21}$$

Here, r = 1, 2 … L, $\hat{y}_r$ is the transformed feature vector with reduced dimensions, and W is a transformation matrix. There are three techniques based on CS-LSL that have been proposed and tested.

16.5.1 Implementations

Three linear subspace learning techniques – unsupervised PCA, supervised linear discriminant analysis, and independent component analysis – are used to select fewer features from compressively sampled signals. Moreover, an advanced technique called compressive sampling with correlated principal and discriminant components (CS-CPDC) has been recently introduced (Ahmed and Nandi 2018c).

16.5.1.1 CS-PCA
The compressive sampling and principal component analysis (CS-PCA) method receives a large amount of vibration data as input and produces fewer features as output, which can be used for fault classification of rotating machines (Ahmed et al. 2017a). Given a dataset of rotating machine vibrations $X \in R^{n \times L}$, CS-PCA first uses the CS based on the MMV model to produce compressively sampled signals, i.e. compressed data $Y = \{y_1, y_2, \ldots, y_L\} \in R^m$ where $1 \leq l \leq L$, $m < < n$, and lets each of these signals fit in with one of the c classes of machine health. To find the larger attributes of the compressively sampled vibration signals, CS-PCA uses PCA to compute a

W projection matrix using the scatter matrix, i.e. the covariance matrix C of the compressively sampled data, which can be computed using the following equation:

$$C = \frac{1}{L} \sum_{i=1}^{L} (y_i - \bar{y})(y_i - \bar{y})^T \tag{16.22}$$

Here, $\bar{y}$ is the mean of all samples and $C \in R^{m \times m}$ is the covariance matrix of Y. The transformation matrix W is then computed by finding the eigenvectors of C. In the produced projection matrix W, successive column vectors from left to right correspond to decreasing eigenvalues. We select the $m1$ eigenvectors corresponding to the $m1$ largest eigenvalues. Hence, a new $m1$-dimensional space $\hat{Y}1 \in R^{m1 \times L}$ is produced from $Y \in R^{m \times L}$, where $m1 < < m$.

16.5.1.2 CS-LDA

The compressive sampling and linear discriminant analysis (CS-LDA) method receives a large amount of roller bearing vibration data as input and produces fewer features as output, which can be used for fault classification of rotating machines (Ahmed et al. 2017a). CS-LDA produce a set of compressively sampled signals $Y \in R^{m \times L}$; here Y can be presented as $Y = [y_1, y_2, ..., y_l]$, where $1 \leq l \leq L$; and let each of these signals fit in with one of the c classes of rotating machine conditions. To compute discriminant attributes from the compressively sampled signals, the CS-LDA method adopts LDA, which considers maximizing the Fisher criterion function J (W), i.e. the ratio of the between the class scatter (S_B) and the within-class scatter (S_w) such that,

$$J(W) = \frac{|W^T S_B W|}{|W^T S_w W|} \tag{16.23}$$

where

$$S_B = \frac{1}{L} \sum_{i=1}^{c} l_i (\mu^i - \mu)(\mu^i - \mu)^T \tag{16.24}$$

$$S_w = \frac{1}{L} \sum_{i=1}^{c} \sum_{j=1}^{l_i} (y_j^i - \mu^i)(y_j^i - \mu^i)^T \tag{16.25}$$

Here, μ^i is the mean vector of class i, $y \in R$ of size $L * m$ is the training dataset, y_1^i represents the dataset belong to the cth class, n_i is the number of measurements of the ith class, μ^i is the mean vector of class i, and μ is the mean vector of the training dataset. LDA projects the space of the compressively sampled data onto a $(c - 1)$ dimension space by finding the optimal projection matrix W by maximizing J (W). Now W is composed of the selected eigenvectors $(\hat{w}_1, ..., \hat{w}_{m2})$ with the first $m2$ largest eigenvalues $(m2 = c - 1)$. Consequently, a new $m2$-dimensional space of discriminant attributes $\hat{Y}2 \in R^{m2 \times L}$ is produced from $Y \in R^{m \times L}$, where $m2 < < m$. With these fewer learned features using CS-PCA and CS-LDA, a classifier is used to classify rotating machine health conditions. For example, in (Ahmed et al. 2017a), both CS-PCA and CS-LDA utilise the multinomial LRC described in Section 11.2 to classify roller bearing health conditions.

16.5.1.3 CS-CPDC

The compressive sampling with correlated principal and discriminant components (CS-CPDC) method has been proposed recently by Ahmed and Nandi (2018b, c). It is a three-stage hybrid method for classification of bearing faults. In the first stage (Figure 16.8a), CS-CPDC uses MMV-CS to obtain compressively sampled raw vibration signals. In the second stage (Figure 16.8b), it employs a multistep approach of PCA, LDA, and CCA to extract features from the obtained compressively sampled signals. In the third stage (Figure 16.8c), it applies SVM to classify bearing health conditions using the learned features from the previous stage.

In the first stage, with the intention of reducing the amount of data and improving analysis effectiveness, CS-CPDC acquires compressively sampled signals using the MMV-CS framework described in Algorithm 16.1. While the CS projections obtained in the first stage help to recover the original signal from low-dimensional features, they may not be the best from a discriminant point of view. Furthermore, the size of the CS projections may still represent a large amount of data collected in real operating conditions. Consequently, techniques to extract fewer features of the CS projections are required. Accordingly, PCA and LDA are commonly used. However, while an individual set of features (e.g. either PCA or LDA) can be good for representations, it may not be good for classifications. Thus, the aim of the second stage is to generate features for superior classification accuracy. The second stage consists of three steps, as shown in Figure 16.8b.

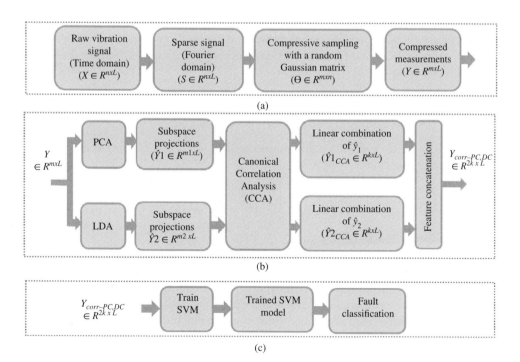

Figure 16.8 Training of the CS-CPDC method: (a) first stage, (b) second stage (c) third stage (Ahmed and Nandi 2018c).

In the first step of the second stage, CS-CPDC finds two feature representations from the compressively sampled signals using PCA and LDA, respectively. Hence, we transform the characteristic space of the compressively sampled signal into a low-dimensional space defined by those basis vectors corresponding to larger eigenvalue components (PCA). Furthermore, we augment these basis vectors with discriminant attributes learned through supervised learning (LDA). Let us consider a set of compressively sampled signals $Y \in R^{m \times L}$; here Y can be presented as $Y = [y_1, y_2, ..., y_l]$, where $1 \leq l \leq L$; and let each of these signals fit in with one of the c classes of machine conditions. To extract feature representations of these signals, CS-CPDC performs a linear transformation to map the m-dimensional space of the compressively sampled vibrations to a lower-dimensional feature space, using Eq. (16.20).

To find the larger attributes of the compressively sampled vibration signals, CS-CPDC uses PCA to compute the W projection matrix as described in Section 16.5.1.1. Hence, a new $m1$-dimensional space $\hat{Y}1 \in R^{m1 \times L}$ is produced from $Y \in R^{m \times L}$, where $m1 < < m$. Furthermore, we employed LDA to compute discriminant attributes from the compressively sampled signals as described in Section 16.5.1.2. Consequently, a new $m2$-dimensional space of discriminant attributes $\hat{Y}2 \in R^{m2 \times L}$ is produced from $Y \in R^{m \times L}$, where $m2 < < m$. These different feature representations extracted from the same dataset always reflect different characteristics of the original signals. The best combination of them retains multiple features of the integration that can be used effectively for classification. We propose CCA (Hardoon et al. 2004) to combine PCA and LDA features to obtain superior classification.

The second step of the multistep procedure of CS-CPDC utilises CCA to combine the different feature representations $\hat{Y}1$ and $\hat{Y}2$ by forming the relationship between them, i.e. maximising the overlapping variance between $\hat{Y}1$ and $\hat{Y}2$. The main idea is to find linear combinations of $\hat{Y}1$ and $\hat{Y}2$ that can maximize the correlation between them based on the following objective function,

$$(W_1, W_2) = \arg \max_{W_1, W_2} W_1 C_{\hat{Y}1\hat{Y}2}$$

$$s.t \quad W_1 C_{\hat{Y}1\hat{Y}1} W_1 = 1, W_2 C_{\hat{Y}2\hat{Y}2} W_2 = 1 \tag{16.26}$$

where $C_{\hat{Y}1\hat{Y}2}$ is the cross-covariance matrix of $\hat{Y}1$ and $\hat{Y}2$ that can be computed using the following equation:

$$C(\hat{Y}1, \hat{Y}2) = \hat{E}\left[\begin{pmatrix}\hat{Y}1\\\hat{Y}2\end{pmatrix}\begin{pmatrix}\hat{Y}1\\\hat{Y}2\end{pmatrix}'\right] = \begin{bmatrix}C_{\hat{Y}1\hat{Y}1} & C_{\hat{Y}1\hat{Y}2}\\C_{\hat{Y}2\hat{Y}1} & C_{\hat{Y}2\hat{Y}2}\end{bmatrix} \tag{16.27}$$

The resulting linear combinations of $\hat{Y}1$ ($\hat{Y}1_{CCA} = W_1 * \hat{Y}1$) and $\hat{Y}2$ ($\hat{Y}2_{CCA} = W_2 * \hat{Y}2$) will maximize their correlation. Finally, in the third step, the learned features $\hat{Y}1_{CCA}$ and $\hat{Y}2_{CCA}$ are concatenated to obtain a vector ($Y_{corr-PC,DC} \in R^{L \times 2k}$) that comprises highly correlated representations of principal and discriminative components, where k is equal to the minimal dimension size of $m1$ and $m2$. These procedures are summarised in Algorithm 16.4. Figure 16.9 shows an illustration of the training process of the first and second stage of our proposed method.

In the third stage, CS-CPDC utilises the multi-class SVM classifier described in Chapter 13 to classify rotating machine health.

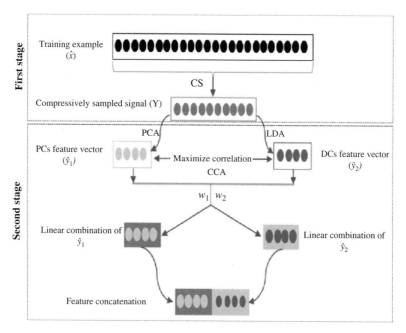

Figure 16.9 Illustration of the training process of the first and second stage of the proposed method (Ahmed and Nandi 2018c). (See insert for a colour representation of this figure.)

Algorithm 16.4   Feature Learning Stage

Input: $Y \in R^{m \times L}$, $y \in R^{1 \times L}$:label information vector for each data points, c: number of classes, $m1$:selected number of principal components

Output: $Y_{corr-PC, DC} \in R^{L \times 2k}$

$PCA(Y) \rightarrow E1 \in R^{m \times m1}$

$\widehat{Y}1 = Y^T * E1$

$LDA(Y, y) \rightarrow E2 \in R^{m \times m2}; m2 = c - 1.$

$\widehat{Y}2 = Y^T * E2$

$CCA(\widehat{Y}1, \widehat{Y}2) \rightarrow w_1, w_2 \in R^{L \times k}; k = \min(m1, m2).$

$\qquad \widehat{Y}1_{CCA} = w_1 * \widehat{Y}1, \widehat{Y}2_{CCA} = w_2 * \widehat{Y}2$

$\qquad Y_{corr-PC,LD} = [\widehat{Y}1_{CCA} \ \widehat{Y}2_{CCA}]$

16.6   CS and Nonlinear Subspace Learning-Based Framework for Fault Diagnosis

It has previously been demonstrated that random linear projections can efficiently preserve the structure of manifolds (Baraniuk and Wakin 2009). Hence, further exploration of the use of the other nonlinear variants of PCA and LDA such as kernel-PCA (KPCA) and other manifold learning techniques in combination with CS may be beneficial to be examined for machine fault diagnosis.

16.6.1 Implementations

Four nonlinear subspace learning techniques have been employed to learn fewer features from compressively sampled signals: KPCA, KLDA, multidimensional scaling (MDS), and stochastic proximity embedding (SPE). Furthermore, a three-stage method combining MMV-CS, geodesic minimal spanning tree (GMST), SPE, and neighbourhood component analysis (NCA) is used to estimate and further reduce the dimensionality of the compressively sampled signals. With these reduced features, a multiclass SVM classifier is used to classify machine health.

16.6.1.1 CS-KPCA

The compressive sampling and kernel principal component analysis (CS-KPCA) method receives a large amount of vibration data as input and produces fewer features as output, which can be used for fault classification of rotating machines. Given a dataset of rotating machine vibrations $X \in R^{n \times L}$, CS-KPCA first uses CS based on the MMV model to produce compressively sampled signals, i.e. compressed data $Y = \{y_1, y_2, ..., y_L\} \in R^m$ where $1 \leq l \leq L$, $m < < n$; and let each of these signals fit in with one of the c classes of machine health conditions. Then, KPCA is used to obtain a further low-dimensional representation of the compressively sampled signals. As described in Section 8.2 in Chapter 8, the KPCA algorithm first maps the compressively sampled data $Y \in R^m$ into a feature space $\hat{Y} \in F$ and then employs a linear PCA in $\hat{Y}$. Assume the compressed signals $Y = \{y_1, y_2, ..., y_L\} \in R^m$ are the training samples for the KPCA algorithm. First, KPCA extends the compressively sampled signals into the hyper-dimension feature space using a nonlinear mapping $\varnothing$ such that,

$$\varnothing : Y \in R^m \rightarrow \varnothing(Y) = \hat{Y} \in F \tag{16.28}$$

Then, to find the larger attributes of $\hat{Y}$, KPCA computes the W projection matrix using the scatter matrix, i.e. the covariance matrix $C_\varnothing$ of $\hat{Y}$, which can be computed using the following equation:

$$C_\varnothing = \frac{1}{L} \sum_{i=1}^{L} (\hat{y}_i - \bar{y})(\hat{y}_i - \bar{y})^T \tag{16.29}$$

Here, $\bar{y} = \sum_{i=1}^{L} \hat{y}_i / L$ is the mean of all samples in $\hat{Y}$. The eigenvector can be computed by solving the following problem,

$$C_\varnothing v = \lambda v \tag{16.30}$$

where v is the eigenvector and λ is the eigenvalue of $C_\varnothing$. Here, v can be represented using the following equation:

$$v = \sum_{i=1}^{L} a_i \varnothing(y_i) = \sum_{i=1}^{L} \alpha_i \hat{y} \tag{16.31}$$

To compute the coefficients α_i, a kernel matrix $K \in R^{L \times L}$ is used, e.g. sigmoid kernel, such that,

$$L\lambda K a = K^2 a \tag{16.32}$$

where $a = (a_1, ..., a_L)^T$ is the normalised eigenvector.

16.6.1.2 CS-KLDA

The compressive sampling and kernel linear discriminant analysis (CS-KLDA) method receives a large amount of vibration data as input and produces fewer features as output, which can be used for fault classification of rotating machines. Given a dataset of rotating machine vibrations $X \in R^{n \times L}$, CS-KLDA first uses CS based on the MMV model to produce compressively sampled signals, i.e. compressed data $Y = \{y_1, y_2, ..., y_L\} \in R^m$ where $1 \leq l \leq L$, $m < < n$; and let each of these signals fit in with one of the c classes of machine health conditions. Then, KLDA is used to obtain a further low-dimensional representation of the compressively sampled signals. Let the compressed signals $Y = \{y_1, y_2, ..., y_L\} \in R^m$ be the training samples for the KLDA algorithm and $L_1, L_2, ..., L_C$ be the number of samples for classes of machine health conditions, respectively ($L = L_1 + L_2 + ... + L_C$ is the total number of observations). First, KLDA extends the compressively sampled signals Y into the hyper-dimension feature space using a nonlinear mapping $\varnothing$ such that,

$$\varnothing : Y \in R^m \rightarrow \varnothing(Y) = \hat{Y} \in F \tag{16.33}$$

Then, a linear LDA is performed in F space. To compute discriminant attributes from $\hat{Y}$, the CS-KLDA method adopts LDA, which considers maximizing the Fisher criterion function $J(W)$, i.e. the ratio between the class scatter (S_B) and the within-class scatter (S_w) such that,

$$J(W) = \frac{|W^T S_B W|}{|W^T S_w W|} \tag{16.34}$$

where

$$S_B = \frac{1}{L} \sum_{i=1}^{c} l_i (\mu^i - \mu)(\mu^i - \mu)^T \tag{16.35}$$

$$S_w = \frac{1}{L} \sum_{i=1}^{c} \sum_{j=1}^{L_i} (\hat{y}_j^i - \mu^i)(\hat{y}_j^i - \mu^i)^T \tag{16.36}$$

Here, μ^i is the mean vector of class i, $\hat{y}_1^i$ represents the dataset belong to the cth class, L_i is the number of measurements of the ith class, and μ is the mean vector of all training dataset in F space. LDA projects the space of the compressively sampled data onto a $(c-1)$ dimension space by finding the optimal projection matrix W by maximizing $J(W)$. The projection matrix W can be represented by the following equation,

$$w = \Phi a \tag{16.37}$$

where $\Phi = \sum_{i=1}^{L} \phi(y_i)$ and $a = (a_1, ..., a_L)^T$. Hence, Eq. (16.34) can be rewritten as Eq. (16.38),

$$J(W) = \frac{|(\Phi a)^T S_B (\Phi a)|}{|(\Phi a)^T S_w (\Phi a)|} \triangleq \frac{|a^T Pa|}{|a^T Qa||} \tag{16.38}$$

Here, $P = \Phi^T S_B \Phi$ and $Q = \Phi^T S_w \Phi$ can be represented by the kernel function (Wang et al. 2008).

16.6.1.3 CS-CMDS

The compressive sampling and classical multidimensional scaling (CS-CMDS) method receives a large amount of vibration data as input and produces fewer features as output, which can be used for fault classification of rotating machines. Given a dataset of rotating machine vibrations $X \in R^{nxL}$, CS-CMDS first uses CS based on the MMV model to produce compressively sampled signals, i.e. compressed data $Y = \{y_1, y_2, \ldots, y_L\} \in R^m$ where $1 \leq l \leq L$, $m < < n$; and let each of these signals fit in with one of the c classes of machine health conditions. Then, CMDS is used to select optimal features of the compressively sampled signals. With these selected features, a classifier can be used to classify the faults of rotating machines. CMDS is a nonlinear optimisation technique to yield a lower-dimensional representation of the compressively sampled data. Let $D = (d_{ij})$ be the pairwise distance matrix of $Y = \{y_1, y_2, \ldots, y_L\} \in R^m$ such that,

$$d_{ij} = \sqrt{\sum_{k=1}^{p} (y_{ik} - y_{jk})^2} \qquad (16.39)$$

Assume the reduced dimension is p ($p \ll m$). Given D and p, MDS reduces Y to $\hat{Y} = \{\hat{y}_1, \hat{y}_2, \ldots, \hat{y}_L\} \in R^p$. CMDS compute $\hat{Y}$ that perfectly preserves D such that,

$$\hat{Y} = \text{argmin} \sqrt{\sum_{k=1}^{p} (y_{ik} - y_{jk})^2} \qquad (16.40)$$

16.6.1.4 CS-SPE

The compressive sampling and stochastic proximity embedding (CS-SPE) method receives a large amount of vibration data as input and produces fewer features as output, which can be used for fault classification of rotating machines. Given a dataset of rotating machine vibrations $X \in R^{nxL}$, CS-SPE first uses CS based on the MMV model to produce compressively sampled signals, i.e. compressed data $Y = \{y_1, y_2, \ldots, y_L\} \in R^m$ where $1 \leq l \leq L$, $m < < n$; and let each of these signals fit in with one of the c classes of machine health conditions. Then, SPE is used to select optimal features of the compressively sampled signals. With these selected features, a classifier can be used to classify the faults of rotating machines. SPE uses a self-organizing iterative scheme to embed m-dimensional data into p dimensions, such that the geodesic distances in the original m dimensions are preserved in the embedded d dimension. To compute a reduced dimension from the compressively sampled signals using SPE, the following steps are performed:

1. Initialize the coordinates y_i. Select an initial learning rate β.
2. Select a pair of points, i and j, at random, and compute their distance: $d_{ij} = \|y_i - y_j\|$. If $d_{ij} \neq r_{ij}$ (r_{ij} is the distance of the corresponding proximity), update the coordinates y_i and y_j using the following equations,

$$y_i \leftarrow y_i + \beta \frac{1}{2} \frac{r_{ij} - d_{ij}}{d_{ij} + \upsilon} (y_i - y_j) \qquad (16.41)$$

and

$$y_j \leftarrow y_j + \beta \frac{1}{2} \frac{r_{ij} - d_{ij}}{d_{ij} + \upsilon} (y_j - y_i) \qquad (16.42)$$

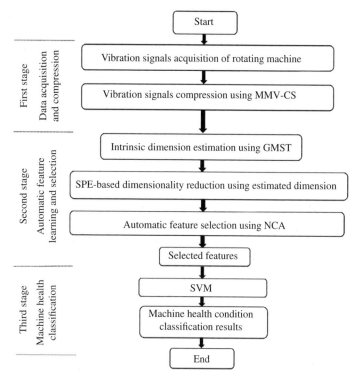

Figure 16.10 Training of the CS-GSN method.

Here, v is a small number to avoid division by zero. For a given number of iterations, this step will be repeated for a prescribed number of steps, and β will be decreased by a suggested decrement $\delta\beta$.

Based on the CS-SPE technique, Ahmed and Nandi proposed a new three-stage method for monitoring rotating machine health. In the first stage of the proposed method, the MMV-CS model is used to obtain compressively sampled signals from the acquired raw vibration signals. In the second stage, a process combining GMST, SPE, and NCA is used to estimate and further reduce the dimensionality of the compressively sampled signals. In the third stage, with these reduced features, a multiclass SVM classifier is used to classify machine health (Ahmed and Nandi 2018d); this method will be referred as CS-GSN. In this method (see Figure 16.10), first, the global dimension estimator GMST is used to define the intrinsic dimensionality, i.e. define the minimal number of features required to represent the compressively sampled data. GMST computes the geodesic graph G from which the intrinsic dimension (p) is estimated by computing multiple minimum spanning trees (MSTs) in which each data sample x_i is linked to its k nearest neighbours such that

$$p(Y) = min \sum_{e \in T} D_{Eucl} \tag{16.43}$$

Here, T represents the set of all the subtrees of G, e is an edge in T, and D_{Eucl} is the Euclidean distance of e.

Second, having defined the minimal number of features p $(p < m)$, the compressively sampled data can be transformed into a reduced-dimensionality space of significant representation. Various linear and nonlinear techniques have been proposed and used to reduce data dimensionality. In this stage of this method, a combination of SPE and NCA to automatically select far fewer features of the compressively sampled data is used. SPE is a nonlinear technique that has several attractive features: (i) it is simple to implement, (ii) it is very fast, (iii) it scales linearly with the size of the data in both time and memory, and (iv) it is relatively insensitive to missing data. So, it was decided that SPE is a suitable technique to adopt for this investigation. With fewer selected features, a classifier can be used for machine fault diagnosis.

16.7 Applications

Two case studies of bearing datasets (real and not simulated) are used to demonstrate how all the presented techniques in this chapter work and validate their efficiency compared with other state-of-the-art fault-diagnosing techniques.

16.7.1 Case Study 1

The vibration dataset used in this case study is acquired from experiments on a test rig that simulates an environment with running roller bearings. In these experiments, several interchangeable faulty roller bearings are inserted in the test rig to represent the types of faults that can normally happen in roller bearings. Six health conditions of roller bearings have been recorded: two normal conditions, i.e. brand new (NO) and worn but undamaged (NW); and four faulty conditions, i.e. IR fault, OR fault, RE fault, and cage fault (CA). The sampling rate was 48 kHz. A good description of the test setup can be found in chapter 6 and the article (Ahmed and Nandi 2018a). A total of 960 examples with 6000 data points were recorded. Figure 16.11 illustrates some typical time-series plots for the six different conditions. As shown in Figure 16.11, each fault modulates the vibration signals with its own unique patterns. For instance, based on the level of damage to the RE and the loading of the bearing, IR and OR fault conditions have a fairly periodic signal, RE fault condition may or may not be periodic, and CA fault conditions generate a random distortion.

First, 50 % of the total observations were randomly selected for training and the other 50% for testing. Then, the selection of the compressed sampling rate (α) using different values (0.1 and 0.2) to generate compressively sampled vibration signals was examined. To ensure that our CS model generates enough samples for the purpose of bearing fault classification, we used the generated compressively sampled signals in each of the sparse representation methods to reconstruct the original signal X by applying the CoSaMP algorithm. The reconstruction errors were measured by root mean squared error (RMSE). For example, by using thresholded WT-based CS with $\alpha = 0.1$, the RMSE for the six bearing conditions are 8.5% (NO), 24.6% (NW), 15.23% (IR), 12.71% (OR), 11.87% (RE), and 5.29% (CA); this has been studied in detail in (Ahmed et al. 2018b). For FFT-based CS using the same sampling rate $\alpha = 0.1$, the RMSE values are 4.8% (NO), 8.9% (NW), 6.3% (IR), 5.6% (OR), 4.7% (RE), and 3.6% (CA), which indicates good signal reconstruction.

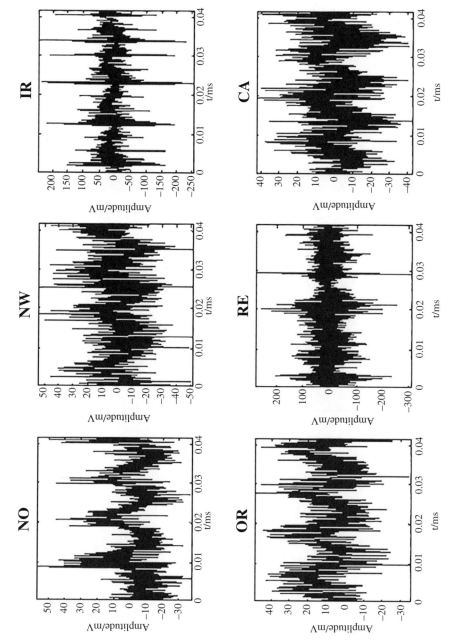

Figure 16.11 Typical time domain vibration signals for the six different conditions (Ahmed and Nandi 2018a).

The following is a brief description of several experiments that have been conducted to verify the capability of the CS-SL framework for diagnosing rotating machine faults.

16.7.1.1 The Combination of MMV-CS and Several Feature-Ranking Techniques

This part of the chapter investigates the combination of MMV-CS and several feature-ranking techniques – i.e. FS, LS, Relief-F, PCC, and Chi-2 – to learn fewer features from a large amount of vibration data from which bearing health conditions are classified. Three classification algorithms – LRC, ANN, and SVM – are used to classify bearing health conditions. The vibration dataset of roller bearing in this case study is used in this investigation with the aim of (i) validating the proposed framework for diagnosing bearing health conditions, and (ii) observing the best combinations of MMV-CS, feature-selection techniques, and classifiers with reduced complexity and improved classification accuracy. Tables 16.1 and 16.2 present testing classification results for five CS-FR based techniques using FFT and WT sparsifying transforms, respectively. The compressively sampled signals were obtained using the aforementioned compressive sampling rates, i.e. $\alpha = 0.1$ and 0.2. The classification results were obtained using LRC, ANN, and SVM with 10-fold cross-validation and by averaging the results of 20 trials for each classifier and for each experiment.

From the results in Tables 16.1 and 16.2, we can see that among the various proposed combinations of CS with FFT, feature-selection techniques, and classifiers, most of the combinations with LRC and ANN achieved better results than SVM with the fitcecoc

Table 16.1 Classification accuracy of roller bearing health conditions for the FFT sparsifying transform with different combinations of MMV-CS and feature ranking and selection techniques (all classification accuracies $\geq$ 99% in bold).

With FFT algorithm	k	SVM		LRC		ANN	
		$\alpha = 0.1$	$\alpha = 0.2$	$\alpha = 0.1$	$\alpha = 0.2$	$\alpha = 0.1$	$\alpha = 0.2$
CS-FS	60	85.8 ± 8.4	89.5 ± 5.1	94.3 ± 5.6	**98.7 ± 0.7**	93.8 ± 6.3	**98.3 ± 0.8**
	120	89.6 ± 11.4	92.3 ± 6.2	**99.7 ± 0.4**	**99.8 ± 0.3**	**99.2 ± 1.3**	**99.7 ± 0.7**
	180	97.4 ± 5.2	**99.1 ± 0.9**	**99.9 ± 0.1**	**99.9 ± 0.2**	**99.8 ± 0.5**	**99.8 ± 0.3**
CS-LS	60	92.7 ± 7.0	92.1 ± 5.1	95.8 ± 0.9	93.4 ± 1.2	98.6 ± 0.9	97.3 ± 2.1
	120	98.7 ± 1.4	98.9 ± 1.1	**99.5 ± 0.3**	**99.8 ± 0.6**	**99.9 ± 0.2**	**99.4 ± 0.7**
	180	**99.2 ± 0.8**	**99.2 ± 0.7**	**99.8 ± 0.2**	**99.9 ± 0.1**	**99.9 ± 0.1**	**99.9 ± 0.2**
CS-Relief-F	60	85.8 ± 13.7	77.9 ± 1.5	**99.5 ± 0.3**	**99.3 ± 0.6**	**99.7 ± 0.6**	**99.7 ± 0.3**
	120	89.3 ± 8.4	90.9 ± 7.8	**99.8 ± 0.3**	**99.9 ± 0.2**	**99.9 ± 0.2**	**99.7 ± 0.4**
	180	95.6 ± 5.8	96.2 ± 4.8	**99.9 ± 0.1**	**99.9 ± 0.1**	**99.9 ± 0.3**	**100 ± 0.0**
CS-PCC	60	78.3 ± 14.0	73.2 ± 7.4	98.5 ± 1.3	**99.3 ± 0.5**	**99.4 ± 0.3**	**99.4 ± 0.6**
	120	93.2 ± 7.5	93.8 ± 7.9	**99.8 ± 0.3**	**99.5 ± 0.2**	**99.8 ± 0.3**	**99.7 ± 0.4**
	180	97.7 ± 5.1	96.7 ± 5.7	**99.9 ± 0.1**	**99.9 ± 0.2**	**99.7 ± 0.5**	**100 ± 0.0**
CS-Chi-2	60	64.2 ± 7.1	68.3 ± 7.6	98.0 ± 2.2	98.7 ± 1.4	96.9 ± 2.0	95.2 ± 3.9
	120	83.2 ± 9.0	76.9 ± 9.3	**99.5 ± 0.5**	**99.5 ± 0.4**	**99.2 ± 0.9**	**99.4 ± 0.8**
	180	95.2 ± 5.2	94.9 ± 6.3	**99.9 ± 0.1**	**99.9 ± 0.1**	**99.9 ± 0.3**	**99.9 ± 0.3**

Table 16.2 Classification accuracy of roller bearing health conditions for the WT sparsifying transform with different combinations of MMV-CS and feature ranking and selection techniques (all classification accuracies ≥ 99% in bold).

With WT algorithm	k	SVM $\alpha = 0.1$	SVM $\alpha = 0.2$	LRC $\alpha = 0.1$	LRC $\alpha = 0.2$	ANN $\alpha = 0.1$	ANN $\alpha = 0.2$
CS-FS	60	91.5 ± 8.4	95.2 ± 4.7	92.2 ± 7.5	96.4 ± 3.9	91.6 ± 8.3	97.1 ± 2.8
	120	94.9 ± 4.9	94.8 ± 5.2	96.3 ± 3.8	96.2 ± 4.0	97.2 ± 2.8	98.3 ± 1.7
	180	95.8 ± 3.9	96.3 ± 3.8	98.9 ± 1.2	97.8 ± 2.2	98.2 ± 1.9	$\mathbf{99.1 \pm 0.9}$
CS-LS	60	92.6 ± 7.4	92.2 ± 7.7	91.9 ± 8.2	93.3 ± 6.9	92.9 ± 7.0	93.7 ± 6.3
	120	91.4 ± 6.9	93.5 ± 6.4	92.8 ± 7.2	94.2 ± 5.7	95.3 ± 4.8	95.2 ± 4.8
	180	94.4 ± 5.4	93.2 ± 6.9	95.4 ± 4.6	95.5 ± 4.4	96.8 ± 2.9	95.9 ± 4.1
CS-Relief-F	60	94.7 ± 5.3	94.2 ± 5.9	94.1 ± 5.1	93.7 ± 6.4	94.6 ± 5.4	94.4 ± 5.7
	120	95.3 ± 4.7	95.5 ± 4.4	96.4 ± 3.2	95.9 ± 4.2	97.2 ± 2.8	97.0 ± 3.1
	180	97.7 ± 2.1	98.8 ± 1.2	98.4 ± 1.6	$\mathbf{99.3 \pm 0.7}$	$\mathbf{99.1 \pm 0.9}$	$\mathbf{99.9 \pm 0.1}$
CS-PCC	60	71.5 ± 17.8	79.9 ± 17.8	94.1 ± 5.7	94.3 ± 4.6	94.9 ± 5.3	95.7 ± 4.3
	120	78.2 ± 16.2	82.2 ± 15.8	95.0 ± 4.9	95.0 ± 4.9	96.4 ± 3.6	96.3 ± 3.8
	180	83.5 ± 13.7	87.4 ± 11.6	98.8 ± 1.2	$\mathbf{99.0 \pm 1.1}$	$\mathbf{99.3 \pm 0.7}$	$\mathbf{99.6 \pm 0.3}$
CS-Chi-2	60	94.3 ± 3.8	95.2 ± 3.2	95.1 ± 4.5	96.7 ± 2.7	95.8 ± 3.9	96.3 ± 3.7
	120	96.2 ± 3.7	97.1 ± 3.4	96.8 ± 2.9	97.9 ± 3.2	98.8 ± 0.4	98.1 ± 2.9
	180	$\mathbf{99.6 \pm 0.3}$	$\mathbf{99.6 \pm 0.2}$	$\mathbf{99.5 \pm 0.4}$	$\mathbf{99.3 \pm 0.7}$	$\mathbf{99.2 \pm 0.7}$	$\mathbf{99.9 \pm 0.1}$

function. In particular, results from CS-Chi-2 and CS-Relief-F for all values of the sampling rate (α) and the number of selected features (k) with both LRC and ANN are better than with SVM. Also, all the combinations of CS with FFT and the considered feature-selection techniques with LRC and ANN achieved high classification accuracies (all above 99%) for all values of α with $k = 120$. Moreover, Ahmed and Nandi have demonstrated that all classification accuracies are above 99% for all the classifiers considered with CS-FS, CS-Relief-F, CS-PCC, and CS-Chi-2 with both WT and FFT sparse representation techniques using $\alpha = 0.4$ and $k = 180$. Also, they have demonstrated that for CS-LS, all considered classifiers achieved accuracy results above 99% using $\alpha = 0.4$ and $k = 180$ with FFT only. All these results and more are reported in (Ahmed and Nandi 2018b).

In summary, these results show that the CS-FR framework with various methods presented in this chapter has the ability to classify bearing health conditions with high classification accuracy, with the following comments:

1. FFT as a sparsifying transform method for our proposed MMV-based CS model can achieve better results than thresholded WT.
2. LRC and ANN have the ability to achieve high classification accuracy with different values of the sampling rate (α) and number of selected features (k) for all the considered CS and feature-selection technique combinations.
3. SVM has the ability to achieve good classification accuracy with a larger number of selected features, i.e. $k = 180$, and larger values of α, e.g. $\alpha = 0.4$, for certain combinations (Ahmed and Nandi 2018b).

4. With the larger number of selected features, all the proposed methods achieved high classification accuracy. Thus, for the application of the proposed methods in fault diagnosis, it is recommended to select a larger number of features from compressively sampled signals.

16.7.1.2 The Combination of MMV-CS and Several Linear and Nonlinear Subspace Learning Techniques

In this part of this chapter, we describe various experiments that have been conducted to verify the ability of the CSL-LSL and CS-NLSL frameworks, described in Sections 0 and 0, respectively, for rotating MCM. This achieved by investigating the CS-PCA, CS-LDA, CS-CPDC, CS-KPCA, CS-KLDA, CS-CMDS, CS-SPE, and CS-GSN methods to learn fewer features of a large amount of the collected vibration data for diagnosing roller bearing faults. In these techniques, first MMV-CS with the FFT sparsifying transform is used to obtain the compressively sampled data from which a subset feature can be learned. With these learned features of each technique, averaging the results of 20 trials from SVM with 10-fold cross-validation is employed to deal with the classification. To evaluate the performance of these methods, we first compressively sampled each testing signal using the same values of α used for the sample training set, and then the trained algorithm is used to obtain the learned features of the testing set. Once the features were learned, the trained SVM is used to classify the testing signals. The overall results are shown in Table 16.3, where the classification accuracy is the average of 20 trials for each experiment.

It can be clearly seen that the larger the value of α is, the better the classification accuracy of all CS-SL based techniques studied here. However, high levels of classification accuracy are achieved with less than 15% of the original data samples. In particular, accuracies from CS-CPDC, CS-KLDA, and CS-MDS methods are 99.8%, 99.3%, and 99.1%, respectively for only 10% of the whole dataset.

For further verification of the efficiency of the CS-SL based techniques, complete comparison results of the classification accuracy using different combinations based on the CS-LSL and CS-NLSL frameworks are compared with some recently published results

Table 16.3 Classification accuracy of roller bearing health conditions for the FFT sparsifying transform with different combinations of MMV-CS and subspace learning techniques (all classification accuracies ≥ 99% in bold).

Algorithm	Compressive sampling rate (α)	
	0.1	0.2
CS-PCA	95.0 ± 4.8	97.2 ± 2.5
CS-LDA	96.8 ± 2.1	98.4 ± 1.3
CS-CPDC	$\mathbf{99.8 \pm 0.2}$	$\mathbf{99.9 \pm 0.1}$
CS-KPCA	98.6 ± 2.5	$\mathbf{99.3 \pm 0.8}$
CS-KLDA	$\mathbf{99.3 \pm 1.1}$	$\mathbf{100.0 \pm 0.0}$
CS-MDS	$\mathbf{99.1 \pm 1.2}$	$\mathbf{99.6 \pm 0.7}$
CS-SPE	96.2 ± 3.7	98.5 ± 1.6
CS-GSN	98.8 ± 2.4	$\mathbf{99.4 \pm 0.5}$

Table 16.4 A comparison with the classification results from the literature on the vibration bearing dataset described in Case study 1.

	Classification accuracy (%)
Raw vibration with entropic features (Wong et al. 2015)	98.9 ± 1.2
Compressed sampled with $\alpha = 0.5$ followed by signal reconstruction (Wong et al. 2015)	92.4 ± 0.5
Compressed sampled with $\alpha = 0.25$ followed by signal reconstruction (Wong et al. 2015)	84.6 ± 3.4
FMM-RF (SampEn + PS) (Seera et al. 2017)	99.81 ± 0.41
GP generated feature sets (un-normalised) (Guo et al. 2005)	
ANN	96. 5
SVM	97.1
CS-LSL and CS-NLSL based methods with FFT, $\alpha = 0.2$, and SVM classifier.	
CS-PCA	97.2 ± 2.5
CS-LDA	98.4 ± 1.3
CS-CPDC	99.9 ± 0.1
CS-KPCA	99.3 ± 0.8
CS-KLDA	100.0 ± 0.0
CS-MDS	99.6 ± 0.7
CS-SPE	98.5 ± 1.6
CS-GSN	99.4 ± 0.5

using the same vibration dataset. For instance, (Wong et al. 2015) reported results for three methods: one method uses all the original vibration data from which entropic features are extracted, and the other two use compressed measurements to recover the original vibration signals; from the recovered signals, entropic features are extracted. With the extracted features, SVM is used to classify bearing health conditions. Moreover, a hybrid model consisting of the fuzzy min-max (FMM) neural network and random forest (RF) with sample entropy (SampEn) and power spectrum (PS) features is used to classify bearing health conditions; the results are reported in (Seera et al. 2017). In (Guo et al. 2005), a genetic programming (GP) based approach is proposed for feature extraction from raw vibration data; with extracted features, SVM and ANN are used to classify bearing health conditions. Table 16.4 presents classification results of bearing health conditions using our proposed methods with $\alpha = 0.2$ and the reported results in the studies mentioned using the same dataset used in this case study.

It can be clearly seen that classification results of CS-LSL and CS-NLSL based methods listed in Table 16.4 are better than those reported in (Wong et al. 2015) and (Guo et al. 2005). Also, our results from CS-LDA, CS-CPDC, CS-KPCA, CS-KLDA, CS-MDS, CS-SPE, and CS-GSN are as good as, if not better than, results reported in (Seera et al. 2017); although we are using only 20% ($\alpha = 0.2$) of the original vibration data, our results are not matched by the method in (Seera et al. 2017) using all of the raw vibration data.

This section has validated the CS-SL based techniques presented in this chapter and has shown that the many combinations of CS and subspace learning methods achieved high classification accuracies for bearing faults. The next section of this chapter will validate the usage of these methods using a publicly available bearing vibration dataset. The advantage of the shared dataset is that we can easily compare the results of other researchers.

16.7.2   Case Study 2

The bearing datasets used in this case are provided by Case Western Reserve University (http://csegroups.case.edu/bearingdatacenter/home). The data were acquired from a motor driving mechanical system where the faults were seeded into the drive-end bearing of the motor. The bearing datasets were collected under normal condition (NO), with an IR fault (IR), with a roller element fault (RE), and with an OR fault (OR). The datasets are further categorised by fault width (0.18–0.53 mm) and motor load (0–3 hp). The sampling rates used were 12 kHz for some of the sampled data and 48 kHz for the rest of the sampled data. The bearing vibration signals were acquired under normal NO, IR, OR, and RE conditions for four different speeds. At each speed, 100 time-series were taken for each condition per load. For the IR, OR, and RE conditions, vibration signals for four different fault widths (0.18, 0.36, 0.53, and 0.71 mm) were separately recorded. In this book, of these acquired vibration signals, three groups of datasets were organised to be used in the evaluation of the methods presented in this chapter and Chapter 17.

The first group of datasets is chosen from the data files of the vibration signals sampled at 48 kHz with fault width (0.18, 0.36, and 0.53) and fixed loads including 1, 2, and 3 hp; and the number of examples chosen is 200 examples per condition. This gave three different datasets A, B, and C with 2000 total examples and 2400 data points for each signal. The second group of datasets is chosen from the vibration signals sampled at 12 kHz with fault width (0.18, 0.36, and 0.53) and variable-load motor operating conditions including 0, 1, 2, and 3 hp; and the number of examples chosen is 400 per condition. This gave a dataset D with 4000 total examples and 1200 data points for each signal. The third type of bearing dataset is chosen from the data files of vibration signals sampled at 12 kHz with fault size (0.18, 0.36, 0.53, and 0.71) and load 2 hp; and the number of examples chosen is 60 examples per condition. This gave a dataset E with 720 total examples with 2000 data points for each signal. A detailed description of these datasets can be found in the bearing data centre of the Case Western Reserve University website and/or in (Ahmed and Nandi 2018a).

The following is a brief description of several experiments have been conducted to verify the capability of the CS-SL framework for diagnosing rotating machine faults.

16.7.2.1   The Combination of MMV-CS and Several Feature-Ranking Techniques

To apply the CS-FR framework in dataset E, 240 examples are randomly selected for training and 480 examples are used for testing. We applied the MMV-CS model with the FFT to obtain compressively sampled signals from the raw vibration signals using α equal to 0.1 and the same feature-selection methods as in the case study 1 to select fewer features of these compressively sampled signals. With these fewer selected features, we employed LRC, ANN, and SVM to deal with the classification problem. The classification accuracies are achieved by averaging the results of 20 trials for each classifier and for each experiment. The performance comparisons are carried out between CS-FR based algorithms and other state-of-the-art fault-diagnosis algorithms introduced in (Yu et al.

2018): feature selection by adjunct Rand index and standard deviation ratio (FSAR) to select features from the original feature set (OFS). Other methods use PCA, LDA, local Fisher discriminant analysis (LFDA), and support margin LFDA (SM-LFDA) to reduce the dimension of selected features using FSAR. With the selected features, SVM is used for the purpose of classification. Table 16.5 presents the comparisons with some recently published results in (Yu et al. 2018). It is clear that all the results from CS-FR based algorithms outperform results reported in (Yu et al. 2018).

16.7.2.2 The Combination of MMV-CS and Several Linear and Nonlinear Subspace Learning Techniques

To apply CS-SL based methods, experiments in dataset D were conducted for $\alpha = 0.1$ with a training size of 40%, and 20 trials for each experiment. The results are compared to some recently published results using the same vibration dataset, i.e. dataset D. For instance (Zhang et al. 2015b) results are reported for a hybrid model that integrates permutation entropy (PE), ensemble empirical mode decomposition (EEMD), and an SVM optimised by inter-cluster distance in the feature space. Moreover, results are reported for a deep neural network (DNN) based on a stacked denoising autoencoder (Xia et al. 2017). More recently, results are reported for a convolutional neural network (CNN) based approach in (Xia et al. 2018). Table 16.6 presents classification results for bearing health conditions using the CS-SL based methods introduced in this chapter with $\alpha = 0.1$ and the reported results from the studies mentioned, using the same dataset used in this case study.

It can be clearly seen that classification results from the CS-LSL and CS-NLSL based methods listed in Table 16.6 are better than those reported in (Zhang et al. 2015b) and (Xia et al. 2017). Also, our results from CS-CPDC, CS-KPCA, CS-KLDA, CS-MDS, CS-SPE, and CS-GSN are as good as, if not better than, results reported in (Xia et al. 2018); although we are using only 10% ($\alpha = 0.1$) of the original vibration data our results are not matched by the method in (Xia et al. 2018) using all the raw vibration data.

16.8 Discussion

A recently proposed compressive sampling and subspace learning framework, CS-SL, has been described in this chapter. The framework employs a CS model with three different subspace learning methods. The first method is based on CS and feature ranking and selection techniques (CS-FR), the second method is the CS model and linear subspace learning techniques (CS-LSL), and the third method is based on the CS model and nonlinear subspace learning. CS-SL based methods have the ability to receive a large amount of vibration data as input and produce fewer features as output, which can be used for fault classification of rotating machines. Our validation experiments have demonstrated that small values of α (0.1 and 0.2) can lead to high classification accuracies using different combinations of methods based on the CS-SL framework. The comparison of non-CS-based techniques with the CS-SL based techniques shows that the learned features of CS-SL based techniques achieved better classification results, even though we are using only 10% ($\alpha = 0.1$) or 20% ($\alpha = 0.2$) of the original vibration data. The results in this chapter indicate that the proposed methods have the ability to reduce the original vibration data and yet achieve high classification accuracies. However, the size of the CS

Table 16.5 A comparison with the classification results from the literature on bearing dataset described in case study 2.

Methods		Classification accuracy (%)
OFS-FSAR-SVM (Yu et al. 2018)		
(25 selected features)		91.46
(50 selected features)		69.58
OFS-FSAR-PCA-SVM (Yu et al. 2018)		
(25 selected features)		91.67
(50 selected features)		69.79
OFS-FSAR-LDA-SVM (Yu et al. 2018)		
(25 selected features)		86.25
(50 selected features)		92.7
OFS-FSAR-LFDA-SVM (Yu et al. 2018)		
(25 selected features)		93.75
(50 selected features)		94.38
OFS-FSAR-(SM-LFDA)-SVM (Yu et al. 2018)		
(25 selected features)		94.58
(50 selected features)		95.63
Our proposed framework with FFT, $\alpha = 0.1$, k = 25		
CS-FS	LRC	98.4 ± 1.6
	ANN	99.6 ± 0.5
	SVM	97.4 ± 2.7
CS-LS	LRC	99.1 ± 0.8
	ANN	99.2 ± 0.8
	SVM	98.5 ± 1.6
CS-Relief-F	LRC	99.3 ± 0.6
	ANN	99.2 ± 0.8
	SVM	97.8 ± 2.4
CS-PCC	LRC	99.2 ± 0.8
	ANN	99.5 ± 0.6
	SVM	97.9 ± 1.9
CS-Chi-2	LRC	97.5 ± 2.6
	ANN	99.9 ± 0.1
	SVM	94.7 ± 5.4

Table 16.6 A comparison with the classification results from the literature on bearing dataset D (Ahmed and Nandi 2018b).

Algorithm	Classification accuracy (%)
(Zhang et al. 2015a)	97.91 ± 0.9
(Xia et al. 2017)	97.59
(Xia et al. 2018)	99.44
CS-PCA	98.1 ± 1.8
CS-LDA	98.9 ± 1.1
CS-CPDC	99.9 ± 0.1
CS-KPCA	99.6 ± 0.4
CS-KLDA	99.2 ± 0.7
CS-SPE	99.1 ± 0.9
CS-GSN	99.7 ± 0.3

projections may still represent a large amount of data collected in real operating conditions; consequently, more reduction may be required in some MCM applications. The next chapter, therefore, moves on to present another method for machine fault diagnosis from highly compressed measurements using sparse overcomplete features. The advantages of highly compressed measurements in vibration-based machine condition monitoring will result in more reduction in computational complexity, storage requirements, and bandwidth for transmitting compressed data.

References

Ahmed, H. and Nandi, A. (2018d). Three-stage method for rotating machine health condition monitoring using vibration signals. In: *Proceedings of the 2018 Prognostics and System Health Management Conference (PHM-Chongqing)*, 285–291. IEEE.

Ahmed, H.O.A. and Nandi, A.K. (2017). Multiple measurement vector compressive sampling and fisher score feature selection for fault classification of roller bearings. In: *2017 22nd International Conference on Digital Signal Processing (DSP)*, 1–5. IEEE.

Ahmed, H.O.A. and Nandi, A.K. (2018a). Compressive sampling and feature ranking framework for bearing fault classification with vibration signals. *IEEE Access* 6: 44731–44746.

Ahmed, H.O.A. and Nandi, A.K. (2018b). Intelligent condition monitoring for rotating machinery using compressively-sampled data and sub-space learning techniques. In: *International Conference on Rotor Dynamics*, 238–251. Cham: Springer.

Ahmed, H.O.A. and Nandi, A.K. (2018c). *Three-Stage Hybrid Fault Diagnosis for Rolling Bearings with Compressively-Sampled Data and Subspace Learning Techniques*, vol. 66(7), 5516–5524. Institute of Electrical and Electronics Engineers.

Ahmed, H.O.A., Wong, M.D., and Nandi, A.K. (2017a). Compressive sensing strategy for classification of bearing faults. In: *2017 IEEE International Conference on Acoustics, Speech and Signal Processing (ICASSP)*, 2182–2186. IEEE.

Ahmed, H.O.A., Wong, M.D., and Nandi, A.K. (2017b). Classification of bearing faults combining compressive sampling, laplacian score, and support vector machine. In: *Industrial Electronics Society, IECON 2017-43rd Annual Conference of the IEEE*, 8053–8058. IEEE.

Bao, Y., Beck, J.L., and Li, H. (2011). Compressive sampling for accelerometer signals in structural health monitoring. *Structural Health Monitoring* 10 (3): 235–246.

Baraniuk, R.G. and Wakin, M.B. (2009). Random projections of smooth manifolds. *Foundations of Computational Mathematics* 9 (1): 51–77.

Candes, E.J. and Tao, T. (2006). Near-optimal signal recovery from random projections: universal encoding strategies. *IEEE Transactions on Information Theory* 52 (12): 5406–5425.

Candès, E.J. and Wakin, M.B. (2008). An introduction to compressive sampling. *IEEE Signal Processing Magazine* 25 (2): 21–30.

Chang, S.G., Yu, B., and Vetterli, M. (2000). Adaptive wavelet thresholding for image denoising and compression. *IEEE Transactions on Image Processing* 9 (9): 1532–1546.

Chen, J. and Huo, X. (2006). Theoretical results on sparse representations of multiple-measurement vectors. *IEEE Transactions on Signal Processing* 54 (12): 4634–4643.

Chen, X., Du, Z., Li, J. et al. (2014). Compressed sensing based on dictionary learning for extracting impulse components. *Signal Processing* 96: 94–109.

Dai, W. and Milenkovic, O. (2009). Subspace pursuit for compressive sensing signal reconstruction. *IEEE Transactions on Information Theory* 55 (5): 2230–2249.

Davenport, M.A., Wakin, M.B., and Baraniuk, R.G. (2006). Detection and estimation with compressive measurements. Dept. of ECE, Rice University, technical report.

Donoho, D.L. (2006). Compressed sensing. *IEEE Transactions on information theory* 52 (4): 1289–1306.

Donoho, D.L., Tsaig, Y., Drori, I., and Starck, J.L. (2006). Sparse solution of underdetermined linear equations by stagewise orthogonal matching pursuit. *IEEE Trans. on Information theory* 58 (2): 1094–1121.

Eldar, Y.C. (2015). *Sampling Theory: Beyond Bandlimited Systems*. Cambridge University Press.

Guo, H., Jack, L.B., and Nandi, A.K. (2005). Feature generation using genetic programming with application to fault classification. *IEEE Transactions on Systems, Man, and Cybernetics, Part B (Cybernetics)* 35 (1): 89–99.

Hardoon, D.R., Szedmak, S., and Shawe-Taylor, J. (2004). Canonical correlation analysis: An overview with application to learning methods. *Neural Comput.* 16 (12): 2664–2699.

Holland, D.J., Malioutov, D.M., Blake, A. et al. (2010). Reducing data acquisition times in phase-encoded velocity imaging using compressed sensing. *Journal of Magnetic Resonance* 203 (2): 236–246.

Li, J., Cheng, K., Wang, S. et al. (2017). Feature selection: a data perspective. *ACM Computing Surveys (CSUR)* 50 (6): 94.

Li, X.F., Fan, X.C., and Jia, L.M. (2012). Compressed sensing technology applied to fault diagnosis of train rolling bearing. *Applied Mechanics and Materials* 226: 2056–2061.

Merlet, S.L. and Deriche, R. (2013). Continuous diffusion signal, EAP and ODF estimation via compressive sensing in diffusion MRI. *Medical Image Analysis* 17 (5): 556–572.

Needell, D. and Tropp, J.A. (2009). CoSaMP: iterative signal recovery from incomplete and inaccurate samples. *Applied and Computational Harmonic Analysis* 26 (3): 301–321.

Qaisar, S., Bilal, R.M., Iqbal, W. et al. (2013). Compressive sensing: from theory to applications, a survey. *Journal of Communications and Networks* 15 (5): 443–456.

Rao, K.R. and Yip, P.C. (2000). *The Transform and Data Compression Handbook*, vol. 1. CRC Press.

Rossi, M., Haimovich, A.M., and Eldar, Y.C. (2014). Spatial compressive sensing for MIMO radar. *IEEE Transactions on Signal Processing* 62 (2): 419–430.

Rudelson, M. and Vershynin, R. (2008). On sparse reconstruction from Fourier and Gaussian measurements. *Communications on Pure and Applied Mathematics* 61 (8): 1025–1045.

Seera, M., Wong, M.D., and Nandi, A.K. (2017). Classification of ball bearing faults using a hybrid intelligent model. *Applied Soft Computing* 57: 427–435.

Shao, H., Jiang, H., Zhang, H. et al. (2018). Rolling bearing fault feature learning using improved convolutional deep belief network with compressed sensing. *Mechanical Systems and Signal Processing* 100: 743–765.

Sun, L., Liu, J., Chen, J., and Ye, J. (2009). Efficient recovery of jointly sparse vectors. In: *Advances in Neural Information Processing Systems*, 1812–1820.

Tang, G., Hou, W., Wang, H. et al. (2015a). Compressive sensing of roller bearing faults via harmonic detection from under-sampled vibration signals. *Sensors* 15 (10): 25648–25662.

Tang, G., Yang, Q., Wang, H.Q. et al. (2015b). Sparse classification of rotating machinery faults based on compressive sensing strategy. *Mechatronics* 31: 60–67.

Tropp, J.A. and Gilbert, A.C. (2007). Signal recovery from random measurements via orthogonal matching pursuit. *IEEE Transactions on Information Theory* 53 (12): 4655–4666.

Wang, L., Chan, K.L., Xue, P., and Zhou, L. (2008). A kernel-induced space selection approach to model selection in KLDA. *IEEE Transactions on Neural Networks* 19 (12): 2116–2131.

Wong, M.L.D., Zhang, M., and Nandi, A.K. (2015). Effects of compressed sensing on classification of bearing faults with entropic features. *IEEE Signal Processing Conference (EUSIPCO)*: 2256–2260.

Xia, M., Li, T., Liu, L. et al. (2017). Intelligent fault diagnosis approach with unsupervised feature learning by stacked denoising autoencoder. *IET Science, Measurement & Technology* 11 (6): 687–695.

Xia, M., Li, T., Xu, L. et al. (2018). Fault diagnosis for rotating machinery using multiple sensors and convolutional neural networks. *IEEE/ASME Transactions on Mechatronics* 23 (1): 101–110.

Yu, X., Dong, F., Ding, E. et al. (2018). Rolling bearing fault diagnosis using modified LFDA and EMD with sensitive feature selection. *IEEE Access* 6: 3715–3730.

Yuan, H. and Lu, C. (2017). Rolling bearing fault diagnosis under fluctuant conditions based on compressed sensing. *Structural Control and Health Monitoring* 24 (5).

Zhang, X., Hu, N., Hu, L. et al. (2015a). A bearing fault diagnosis method based on the low-dimensional compressed vibration signal. *Advances in Mechanical Engineering* 7 (7): 1–12.

Zhang, X., Liang, Y., and Zhou, J. (2015b). A novel bearing fault diagnosis model integrated permutation entropy, ensemble empirical mode decomposition and optimized SVM. *Measurement* 69: 164–179.

17

Compressive Sampling and Deep Neural Network (CS-DNN)

17.1 Introduction

In cases where high levels of data compression are required to address the challenges of learning from high-dimensional data in applications of machine condition monitoring, compressive sampling (CS) projections may still represent a large amount of vibration data collected in real operating conditions. One way to overcome this issue is to start by learning a subset of features from the CS projections (Chapter 16) and then apply classification to this learned subset of features. In this chapter, we present another approach that has been proposed recently through the design of an intelligent fault-classification method from highly compressed measurements using sparse-overcomplete features and training a deep neural network through a sparse autoencoder (CS-SAE-DNN). This method includes the extraction of overcomplete sparse representations from highly compressed measurements. It involves unsupervised feature learning with a sparse autoencoder (SAE) algorithm for learning feature representations in multiple stages of nonlinear feature transformation based on a deep neural network (DNN) (Ahmed et al. 2018).

17.2 Related Work

As discussed in Section 16.3, several studies have been undertaken to use CS in machine condition monitoring and fault diagnosis. Moreover, a number of studies have begun to examine DNNs using an autoencoder (AE) algorithm for the purpose of machinery fault diagnosis. For example, Tao et al. (2015) proposed a DNN algorithm framework for diagnosing bearing faults based on a stacked autoencoder and softmax regression. Jia et al. (2016) showed the effectiveness of a proposed DNN-based intelligent method in the classification of different datasets from rolling element bearings and planetary gearboxes with massive samples using an AE as a learning algorithm. In a recent paper by Sun et al. (2016), a SAE-based DNN approach with the help of the denoising coding and dropout method using one hidden layer was proposed for induction motor fault classification with 600 data samples and 2000 features from each induction motor working condition. The results of these investigations validate the effectiveness of DNN based on an AE learning algorithm in machinery fault classification.

Condition Monitoring with Vibration Signals: Compressive Sampling and Learning Algorithms for Rotating Machines,
First Edition. Hosameldin Ahmed and Asoke K. Nandi.
© 2020 John Wiley & Sons Ltd. Published 2020 by John Wiley & Sons Ltd.

Moving on, we will now consider other types of deep learning architectures, e.g. convolutional deep neural networks (CNNs), deep belief networks (DBNs), and recurrent neural networks (RNNs), which have been used for machine fault diagnosis. For example, unlike a standard neural network (NN), the architecture of a CNN is usually composed of a convolutional layer and a sub-sampling layer, also called a pooling layer. The CNN learns abstract features from alternating and stacking convolutional layers and pooling operations. The convolutional layers convolve multiple local filters with raw input data and generate invariant local features, and the pooling layers extract the most-significant features (Ahmed et al. 2018). Many studies have used CNNs for diagnosing bearing faults (Guo et al. 2016; Fuan et al. 2017; Zhang et al. 2018; and Shao et al. 2018).

DBNs are generative neural networks that stack multiple restricted Boltzmann machines (RBMs) that can be trained in a greedy layer-wise unsupervised way; then they can be further fine-tuned with respect to labels of the training data by adding a softmax layer in the top layer. Many researchers have used DBNs for diagnosing bearing faults (Shao et al. 2015; Ma et al. 2016; Tao et al. 2016). RNNs build connections between units from a direct cycle and map from the entire history of previous inputs to target vectors in principal; this allows a memory of previous inputs to be kept in the network state. As is the case with DBNs, RNNs can be trained via backpropagation through time for supervised tasks with sequential input data and target outputs. Examples of studies that used RNN to diagnose bearing faults in rotating machines can be found in (Lu et al. 2017; Jiang et al. 2018). A good overview of these deep learning architectures can be found in (Zhao et al. 2016; Khan & Yairi 2018).

17.3   CS-SAE-DNN

The compressive sampling and SAE-based deep neural network (CS-SAE-DNN) method has been proposed recently by Ahmed et al. (2018). It uses CS for the sparse time-frequency representation model described in Section 16.2.3 to produce highly compressed vibration measurements from the high-dimensional vibration data collected for the purpose of machine condition monitoring (MCM). To predict the status of rotating machines, CS-SAE-DNN learns sparse-overcomplete feature representations from these CS-based highly compressed measurements, which can be used for machine fault classification. Thus, it allows high compression of data that results in reduced computations and reduced transmission costs.

17.3.1   Compressed Measurements Generation

To obtain the compressed measurements from the very large vibration data of a rotating machine, CS-SAE-DNN uses the CS model described in Section 16.2.3. It applies the Haar wavelet basis as a sparse representation of the vibration signals. In fact, CS-SAE-DNN applies the thresholded Haar wavelet transform (WT) with five decomposition levels as a sparsifying transform to obtain the sparse representation of the vibration signal. Based on CS theory, the obtained compressed measurements must possess the quality of the original signal, i.e. have sufficient information from the original signal. Therefore, we need to test that our CS model generates enough samples for the purpose of machine fault diagnosis. One approach to test the CS model is the flip test, which has the ability to test the efficiency of any sparsity model, any signal, any sampling rate, and any recovery algorithm toward an accurate CS model.

17.3.2 CS Model Testing Using the Flip Test

The *flip test* was designed by Adcock et al. (2017) to test the CS model. The basic idea of this test is to flip the sparsity basis coefficients that represent the sparse representations and then perform a reconstruction with a measurements matrix using the sampling operator and the recovery algorithm. If the sparse vector(s) is not recovered within a low tolerance, then we decrease the thresholding level to obtain a sparse signal and repeat until x is recovered exactly. The procedure of the flip test can be summarised as follows:

1. Obtain the sparse signal $s \in R^{1 \times n}$ of the original signal $x \in R^{1 \times n}$ using the sparsifying transform $\psi \in R^{n \times n}$, where, in our case, the Haar wavelet is used as sparsifying transform. Then, threshold s so that it is perfectly sparse.
2. Obtain the compressed measurements $y \in R^{1 \times m}$ where $m \ll n$ using the CS model of interest.
3. Perform a reconstruction using the compressed measurements y, a measurement matrix ($\theta = \phi * \psi$), and a reconstruction algorithm. If x is not recovered exactly (i.e. within a low tolerance), then decrease the thresholding level for s in step 1 to obtain a new sparse s, and repeat until x is perfectly reconstructed.

17.3.3 DNN-Based Unsupervised Sparse Overcomplete Feature Learning

In this method, we impose some flexible constraints for regularisation on the hidden units of the SAE. These include a sparsity constraint that can be controlled by different parameters such as a sparsity parameter, a weight-decay parameter, and the weight of the sparsity penalty parameter. For the purpose of learning sparse overcomplete representations of our highly compressed measurements, the number of hidden units in each hidden layer is set to be greater than the number of input samples, and we used the encoder part of our unsupervised learning algorithm (i.e. the SAE). One important aspect of this method is to pretrain the DNNs in an unsupervised manner using the SAE, as described in Section 14.2, and then to fine-tune it with a backpropagation (BP) algorithm for classification. The difficulty of multilayer training can be overcome with an appropriate set of features. One of the advantages of pretraining and fine-tuning in this approach is the power to mine fault features flexibly from highly compressed signals. Thus, the proposed approach is expected to achieve better classification accuracy compared with methods based on undercomplete feature representations. Consequently, the efficiency of compressive sensing in machine fault classification is expected to be improved.

Previous research has shown that sparse representations of signals are able to signify the diagnosis features for machinery faults (Liu et al. 2011; Tang et al. 2014; Zhu et al. 2015; Fan et al. 2015; Ren et al. 2016). The advantage of sparse-overcomplete representations is that the number of obtained features is greater than the number of input samples; this has been studied by Lewicki and Sejnowski (2000), who concluded that overcomplete bases can produce a better approximation of the underlying statistical distribution of the data. Olshausen and Field (1997) and Doi et al. (2006) identify several advantages of an overcomplete basis set: for example, their robustness to noise and their ability to improve classification performance. Sparse feature-learning methods normally contain two stages: (i) produce a dictionary W that represents the data $\{x_i\}_{i=1}^N$ sparsely using a learning algorithm, e.g. training an artificial neural network (ANN) with sparsity penalties; and (ii) obtain a feature vector from a new input vector using an encoding algorithm.

Various recent studies investigating sparse feature representation have been carried out. These include SAE, sparse coding (Liu et al. 2011), and RBMs (Lee et al. 2008). The SAE approach has a number of attractive features: (i) it is simple to train, (ii) the encoding stage is very fast, and (iii) it has the ability to learn features when the number of hidden units is greater than the number of input samples. Therefore it was decided that SAE is an appropriate method to adopt for our investigation. In an analysis of AEs, Bengio et al. (2007) found that they can be used as building blocks of DNNs, using greedy layer-wise pretraining.

CS-SAE-DNN applies a learning algorithm in multiple stages of nonlinear feature transformation, where each stage is a kind of feature transformation. One way to do this is by using a DNN with multiple hidden layers; each layer is connected to the layers below it in a nonlinear combination. In the pretraining stage, the SAE is used to train the DNN, and the encoder part of the SAE with a sigmoid activation function is used to learn the overcomplete feature representations.

As expressed in Figure 17.1, the proposed method produces overcomplete representations from the input compressed measurements (obtained using the CS model described in Section 16.2.3) by setting the number of hidden units (d_i) to be greater than the number of input samples (m), i.e. $d_i > m$ in each hidden layer (i), where $d_{i+1} > d_i$ for $i = 1, 2, 3, \ldots, n$. Input (n) represents the output of Encoder ($n - 1$), and d_n is the number of hidden layers in Encoder (n). As shown in Figure 17.1, DNN training includes two levels of training: pretraining using an unsupervised learning algorithm, and retraining using a backpropagation algorithm. In the pretraining stage, the unlabelled bearing compressed measurements (y) are first used to train the DNN by setting the parameters in each hidden layer and to compute the sparse-overcomplete feature representations. In fact, in the DNN based on SAE, we are making use of the SAE algorithm applied multiple times through the network. Therefore, the output overcomplete feature vector from the first encoder is the input for the second encoder.

Finally, the fault classification is achieved using two stages: (i) pretraining classification based on a stacked AE and softmax regression layer, which is the deep net stage (the first stage); and (ii) retraining classification based on the BP algorithm, which is the fine-tuning stage (the second stage). Figure 17.2 shows an illustration of the pretraining process using two hidden layers. With enforcement of sparsity constraints and by setting the number of units in each hidden layer to be greater than the input samples, each AE learns useful features of the compressed unlabelled training samples. The training process is performed by optimising the cost function $CF_{sparse}(W, b)$ in the following equation,

$$CF_{sparse}(W, b) = \frac{1}{2m} \sum_{i=1}^{m} \|\tilde{y}_i - y_i\|^2 + \lambda \|W\|^2 + \beta \sum_{j=1}^{d} KL(\rho \| \hat{\rho}) \qquad (17.1)$$

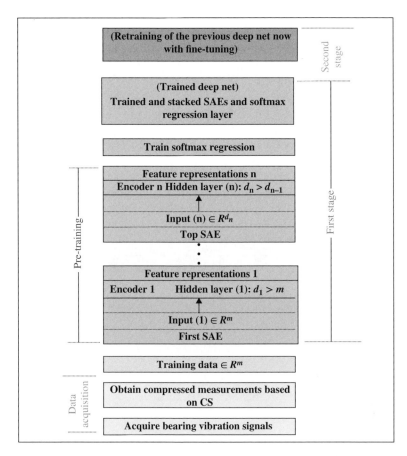

Figure 17.1 Training the CS-SAE-DNN method (Ahmed et al. 2018). (See insert for a colour representation of this figure.)

where m is the input size, i.e. the highly compressed measurements size, d is the hidden layer size, λ is the weight decay parameter, β is the weight of the sparsity penalty term, ρ is a sparsity parameter, and $\hat{\rho}$ is the average threshold activation of hidden units.

The optimization is performed using the scaled conjugate gradient (SCG), which is a member of the conjugate gradient (CG) methods (Møller 1993). In the first learning stage, the encoder part of the first SAE with sigmoid functions for the range of unit activation function values [0, 1] is used to learn features from compressed vibration signals of length m, where the number of hidden units $d_1 > m$ and the extracted overcomplete features (v_1) are used as the input signals for the second learning stage. Then Encoder 2 of the second SAE with a number of hidden units d_2 is used to extract overcomplete features (v_2) from (v_1). Finally, softmax regression is trained using (v_2) to classify bearing health conditions.

The full network formed, i.e. the trained deep net (see Figure 17.1), which comprises the trained and stacked SAEs and the softmax regression layer, is used to compute the classification results of the first stage on the test set.

The pretraining process can be described in the following Algorithm 17.1.

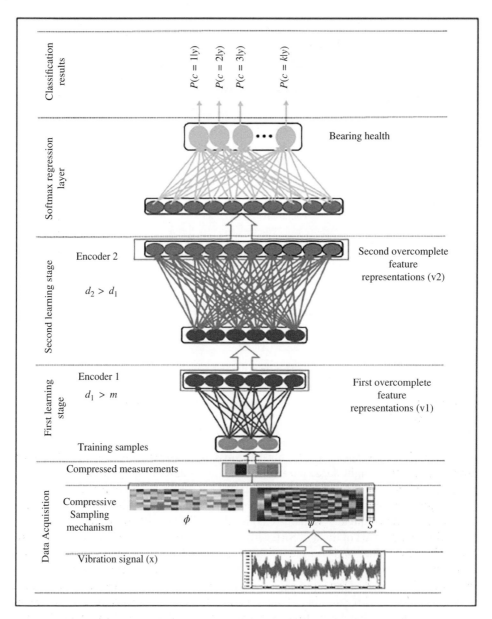

Figure 17.2 Illustration of the proposed method using two hidden layers. Data flow from the bottom to the top (Ahmed et al. 2018). (See insert for a colour representation of this figure.)

Algorithm 17.1   Pretraining

Given a DNN of n hidden layers, the pretraining process using SAE to learn overcomplete features will be conducted on each layer.

(1) Initialization, m is the number of input samples, n is the number of hidden layers,

$$d_0 = m, i = 1, 2, \ldots, n$$

(2) For $i = 1$ *to* n,

(3) Set up the number of hidden units (d_i) to be greater than the input samples (m), i.e. $d_i > d_{i-1}$.

(4) Set up a sparsity parameter, a weight decay parameter, the weight of the sparsity penalty term, and the maximum training epoch that achieves the lowest possible reconstruction error E $(x_i, \tilde{x}_i)$.

(5) Use the SCG algorithm for network training. By utilising SCG, the learning rate is automatically adopted at each epoch and the average threshold activation of hidden units ($\hat{\rho}$) can be computed using Eq. (14.8).

(6) Based on Eq. (14.9), compute the cost function.

(7) Using the encoder part of the SAE, calculate the output overcomplete feature vector v_i and use it as the input of the following hidden layer.

(8) $i = i + 1$.

(9) Repeat steps 2–8 until $i = n$.

(10) Use the overcomplete feature vector of the last hidden layer v_n to be the input of the softmax regression layer.

17.3.4   Supervised Fine Tuning

The classification results of the first stage can be improved by performing backpropagation, also called fine-tuning, on the formed deep net. The supervised fine tuning in our CS-SAE-DNN is performed by retraining the trained deep net from the first stage on the training data using the class labels of the machine health conditions. The retrained deep net is then can be used to compute the classification results of the second stage on the test set.

17.4   Applications

Two case studies of bearing datasets (real and not simulated) described in Sections 16.7.1 and 16.7.2 are used to demonstrate how CS-SAE-DNN works and to validate its efficacy compared with other state-of-the-art fault-diagnosing techniques.

17.4.1   Case Study 1

The vibration data used in this case study were recorded under six roller bearing health conditions with 160 examples of each condition and a total of 960 raw data files. This dataset has been described in more detail in Section 16.7.1. In order to verify the validity of the proposed method, several experiments were carried out to learn overcomplete features of various highly compressed bearing datasets obtained using the CS framework with different compressed sampling rates. Fifty percent of these compressed samples are randomly selected for the pretraining stage of DNN, and then these samples are used to retrain the deep net; the other 50% of the samples are used for testing performance. Then, the obtained overcomplete features are used for classification using different DNN settings. Finally, CS-SAE-DNN is compared with several existing methods.

We began by obtaining the compressed vibration signal from the large data of rolling element bearing vibration signals using thresholded Haar WT, described in Section 16.2.3, and a random Gaussian matrix. First, we used the Haar wavelet basis with five decomposition levels as a sparsifying transform, where the wavelet coefficients are thresholded using the penalised hard threshold to obtain sparse representations of the original vibration signals. Figure 17.3a shows the wavelet coefficients of the vibration. After applying the penalised hard threshold, the wavelet coefficients are sparse in the Haar wavelet domain, as shown in Figure 17.3b, where only 216 are nonzero (nnz) in the NO wavelet coefficients: that is, 95.8% of the 5120 coefficients are zeros. The other conditions – NW, IR, OR, RE, and CA – have 276, 209, 298, 199, and 299 nonzero elements, respectively; and 94.6%, 95.9, 94.2, 96.1, and 94.2% of 5120 coefficients are zeros.

The CS model was applied using different sampling rates (α) (0.0016, 0.003, 0.006, 0.013, 0.025, 0.05, and 0.1) with 8, 16, 32, 64, 128, 256, and 512 compressed measurements of the original vibration signal using a random Gaussian matrix. The size of the Gaussian matrix is m by N, where N is the length of the original vibration signal measurements and m is the number of compressed signal elements (i.e. $m = \alpha*N$). Based on the CS framework, multiplying this matrix by the signal sparse representations generates different sets of compressed measurements of the vibration signal. The obtained compressed measurements must possess the quality of the original signal, i.e. have sufficient information from the original signal. Thus, we need to test that our CS model generates enough samples for the purpose of bearing fault classification. Roman et al. (2014) proposed a generalised flip test for the CS model that has the ability to test the efficiency of any sparsity model, any signal class, any sampling operator, and any recovery algorithm toward the accurate CS model. The basic idea of this test is to flip sparsity basis coefficients that represent the sparse representations and then perform a reconstruction with a measurements matrix using the sampling operator and the recovery algorithm. If the sparse vector(s) is not recovered within a low tolerance, then we decrease the thresholding level to obtain a sparse signal and repeat until s is recovered exactly. More details of the original flip test can be found in (Roman et al. 2014; Adcock et al. 2017).

Following the idea of the generalised flip test, we tried different sampling rates (0.05, 0.1, 0.15, and 0.2) and tested the efficiency of our CS model by thresholding the wavelet coefficients to obtain a sparse signal (s) and then by reconstructing s from the obtained compressed measurements using a random Gaussian matrix. The CoSaMP algorithm is used to reconstruct the sparse signal. The reconstruction errors measured by root mean squared error (RMSE) for the six bearing conditions are presented in Table 17.1. The second column depicts the reconstruction errors compared to the original thresholded coefficients using 5% of the original signal for the six conditions; the third column shows the reconstructions errors using 10% of the original signal; the fourth is for 15% of the original signal; and the fifth column is for 20% of the original signal.

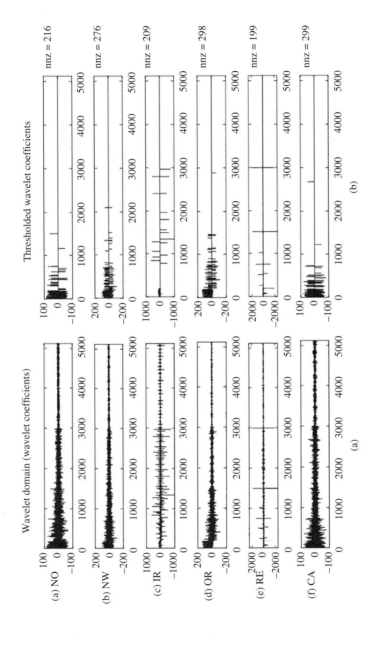

Figure 17.3 (a) Wavelet coefficients and (b) corresponding thresholded wavelet coefficients for each condition signal (nnz refers to the number of nonzero elements) (Ahmed et al. 2018).

Table 17.1 Results of root mean square error (RMSE) for various sampling rates.

	$\alpha = 0.05$	$\alpha = 0.10$	$\alpha = 0.15$	$\alpha = 0.20$
NO	9.37	8.45	0.04	0.03
NW	11.29	24.64	5.8	0.06
IR	20.82	15.23	0.16	0.09
OR	14.38	12,71	8.18	0.07
RE	17.02	11.87	0.35	0.12
CA	8.3	5.29	0.14	0.04

It is clear that as α increases, RMSE decreases, indicating better signal reconstruction. Better signal reconstruction indicates that the compressed measurements possess the quality of the original signal.

Fifty percent of these compressed samples are randomly selected for the pretraining stage of the DNN; then these samples are used to retrain the deep net. The other 50% of the samples are used for testing performance. Then, the obtained overcomplete features are used for classification using different DNN settings. Finally, we compare our proposed method, using our highly compressed datasets, with several existing methods.

To learn features from these compressed measurements, we used a SAE neural network with a limited number of hidden layers (two, three, and four hidden layers). The structures of these different hidden layers are chosen to be in the form of overcomplete feature learning (expansion), where the number of neurons in different hidden layers of the network structure is twice the number of neurons in the preceding layer: for example, if the number of input samples in the input layer is z, then the number of nodes in the first hidden layer is $2z$, $4z$ in the second hidden layer, and so on. The number of nodes in the output layer is limited by the number of bearing conditions (six conditions). A bidirectional deep architecture of stacked AEs has been used for the purpose of deep learning; these include feedforward and BP. The parameters that control the effects of using regularisers by the SAE were set as follows: the weight decay (λ) was set to a very small value of 0.002, the weight of the sparsity penalty term (β) was set to 4, and the sparsity parameter (ρ) was set to 0.1. The maximum training epoch is 200.

The overall classification results obtained from these experiments are shown in Table 17.2. The first major comment is that the classification accuracy after the second stage is better than that after the first stage for every dataset at these values of α. The classification in the deep net stage (the first stage) achieved good results for larger numbers of measurements, i.e. for values of m equal to 512, 256, and 128; and high accuracy was achieved by the DNN with two hidden layers using only 64 samples from our signal. Most of the classification accuracies for the DNNs with two, three, and four hidden layers using fine-tuning (the second stage) are 99% or above, and some are 100% for even less than 1% compressed measurements of the original vibration signal, i.e. when $\alpha = 0.006$. The two hidden layers of the DNN achieved high classification accuracy (98%) for α equal to 0.003 and 0.0016 with 16 and 8 compressed measurements. Moreover, the three hidden layers of the DNN achieved 100% with only 16 measurements, i.e. $\alpha = 0.003$. Taken together, these results show that the proposed

Table 17.2 Overall classification accuracies and their related standard deviations.

No. of hidden layers	Classification stage	Compressive sampling rate						
		$\alpha = 0.0016$ (m = 8)	$\alpha = 0.003$ (m = 16)	$\alpha = 0.006$ (m = 32)	$\alpha = 0.013$ (m = 64)	$\alpha = 0.025$ (m = 128)	$\alpha = 0.05$ (m = 256)	$\alpha = 0.1$ (m = 512)
2	Deep net (the first stage)	95.3 ± 1.0	95.9 ± 0.3	96.0 ± 1.3	99.1 ± 0.3	99.8 ± 0.1	100	99.8 ± 0.1
	Fine-tuning (the second stage)	98.0 ± 0.3	98.8 ± 0.3	99.6 ± 0.1	99.6 ± 0.1	99.8 ± 0.1	100	99.8 ± 0.1
3	Deep net (the first stage)	64.8 ± 4.2	94.9 ± 2.4	85.6 ± 4.4	93.6 ± 2.2	99.2 ± 2.6	99.6 ± 0.1	100
	Fine-tuning (the second stage)	82.0 ± 5.1	100	99.5 ± 0.3	99.8 ± 0.1	100	100	100
4	Deep net (the first stage)	36.6 ± 8.3	55.2 ± 6.1	90.5 ± 2.9	95.6 ± 1.2	99.8 ± 0.1	98.6 ± 3.9	100
	Fine-tuning (the second stage)	83.1 ± 1.6	89.4 ± 2.5	96.3 ± 0.8	99.7 ± 0.3	99.8 ± 0.1	99.7 ± 0.3	100

Table 17.3 A comparison with the classification results from literature on the bearing dataset.

		Accuracy (%)
Raw vibration (Wong et al. 2015)		98.9 ± 1.2
Compressed sensed ($\alpha = 0.5$) (Wong et al. 2015) followed by reconstruction		92.4 ± 0.5
Compressed sensed ($\alpha = 0.25$) (Wong et al. 2015) followed by reconstruction		84.6 ± 3.4
This work ($\alpha = 0.5$) (two hidden layers)	Deep net (the first stage)	99.1 ± 1.7
	Fine-tuning (the second stage)	100 ± 0.0
This work ($\alpha = 0.25$) (two hidden layers)	Deep net (the first stage)	99.6 ± 1.2
	Fine-tuning (the second stage)	100 ± 0.0
This work ($\alpha = 0.1$) (two hidden layers)	Deep net (the first stage)	99.8 ± 0.1
	Fine-tuning (the second stage)	99.8 ± 0.1

method has the ability to classify bearing conditions with high accuracy from highly compressed bearing vibration measurements.

The performance comparison of several methods uses the same vibration dataset as in (Wong et al. 2015). One method uses all the original samples. Each of the other two methods uses compressed measurements (for values of α of 0.5 and 0.25) and then reconstructs the original signals. These three have been reported in (Wong et al. 2015). The remaining three are our proposed methods to demonstrate the possibility of sampling the vibration data of roller element bearings at less than the Nyquist rate using CS and to perform fault classification without reconstructing the original signal. Table 17.3 shows the classification results of bearing faults using CS-SAE-DNN with two hidden layers and a sampling rate α of 0.5, 0.25, and 0.1; and the reported results in (Wong et al. 2015) using the same dataset. It is clear that all results from CS-SAE-DNN are better than those in (Wong et al. 2015).

CS-SAE-DNN requires a number of parameters to run. Of these parameters, SAE parameters need to be set, including the sparsity parameter (ρ), weight decay (λ), and weight of the sparsity penalty term (β). To test the influence of these parameter values on bearing fault classification performance, numerous experiments have been carried out using the CS-SAE-DNN method with $\alpha = 0.05$, two hidden layers, and different values of SAE parameterisation. As can be seen from Figure 17.4, while classification accuracy is sensitive to the value of λ, there is a very broad range of values for ρ and β for which classification accuracies are very high and stable.

17.4.2 Case Study 2

Three vibration datasets, A, B, and C, were used in this case study, which were recorded under 10 roller bearing health conditions with 200 examples of each condition and a total of 2000 raw data files. These datasets have been described in more detail in Section 16.7.2. To apply CS-SAE-DNN to each of these datasets (i.e. A, B, and C), first CS model

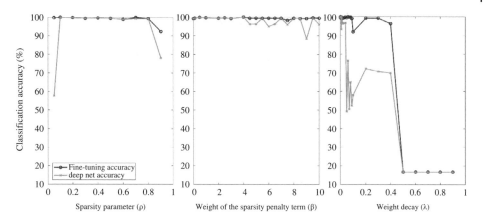

Figure 17.4 Effects of parameterization on classification accuracy (Ahmed et al. 2018). (See insert for a colour representation of this figure.)

is applied to obtain compressed vibration signals using different sampling rates (α: 0.025, 0.05, 0.1, and 0.2) with 60, 120, 240, and 480 compressed measurements. Fifty percent of these compressed samples are randomly selected for the pretraining stage of the DNN, and then these samples are used to retrain the deep net; the other 50% of the samples are used for testing performance. Then, the SAE-based DNN with two hidden layers is used to process the compressed measurements of each dataset. Classification accuracy rates are obtained by averaging the results of 10 experiments for each compressed dataset obtained using different sampling rates. The average accuracies and their corresponding standard deviations of 10 experiments for each dataset are shown in Table 17.4. One of the more significant findings to emerge from the results in Table 17.4 is that classification results after the fine-tuning stage (the second stage) are better than that after the first stage for all datasets A, B, and C with different values of α. Also, it shows that the deep net stage (the first stage) achieved good results, with 99.6% and 99.5% for a = 0.2 with datasets B and C, respectively. Most of the classification accuracy results after the second stage are above 99% for values of α in the range of 0.05–0.2. In particular, results after the second stage of our proposed method for datasets B and C with α equal to 0.2 achieved 100% accuracy for every single run in our investigations; 100% accuracy was also achieved for dataset C with α equal to 0.1. Overall, these results indicate that the proposed method has the ability to classify bearing conditions with high accuracy from highly compressed vibration measurements.

To examine the speed and accuracy of CS-SAE-DNN in several scenarios, three experiments are conducted (all with two hidden layers and fine-tuning) using the three datasets A, B, and C. The results are presented in Table 17.5. The first column refers to the three datasets. The second and third columns describe accuracies and execution times of using a "traditional" AE-based DNN (Jia et al. 2016) with 2400 inputs from the Haar wavelet (with no CS); this approach will be referred as WT-DNN. The fourth and fifth columns describe accuracies and execution times of using our SAE-based DNN with 2400 inputs from the Haar wavelet (with no CS); this approach will be referred as WT-SAE-DNN. Two things are clear for each of the three datasets: (i) our AE-based DNN (even without CS) is much faster than (or requires only 80% of the time of) the "traditional" AE-based DNN, and yet (ii) our classification results are very competitive.

Table 17.4 Classification results for bearing datasets A, B, and C of the second machine.

Datasets	Classification stage	Compressive sampling rate			
		$\alpha = 0.025$ (m = 60)	$\alpha = 0.05$ (m = 120)	$\alpha = 0.1$ (m = 240)	$\alpha = 0.2$ (m = 480)
A	Deep net (the first stage)	89.7 ± 5.4	93.7 ± 2.9	92.5 ± 2.2	98.7 ± 0.8
	Fine-tuning (the second stage)	97.6 ± 0.7	98.4 ± 1.3	99.3 ± 0.6	99.8 ± 0.2
B	Deep net (the first stage)	94.5 ± 1.1	96.2 ± 1.6	98.2 ± 0.6	99.6 ± 0.6
	Fine-tuning (the second stage)	98.9 ± 0.2	99.3 ± 0.4	99.7 ± 0.5	100 ± 0.0
C	Deep net (the first stage)	97.3 ± 0.8	98.5 ± 1.4	98.2 ± 0.4	99.5 ± 0.2
	Fine-tuning (the second stage)	99.4 ± 0.6	99.7 ± 0.3	100 ± 0.0	100 ± 0.0

Table 17.5 A comparison of results to examine speed and accuracy in several scenarios.

	WT-DNN		WT-SAE-DNN		CS-SAE-DNN	
	Accuracy (%)	Time (mins)	Accuracy (%)	Time (mins)	Accuracy (%)	Time (mins)
A	99.0 ± 0.1	41.5 ± 0.3	99.4 ± 0.5	34.1 ± 0.7	99.3 ± 0.6	5.7 ± 0.1
B	99.1 ± 0.6	43.3 ± 0.7	99.5 ± 0.8	32.9 ± 0.3	99.7 ± 0.5	5.9 ± 0.4
C	99.6 ± 0.3	43.1 ± 0.2	99.8 ± 1.1	34.2 ± 0.9	100	5.7 ± 0.2

The sixth and seventh columns describe accuracies and execution times of using our proposed SAE-based DNN with 240 inputs from the Haar wavelet (with CS), i.e. CS-SAE-DNN with $\alpha = 0.1$. Two things are clear for each of the three datasets: (i) our AE-based DNN (even with CS) is significantly faster than (or requires only 15% of the time of) the "traditional" AE-based DNN, and yet (ii) our classification results are as good as, if not better than, the other two scenarios. In summary, the significant reduction in computation time comes from two sources: (i) using our proposed SAE and (ii) using CS. Finally, our complete proposal (using SAE and CS) achieves classification results for all three datasets that are as good as, if not better than, the other two scenarios.

Moreover, comparisons with some recently published results in (De Almeida et al. 2015; Jia et al. 2016) with the same roller bearing datasets A, B, and C are presented in Table 17.6. The second left-hand column presents the classification results of the

Table 17.6 A comparison with the results from literature for bearing datasets A, B, and C of the second dataset.

Method Dataset	(Jia et al. 2016) DNN	(Jia et al. 2016) BPNN	(de Almeida et al., 2015) MLP	CS-SAE-DNN (with two hidden layers) $\alpha = 0.1$
A	99.95 ± 0.06	65.20 ± 18.09	95.7	99.3 ± 0.6
B	99.61 ± 0.21	61.95 ± 22.09	99.6	99.7 ± 0.5
C	99.74 ± 0.16	69.82 ± 17.67	99.4	100 ± 0.0

DNN-based method in (Jia et al. 2016), while the third column shows the classification results of the backpropagation neural network (BPNN) based method in (Jia et al. 2016). In (De Almeida et al. 2015), a generic multilayer perceptron (MLP) was used for classification purposes.

It can be clearly seen that the results from CS-SAE-DNN with fine-tuning (second stage) are very competitive. In particular, results from our fine-tuning method with dataset C achieved 100% accuracy for every single run in our investigations, even though we are using a limited amount (only 10%) of the original data; these results are not matched by any of the other methods using 100% of the original data.

17.5   Discussion

A recently proposed fault-diagnosis method, CS-SAE-DNN, has been described and validated in this chapter. The method employs a CS model and SAE-based DNN. The most obvious finding to emerge from our validation experiments is that, despite achieving fairly high classification accuracy in the first stage, the proposed method is able to achieve even higher classification accuracy in the second stage from highly compressed measurements compared to the existing methods. In particular, most of the classification accuracies for DNNs with two, three, and four hidden layers using fine-tuning (the second stage) are 99% or above, and some are 100% for even less than 1% compressed measurements of the original vibration signal, i.e. when $\alpha = 0.006$. The DNNs with two hidden layers achieved high classification accuracy (98%) for α equal to 0.003 and 0.0016 with 16 and 8 compressed measurements. This provides strong evidence for the advantage of the CS-SAE-DNN technique in cases where high levels of data compression are required to address the challenges of learning from high-dimensional data. Moreover, classification results from our proposed method outperform those achieved by reconstructing the original signals. Additionally, a significant reduction in computation time is achieved using our proposed method compared to another AE-based DNN method, with better classification accuracies.

References

Adcock, B., Hansen, A.C., Poon, C., and Roman, B. (2017). Breaking the coherence barrier: a new theory for compressed sensing. *Forum of Mathematics, Sigma* 5. Cambridge University Press. arXiv: 1302.0561, 2014.

Ahmed, H.O.A., Wong, M.L.D., and Nandi, A.K. (2018). Intelligent condition monitoring method for bearing faults from highly compressed measurements using sparse over-complete features. *Mechanical Systems and Signal Processing* 99: 459–477.

Bengio, Y., Lamblin, P., Popovici, D., and Larochelle, H. (2007). Greedy layer-wise training of deep networks. In: *Advances in Neural Information Processing Systems*, 153–160. http://papers.nips.cc/paper/3048-greedy-layer-wise-training-of-deep-networks.pdf.

de Almeida, L.F., Bizarria, J.W., Bizarria, F.C., and Mathias, M.H. (2015). Condition-based monitoring system for rolling element bearing using a generic multi-layer perceptron. *Journal of Vibration and Control* 21 (16): 3456–3464.

Doi, E., Balcan, D.C., and Lewicki, M.S. (2006). A theoretical analysis of robust coding over noisy overcomplete channels. In: *Advances in Neural Information Processing Systems*, 307–314. http://papers.nips.cc/paper/2867-a-theoretical-analysis-of-robust-coding-over-noisy-overcomplete-channels.pdf.

Fan, W., Cai, G., Zhu, Z.K. et al. (2015). Sparse representation of transients in wavelet basis and its application in gearbox fault feature extraction. *Mechanical Systems and Signal Processing* 56: 230–245.

Fuan, W., Hongkai, J., Haidong, S. et al. (2017). An adaptive deep convolutional neural network for rolling bearing fault diagnosis. *Measurement Science and Technology* 28 (9): 095005.

Guo, X., Chen, L., and Shen, C. (2016). Hierarchical adaptive deep convolution neural network and its application to bearing fault diagnosis. *Measurement* 93: 490–502.

Jia, F., Lei, Y., Lin, J. et al. (2016). Deep neural networks: a promising tool for fault characteristic mining and intelligent diagnosis of rotating machinery with massive data. *Mechanical Systems and Signal Processing* 72: 303–315.

Jiang, H., Li, X., Shao, H., and Zhao, K. (2018). Intelligent fault diagnosis of rolling bearings using an improved deep recurrent neural network. *Measurement Science and Technology* 29 (6): 065107.

Khan, S. and Yairi, T. (2018). A review on the application of deep learning in system health management. *Mechanical Systems and Signal Processing* 107: 241–265.

Lee, H., Ekanadham, C., and Ng, A.Y. (2008). Sparse deep belief net model for visual area V2. In: *Advances in Neural Information Processing Systems*, 873–880.

Lewicki, M.S. and Sejnowski, T.J. (2000). Learning overcomplete representations. *Neural Computation* 12 (2): 337–365.

Liu, H., Liu, C., and Huang, Y. (2011). Adaptive feature extraction using sparse coding for machinery fault diagnosis. *Mechanical Systems and Signal Processing* 25 (2): 558–574.

Lu, C., Wang, Z., and Zhou, B. (2017). Intelligent fault diagnosis of rolling bearing using hierarchical convolutional network-based health state classification. *Advanced Engineering Informatics* 32: 139–151.

Ma, M., Chen, X., Wang, S. et al. (2016). Bearing degradation assessment based on Weibull distribution and deep belief network. In: *International Symposium on Flexible Automation (ISFA)*, 382–385. IEEE.

Møller, M.F. (1993). A scaled conjugate gradient algorithm for fast supervised learning. *Neural Networks* 6 (4): 525–533.

Olshausen, B.A. and Field, D.J. (1997). Sparse coding with an overcomplete basis set: a strategy employed by V1. *Vision Research* 37 (23): 3311–3325.

Ren, L., Lv, W., Jiang, S., and Xiao, Y. (2016). Fault diagnosis using a joint model based on sparse representation and SVM. *IEEE Transactions on Instrumentation and Measurement* 65 (10): 2313–2320.

Roman, B., Hansen, A. and Adcock, B., 2014. On asymptotic structure in compressed sensing. arXiv preprint arXiv: 1406.4178.

Shao, H., Jiang, H., Zhang, H. et al. (2018). Rolling bearing fault feature learning using improved convolutional deep belief network with compressed sensing. *Mechanical Systems and Signal Processing* 100: 743–765.

Shao, H., Jiang, H., Zhang, X., and Niu, M. (2015). Rolling bearing fault diagnosis using an optimization deep belief network. *Measurement Science and Technology* 26 (11): 115002.

Sun, W., Shao, S., Zhao, R. et al. (2016). A sparse auto-encoder-based deep neural network approach for induction motor faults classification. *Measurement* 89: 171–178.

Tang, H., Chen, J., and Dong, G. (2014). Sparse representation based latent components analysis for machinery weak fault detection. *Mechanical Systems and Signal Processing* 46 (2): 373–388.

Tao, J., Liu, Y., and Yang, D. (2016). Bearing fault diagnosis based on deep belief network and multisensor information fusion. *Shock and Vibration* 7 (2016): 1–9.

Tao, S., Zhang, T., Yang, J. et al. (2015). Bearing fault diagnosis method based on stacked autoencoder and softmax regression. In: *2015 34th Chinese Control Conference (CCC)*, 6331–6335. IEEE.

Wong, M.L.D., Zhang, M., and Nandi, A.K. (2015). Effects of compressed sensing on classification of bearing faults with entropic features. *IEEE Signal Processing Conference (EUSIPCO)*: 2256–2260.

Zhang, W., Li, C., Peng, G. et al. (2018). A deep convolutional neural network with new training methods for bearing fault diagnosis under noisy environment and different working load. *Mechanical Systems and Signal Processing* 100: 439–453.

Zhao, R., Yan, R., Chen, Z., et al. (2016). Deep learning and its applications to machine health monitoring: A survey. arXiv preprint arXiv: 1612.07640.

Zhu, H., Wang, X., Zhao, Y. et al. (2015). Sparse representation based on adaptive multiscale features for robust machinery fault diagnosis. *Proceedings of the Institution of Mechanical Engineers, Part C: Journal of Mechanical Engineering Science* 229 (12): 2303–2313.

18

Conclusion

18.1 Introduction

Rotating machine condition monitoring is a key technique for ensuring the efficiency and quality of any production procedure. The correct monitoring system helps engineers to find defects and repair them quickly before machine breakdowns. In an ideal situation, we could predict machine failures in advance and carry out maintenance before they happen such that machines would always run in a healthy condition and provide satisfactory work.

Condition monitoring has been applied in many sensitive applications of rotating machines such as power generation, oil and gas, aerospace and defence, automotive, marine, etc. Condition monitoring systems for rotating machinery involve the practice of monitoring measurable data (e.g. vibration, acoustic, etc.), which can be used individually or in a mixture to identify changes in machine condition. The increased level of complexity in modern rotating machines requires more effective and efficient condition monitoring techniques. In view of that, a growing body of literature has resulted from research and development efforts by many research groups around the world. These publications have a direct impact on the present and future development of machine condition monitoring. In addition, the nature of machine condition monitoring takes multiple directions and motivates continuous contributions from generations of researchers.

Various types of condition monitoring techniques have been investigated. These techniques are based on forms of sensor data acquired from rotating machines. Usually, machine condition monitoring techniques can be categorised as follows: vibration monitoring, acoustic emission monitoring, a fusion of vibration and acoustic, electric motor current monitoring, oil analysis, thermography, visual inspection, performance monitoring, and trend monitoring.

Instead of processing collected raw signals, the common approach is to compute certain attributes of the raw signal that can describe the signal in essence. In the machine learning community, these attributes are referred to as features. Sometimes, multiple features are computed to form a reduced dimensionality set of the original high-dimensional dataset. Depending on the number of features in the set, one may carry out further filtering of the computed features using a feature-selection algorithm. There are various techniques for feature extraction and feature selection that can be used in monitoring machine conditions.

Condition Monitoring with Vibration Signals: Compressive Sampling and Learning Algorithms for Rotating Machines, First Edition. Hosameldin Ahmed and Asoke K. Nandi.
© 2020 John Wiley & Sons Ltd. Published 2020 by John Wiley & Sons Ltd.

The core objective is to categorise the acquired signal correctly into the corresponding machine condition, which is generally a multiclass classification problem. Many machine learning–based classification algorithms can be used to deal with the classification problem in machine fault detection and identification.

18.2 Summary and Conclusion

In this book, we attempted to bring together many techniques in one place and outline a comprehensive guide from the basics of the rotating machine to the generation of knowledge using vibration signals. The contents of this book are divided into five parts. Part I, which included Chapters 1 and 2, provided an introduction to machine condition monitoring using vibration signals. Chapter 1 covered three main topics: (i) importance of machine condition monitoring, (ii) maintenance approaches, and (iii) machine condition monitoring techniques. From the three types of maintenance approaches (see Figure 18.1), condition-based maintenance (CBM) has advantages over the other two types of maintenance, i.e. corrective maintenance and time-based maintenance. In this chapter, we showed that decisions regarding maintenance are made based on the current health of the machine, which can be identified through a condition monitoring system. Then, we presented the various types of condition monitoring techniques that can be used to identify the current health of machines. Of these techniques (see Figure 18.1), vibration condition monitoring has become well-accepted for many CBM programs for rotating machines. Hence, Chapter 2 was concerned with principles of rotating machine vibration and acquisition techniques. The first part of this chapter presented vibration principles and described vibration signals produced by rotating machines, and types of vibration signals. Part II described vibration data acquisition techniques and highlighted the advantages and limitations of vibration signals.

Given the collected vibration signals from different sources of rotating machines, one needs to analyse these signals to reveal useful information that can interpret the signal in essence. Hence, the second part of the book described the three main groups of vibration signal analysis – time domain, frequency domain, and time-frequency domain. Chapter 3 described time domain techniques that extract features from raw vibration signals using statistical parameters as well as advanced techniques (see Figure 18.2). All of the studies presented in this chapter used more than one time domain statistical technique to extract features from raw vibration data; some used more than 10. Of these techniques, kurtosis is used in all the mentioned studies. Additionally, other advanced techniques include time synchronous averaging (TSA), which removes any periodic events not synchronous with the specific sampling frequency or sampling time of a vibration signal, autoregressive model (AR), the autoregressive integral moving average (ARIMA) model used to predict machine faults even if data are only available from a machine in its normal condition, and filter-based methods, which have been used by many researchers to remove noise and isolate signals from raw signals.

Chapter 4 presented techniques for vibration signal processing in the frequency domain that have the capability to reveal frequency characteristics that are not easy to observe in the time domain (see Figure 18.2). Typically, the fast Fourier transform (FFT) is the most widely used technique to transform the time domain signal to the frequency domain. Moreover, various frequency spectrum features can be extracted

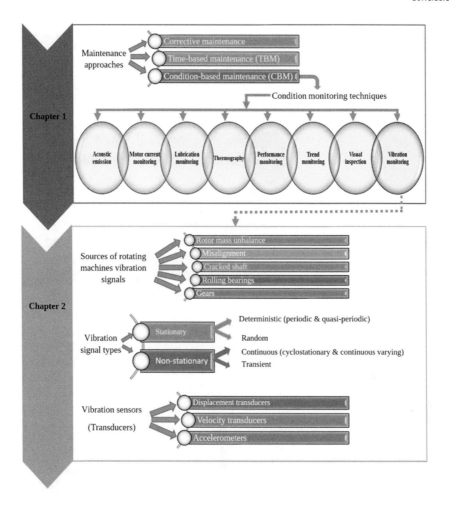

Figure 18.1 Topics covered by Part I of the book. (See insert for a colour representation of this figure.)

and used for diagnosing machine faults. Chapter 5 described various time-frequency domain analysis techniques (see Figure 18.2), which are used for nonstationary waveform signals that are very common when machine faults happen. These include (i) the short-time Fourier transform (STFT), which computes the discrete Fourier transform (DFT) by decomposing a signal into shorter segments of equal length using a time-localised window function. (ii) Wavelet analysis decomposes the signal based on a family of 'wavelets'. Unlike the window used with STFT, the wavelet function is scalable, which makes it adaptable to a wide range of frequencies and time-based resolution. The three main transforms in wavelet analysis are the continuous wavelet transform (CWT), discrete wavelet transform (DWT), and wavelet packet transform (WPT). (iii) Empirical mode decomposition (EMD) decomposes the signal into different scales of intrinsic mode functions (IMFs). (iv) The Hilbert-Huang transform (HHT) can be achieved by decomposing a signal into IMFs using EMD, applying the Hilbert transform (HT) to all IMF components, and computing the instantaneous

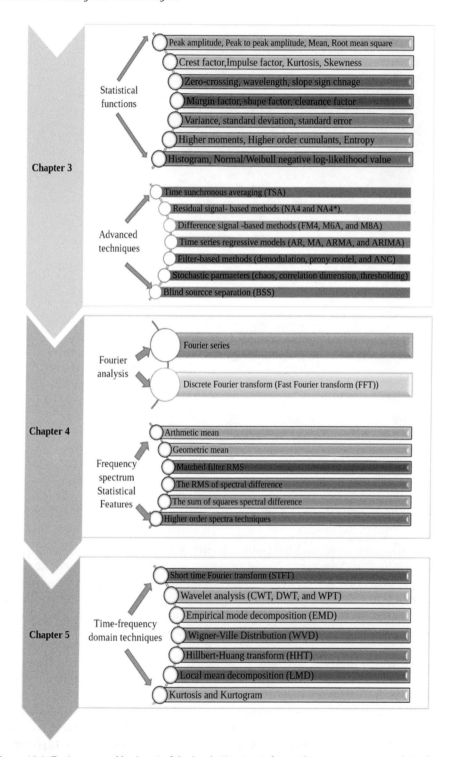

Figure 18.2 Topics covered by Part II of the book. (See insert for a colour representation of this figure.)

frequency to all IMFs. The HT can be achieved by transforming into the frequency domain using a Fourier transform, shifting the phase angle of all components by $\pm\frac{\pi}{2}$, i.e. shifting by $\pm90°$, and then transforming back to the time domain. (v) The Wigner-Ville distribution (WVD) can be derived by generalising the relationship between the power spectrum and the autocorrelation function for nonstationary, time-variant processes. And (vi) spectral kurtosis (SK) can be achieved by decomposing the signal into the time-frequency domain where kurtosis values are defined for each frequency group, while the Kurtogram algorithm (KUR) computes the SK for several window sizes using a bandpass filter.

Part III focused on vibration-based machine condition monitoring processes, highlighted the main problems of learning from vibration data for the purpose of fault diagnosis, and described commonly used methods for data normalisation and dimensionality reduction. Chapter 6 described the general framework for machine fault detection and classification using vibration signals. In addition, several different types of learning that can be applied to vibration signals were described. Furthermore, the main problems of learning from vibration signals and the techniques that can be used to overcome these problems were presented in this chapter (see Figure 18.3). A reasonable approach to tackle the challenges of dealing with high-dimensional data is to learn a subspace from it, i.e. map high-dimensional data into a low-dimensional subspace that are linear or nonlinear combinations of the original high-dimensional data. Chapter 7 described various linear subspace learning techniques that can be used to reduce the dimensionality of vibration signals (see Figure 18.3).

Chapter 8 presented a number of nonlinear subspace learning techniques to achieve a reduced nonlinear subspace feature set from vibration signals with high-dimensional data. Furthermore, Chapter 9 introduced generally applicable methods that can be used to select the most relevant features, which can contribute to the classification accuracy of vibration signals. These include various algorithms for feature ranking, sequential selection algorithms, heuristic-based selection algorithms, and embedded model-based feature-selection algorithms.

The various different techniques for feature learning and feature selection introduced in Part III are often used to obtain a set of features from raw vibration signals, which sufficiently represent the machine's health. With these learned features, the next stage is the classification stage, where a classification algorithm is usually used to correctly categorise the acquired vibration signal into the corresponding machine health condition, which is generally a multiclass classification problem. Therefore, Part IV of this book presented various classification techniques and their applications in machine fault detection and identification based on the features provided (see Figure 18.4). Chapter 10 introduced the basic theory of decision trees, its structure, and the ensemble model that combines decision trees into a decision forest. In addition, this chapter described the application of decision trees and decision forests in various studies of machine fault diagnosis. Chapter 11 presented two probabilistic models for classification: (i) the hidden Markov model (HMM) as a probabilistic generative model and (ii) the logistic regression model (LR) as a probabilistic discriminative model. These classifiers, i.e. HMM and LR, have been widely used in diagnosing machine faults. Therefore, this chapter presented various studies that utilised these classifiers to deal with the classification problem in machine condition monitoring. Chapter 12 introduced the basic concept of artificial neural networks (ANNs). Then, three types of ANN – multilayer perceptron (MLP),

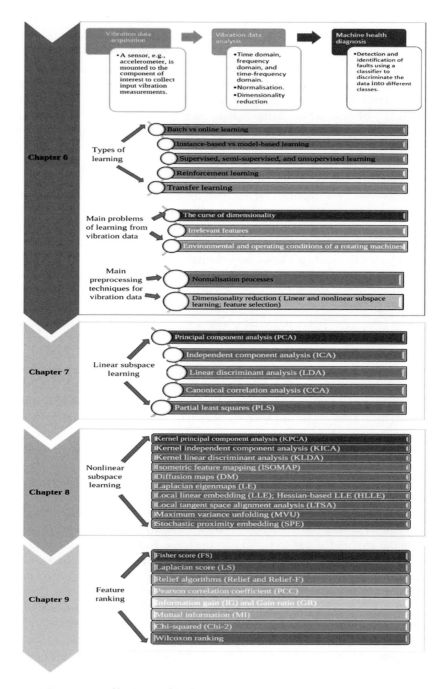

Figure 18.3 Topics covered by Part III of the book. (See insert for a colour representation of this figure.)

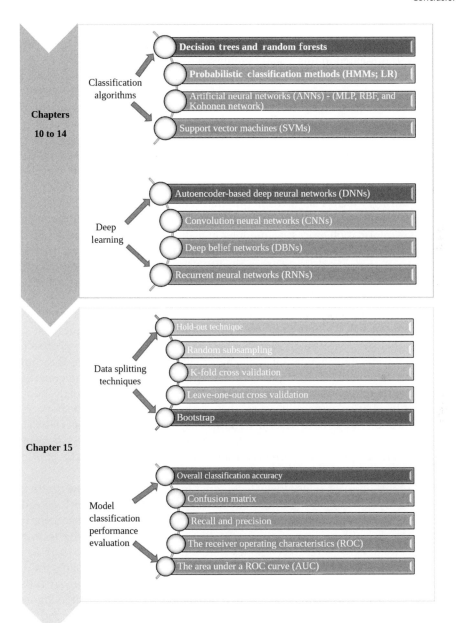

Figure 18.4 Topics covered by Part IV of the book. (See insert for a colour representation of this figure.)

radial basis functions (RBFs), and Kohonen networks – and their application in diagnosing machine faults were described. Most of the studies of machine fault diagnosis using ANNs presented in this chapter employed pre-processing techniques before the application of the ANN-based classifiers. Of these, the genetic algorithm (GA), along with various types of time domain statistical features, frequency domain features, and time-frequency domain features, have been widely used with different types of ANNs. Chapter 13 introduced support vector machines (SVMs), which are one of the most popular machine learning classifiers used in diagnosing machine faults. The pre-processing techniques used by most of the SVM-based studies in machine fault diagnosis are similar to those used by most ANN-based studies.

Chapter 14 introduced recent trends of deep learning in the field of machine condition monitoring, which is another solution that can be used to automatically learn the representations of the data required for feature detection and classification using multiple layers of information processing modules in hierarchical architectures. Additionally, this chapter provided an explanation of commonly used techniques and examples of their applications in diagnosing machine faults. These include autoencoder-based deep neural networks (DNNs), convolutional neural networks (CNNs), deep belief networks (DBNs), and recurrent neural networks (RNNs).

Furthermore, to assess the efficiency of the classification algorithms introduced in this part of the book, Chapter 15 described different validation techniques that can be used to evaluate and verify a classification model's performance before proceeding to its application and implementation in machine fault diagnosis problems. These include the most commonly used data-splitting techniques, the hold-out technique, random subsampling, K-fold cross-validation, leave-one-out cross-validation, and the bootstrap technique. In addition, various measures that can be used to evaluate the performance of a classification model were described in this chapter, including overall classification accuracy, the confusion matrix, the receiver operating characteristic (ROC) graph, the area under a ROC curve (AUC), precision, and recall. Figure 18.4 shows the topics covered by this part of the book.

Part V of this book introduced new techniques for classifying machine health based on compressive sampling (CS) and machine learning algorithms. Chapter 16 introduced a new fault-diagnosis framework called compressive sampling and subspace learning (CS-SL), which combines compressive sampling and subspace learning techniques. Under this framework, several techniques were introduced (see Figure 18.5): (i) compressive sampling and feature ranking (CS-FR), (ii) compressive sampling and linear subspace learning (CS-LSL), and (iii) compressive sampling and nonlinear subspace learning (CS-NLSL). These techniques have the ability to receive a large amount of vibration data as input and produce fewer features as output, which can be successfully used for fault classification of rotating machines.

Chapter 17 presented another approach that has been proposed recently through the design of an intelligent fault-classification method, called (CS-SAE-DNN), from highly compressed measurements using sparse-overcomplete features and training DNNs through a sparse autoencoder (SAE). It has been shown that this method, i.e. CS-SAE-DNN, has an advantage in cases where high levels of data compression are required.

Finally, most of the introduced techniques in this book and links to their publicly accessible software can be found in the book's appendix.

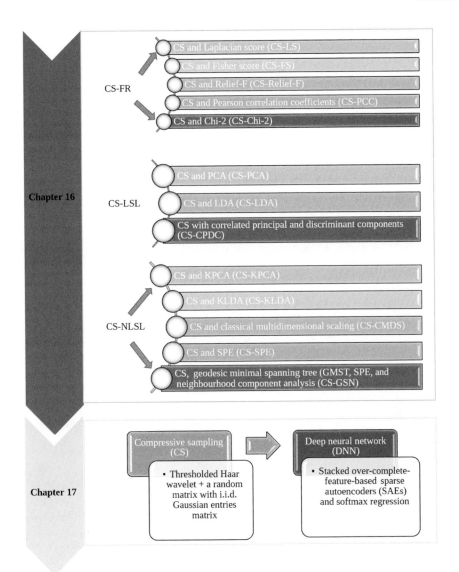

Figure 18.5 Topics covered by Part V of the book. (See insert for a colour representation of this figure.)

Appendix

Machinery Vibration Data Resources and Analysis Algorithms

Machinery vibration data resources:

Dataset	Provider	Link
Drill Bit Dataset Repository	Intelligent Informatics and Automation Laboratory, IIT Kanpur	http://iitk.ac.in/iil/datasets Verma et al. (2015) Verma et al. 2016
Bearing Dataset	Bearing Data Center, Case Western Reserve University	http://csegroups.case.edu/ bearingdatacenter/pages/ download-data-file
Bearing Dataset	Prognostics Data Repository, NASA	https://ti.arc.nasa.gov/tech/dash/ groups/pcoe/prognostic-data- repository
Bearing Dataset	Society For Machinery Failure Prevention Technology	https://mfpt.org/fault-data-sets
Gearbox	PHM Society 2009	https://www.phmsociety.org/ competition/PHM/09

Time-domain analysis algorithms:

Algorithm name	Platform	Package	Function
Max	MATLAB	Signal Processing Toolbox – Measurements and Feature Extraction – Descriptive Statistics	max
Min			min
Mean			mean
Peak-to-peak			Peak2peak
RMS			rms
STD			std
VAR			var
Crest Factor			Peak2rms
Zero Crossing	MATLAB	Bruecker (2016)	crossing
Skewness	MATLAB	Statistics and Machine Learning Toolbox.	skewness

Condition Monitoring with Vibration Signals: Compressive Sampling and Learning Algorithms for Rotating Machines, First Edition. Hosameldin Ahmed and Asoke K. Nandi.
© 2020 John Wiley & Sons Ltd. Published 2020 by John Wiley & Sons Ltd.

Algorithm name	Platform	Package	Function
Kurtosis			kurtosis
TSA	MATLAB	Signal Processing Toolbox	tsa
AR-covariance			arcov
AR-Yule-Walker			aryule
Prony			prony
ARIMA	MATLAB	Econometric Toolbox	arima
ANC	MATLAB	Clemens (2016)	Adapt-filt-tworef
S-ANC	MATLAB	NJJ (2018)	sanc
BSS	R	JADE and BSSasymp (Miettinen et al. 2017)	BSS
BSS	MATLAB	Gang (2015)	YGBSS

Frequency-domain analysis algorithms:

Algorithm name	Platform	Package	Function
Discrete Fourier transform Inverse discrete Fourier transform	MATLAB	Communication System Toolbox	fft ifft
Fast Fourier transform Inverse fast Fourier transform		Fourier Analysis and Filtering	fft ifft
Signal envelope	MATLAB	Signal Processing Toolbox	envelope
Envelope spectrum for machinery diagnosis			envspectrum
Average spectrum vs. order for vibration signal			Orderspectrum
Complex cepstral analysis			cceps
Inverse complex cepstrum			Icceps
Real cepstrum and minimum phase reconstruction			rceps

Time-frequency-domain analysis algorithms:

Algorithm name	Platform	Package	Function
Spectrogram using STFT	MATLAB	Signal Processing Toolbox	spectrogram
Discrete-time analytic signal using Hilbert transform			hilbert
Display wavelet families names	MATLAB	Wavelet Toolbox	Waveletfamilies('f')
Display wavelet families with their corresponding properties			Waveletfamilies('a')
Denoising of a one-dimensional signal using wavelets			wden
The continuous 1-D wavelet transform			cwt
The inverse continuous 1-D wavelet transform			icwt

Algorithm name	Platform	Package	Function
Continuous wavelet transform	Python	Lee et al. (2018)	pywt.cwt
Discrete wavelet transform			pywt.dwt
Fast Kurtogram	MATLAB	Antoni (2016)	Fast_Kurtogram
Visualise spectral kurtosis	MATLAB 2018b		kurtogram
Wigner-Ville distribution and smoothed pseudo Wigner-Ville distribution			wvd
Empirical mode decomposition			Emd
Spectral entropy of signal			pentropy

Linear subspace learning algorithms:

Algorithm name	Platform	Package	Function
Eigenvalues and eigenvectors	MATLAB	Mathematics Toolbox – Linear Algebra	eig
Singular value decomposition			svd
Principal component analysis of raw data		Statistics and Machine Learning Toolbox – Dimensionality Reduction and Feature Extraction	pca
Feature extraction by using reconstruction ICA			rica
Canonical correlation			Canoncorr

Nonlinear subspace learning algorithms;

Algorithm name	Platform	Package	Function
Kernel PCA	MATLAB	Van Vaerenbergh (2016)	km_pca
Kernel CCA			km_cca
MATLAB Toolbox for Dimensionality Reduction	MATLAB	Van der Maaten et al. 2007 The software within this toolbox for some nonlinear methods introduced in this book includes: • LLE • LE • HLLE • LTSA • MVU • KPCA • DM • SPE	compute_mapping(data, method, # of dimensions, parameters)

Feature-selection algorithms:

Algorithm name	Platform	Package	Function
Sequential feature selection	MATLAB	Statistics and Machine Learning Toolbox	sequentialfs
Importance of attributes (predictors) using Relief-F algorithm			relieff
Wilcoxon rank sum test			ranksum
Rank key features by class separability criteria		Bioinformatics Toolbox	rankfeatures
Feature selection library		Giorgio (2018)	relieff laplacian fisher lasso
Find minimum of function using genetic algorithm		Global Optimisation Toolbox	ga

Decision trees and random forests algorithms:

Algorithm name	Platform	Package	Function
Fit binary classification decision tree for multiclass classification	MATLAB	Statistics and Machine Learning Toolbox – Classification – Classification trees	fitctree
Produce sequence of subtrees by pruning			prune
Compact tree			compact
Mean predictive measure of association for surrogate splits in decision tree			surrogateAssociation
Decision tree and Decision forest	MATLAB	Wang (2014)	RunDecisionForest TrainDecisionForest

Probabilistic classification methods:

Algorithm name	Platform	Package	Function
Hidden Markov model posterior state probabilities	MATLAB	Statistics and Machine Learning Toolbox – Hidden Markov models	hmmdecode
Multinomial logistic regression		Statistics and Machine Learning Toolbox – Regression	mnrfit
Maximum likelihood estimates		Statistics and Machine Learning Toolbox – Probability Distributions	mle
Gradient descent optimisation		Allison (2018)	grad_descent

Artificial neural network (ANN) algorithms:

Algorithm name	Platform	Package	Function
Perceptron	MATLAB	Deep Learning Toolbox – Define Shallow Neural Network Architecture	Perceptron
MLP neural network trained by backpropagation		Chen (2018)	mlpReg mlpRegPred
Radial basis function with K means clustering		Shujaat (2014)	RBF
Design probabilistic neural network		Deep Learning Toolbox – Define Shallow Neural Network Architecture	newpnn
Train shallow neural network		Deep Learning Toolbox – Function Approximation and Clustering	train
Self-organising map		Deep Learning Toolbox – Function Approximation and Clustering-Self-Organising maps	selforgmap
Gradient descent backpropagation			net.trainFcn = 'traingd'

Support vector machines algorithms:

Algorithm name	Platform	Package	Function
Train support vector machine classifier	MATLAB	Statistical and Machine Learning Toolbox	svmtrain
Classify using support vector machine (SVM)			svmclassify
Train binary SVM classifier			fitcsvm
Predict labels using SVM classifier			predict
Fit multiclass models for SVM or other classifiers			fitcecoc
LIBSVM: a library for support vector machines	JAVA, MATLAB, OCTAVE, R, PYTHON, WEKA, SCILAB, C#, etc.	Chang 2011	n/a

References

Allison, J. (2018). Simplified gradient descent optimization. Mathworks File Exchange Center. https://uk.mathworks.com/matlabcentral/fileexchange/35535-simplified-gradient-descent-optimization.

Antoni, J. (2016). Fast kurtogram. Mathworks File Exchange Center. https://uk.mathworks.com/matlabcentral/fileexchange/48912-fast-kurtogram.

Bruecker, S. (2016). Crossing. Mathworks File Exchange Center. https://uk.mathworks.com/matlabcentral/fileexchange/2432-crossing.

Chang, C.C. (2011). LIBSVM: a library for support vector machines. *ACM Transactions on Intelligent Systems and Technology*, 2: 27: 1--27: 27, 2011. Software available at http://www.csie.ntu.edu.tw/~cjlin/libsvm.

Chen, M. (2018). MLP neural network trained by backpropagation. Mathworks File Exchange Center. https://uk.mathworks.com/matlabcentral/fileexchange/55946-mlp-neural-network-trained-by-backpropagation.

Clemens, R. (2016). Noise canceling adaptive filter. Mathworks File Exchange Center. https://www.mathworks.com/matlabcentral/fileexchange/10447-noise-canceling-adaptive-filter.

Gang, Y. (2015). A novel BSS. Mathworks File Exchange Center. https://uk.mathworks.com/matlabcentral/fileexchange/50867-a-novel-bss.

Giorgio. (2018). Feature selection library. Mathworks File Exchange Center. https://uk.mathworks.com/matlabcentral/fileexchange/56937-feature-selection-library.

Lee, G.R., Gommers, R., Wohlfahrt, K., et al. (2018). PyWavelets/pywt: PyWavelets v1.0.1 (Version v1.0.1). Zenodo. http://doi.org/10.5281/zenodo.1434616.

Miettinen, J., Nordhausen, K., and Taskinen, S. (2017). Blind source separation based on joint diagonalization in R: the packages JADE and BSSasymp. *Journal of Statistical Software* 76.

NJJ. (2018). sanc(x,L,mu,delta). Mathworks File Exchange Center. https://www.mathworks.com/matlabcentral/fileexchange/65840-sanc-x-l-mu-delta.

PHM Society. (2009). Data analysis competition. https://www.phmsociety.org/competition/PHM/09.

Shujaat, K. (2014). Radial basis function with k mean clustering. Mathworks File Exchange Center. https://uk.mathworks.com/matlabcentral/fileexchange/46220-radial-basis-function-with-k-mean-clustering.

Van der Maaten, L., Postma, E.O., and van den Herik, H.J. (2007). *Matlab Toolbox for Dimensionality Reduction*. MICC, Maastricht University.

Van Vaerenbergh, S. (2016). Kernel methods toolbox. Mathworks File Exchange Center. https://uk.mathworks.com/matlabcentral/fileexchange/46748-kernel-methods-toolbox?s_tid=FX_rc3_behav.

Verma, N.K., Sevakula, R.K., Dixit, S., and Salour, A. (2015). Data driven approach for drill bit monitoring. *IEEE Reliability Magazine* (February): 19–26.

Verma, N.K., Sevakula, R.K., Dixit, S., and Salour, A. (2016). Intelligent condition based monitoring using acoustic signals for air compressors. *IEEE Transactions on Reliability* 65 (1): 291–309.

Wang, Q. (2014). Decision tree and decision forest. Mathworks File Exchange Center. https://uk.mathworks.com/matlabcentral/fileexchange/39110-decision-tree-and-decision-forest.

Index

Condition Monitoring with Vibration Signals: Compressive Sampling and Learning Algorithms for Rotating Machines, First Edition. Hosameldin Ahmed and Asoke K. Nandi.
© 2020 John Wiley & Sons Ltd. Published 2020 by John Wiley & Sons Ltd.